Vitalist Modernism

This book reveals how, when, where, and why Vitalism and its relationship to new scientific theories, philosophies and concepts of energy became seminal from the fin de siècle until the Second World War for such Modernists as Sophie Taeuber-Arp, Hugo Ball, Juliette Bisson, Eva Carrière, Salvador Dalì, Robert Delaunay, Marcel Duchamp, Edvard Munch, Picasso, Yves Tanguy, Gino Severini and John Cage. For them, Vitalism entailed the conception of life as a constant process of metamorphosis impelled by the free flow of energies, imaginings, intuition and memories, unconstrained by mechanistic materialism and chronometric imperatives, to generate what the philosopher Henri Bergson aptly called Creative Evolution.

Following the three main dimensions of Vitalist Modernism, the first part of this book reveals how biovitalism at the fin de siècle entailed the pursuit of corporeal regeneration through absorption in raw nature, wholesome environments, aquatic therapies, electromagnetism, heliotherapy, modern sports, particularly rugby, water sports, the Olympic Games and physical culture to energize the human body and vitalize its life force. This is illuminated by artists as geoculturally diverse as Gustave Caillebotte, Thomas Eakins, Munch and Albert Gleizes. The second part illuminates how simultaneously Vitalism became aligned with anthroposophy, esotericism, magnetism, occultism, parapsychology, spiritism, theosophy and what Bergson called "psychic states", alongside such new sciences as electromagnetism, radiology and the Fourth Dimension, as captured by such artists as Juliette Bisson, Giacomo Balla, Albert Besnard, Umberto Boccioni, Eva Carrière, John Gerrard Keulemans, László Moholy-Nagy, James Tissot, Albert von Schrenck Notzing and Picasso.

During and after the devastation of the First World War, the third part explores how Vitalism, particularly Bergson's theory of becoming, became associated with Dadaist, Neo-Dadaist and Surrealist notions of amorality, atemporality, dysfunctionality, entropy, irrationality, inversion, negation and the nonsensical captured by Hans Arp, Charlie Chaplin, Theo Van Doesburg, Kazimir Malevich, Kurt Schwitters and Vladimir Tatlin alongside Cage's concept of Nothing. After investigating the widespread engagement with Bergson's philosophies and Vitalism and art by Anarchists, Marxists and Communists during and after the First World War, it concludes with the official rejection of Bergson and any form of Vitalism in the Soviet Union under Stalin.

This book will be of vital interest to gallery, exhibition and museum curators and visitors, plus readers and scholars working in art history, art theory, cultural studies, modernist studies, occult studies, European art and literature, health, histories of science, philosophy, psychology, sociology, sport studies, heritage studies, museum studies and curatorship.

Fae Brauer is Professor Emeritus of Art and Visual Culture at the University of East London Centre for Cultural Studies Research; Honorary Professor of Art History and Art Theory at The University of New South Wales and a Commissioning Editor for the Rowman & Littlefield International Radical Cultural Studies Series. She is an elected Fellow of the Royal Society of Arts with an MA and PhD from The Courtauld Institute of Art, University of London.

Science and the Arts since 1750
Series Editors: Barbara Larson, University of West Florida
and Ellen K. Levy, Independent Artist and Writer

This series of monographs and edited volumes explores the arts – painting and sculpture, drama, dance, architecture, design, photography and popular culture materials – as they intersect with emergent scientific theories, agendas, and technologies, from any geographical area from 1750 to now.

Photography, Natural History and the Nineteenth-Century Museum
Exchanging Views of Empire
Kathleen Davidson

Art, Technology and Nature
Renaissance to Postmodernity
Edited by Camilla Skovbjerg Paldam and Jacob Wamberg

The Organic School of the Russian Avant-Garde
Nature's Creative Principles
Isabel Wünsche

Science, Technology, and Utopias
Women Artists and Cold War America
Christine Filippone

Visualizing the Body in Art, Anatomy, and Medicine since 1800
Models and Modeling
Edited by Andrew Graciano

Painting, Science, and the Perception of Coloured Shadows
'The Most Beautiful Blue'
Paul Smith

Vitalist Modernism
Art, Science, Energy and Creative Evolution
Edited by Fae Brauer

For more information about this series, please visit: https://www.routledge.com/Science-and-the-Arts-since-1750/book-series/ASHSER4039

Vitalist Modernism

Art, Science, Energy and Creative Evolution

Edited by Fae Brauer

Routledge
Taylor & Francis Group
NEW YORK AND LONDON

Designed cover image:
Photograph courtesy of the Munch Museum, Oslo and University of Oslo
Photographer: Ove Kvavik

First published 2023
by Routledge
605 Third Avenue, New York, NY 10158

and by Routledge
4 Park Square, Milton Park, Abingdon, Oxon, OX14 4RN

Routledge is an imprint of the Taylor & Francis Group, an informa business

ISBN: 978-0-367-49304-2 (hbk)
ISBN: 978-1-032-42348-7 (pbk)
ISBN: 978-1-003-04559-5 (ebk)

DOI: 10.4324/9781003045595

Typeset in Sabon
by KnowledgeWorks Global Ltd.

In memoriam Virginia Spate (1937–2022),
a pioneering scholar of Modernism.

Contents

Figures

Acknowledgements

This book began its life as the session entitled *Vitalist Modernism* for the 2019 Association of Art Historians Annual Conference at the Universities of Brighton and Sussex from 4 to 6 April 2019 entitled “Art and Visual Culture in an Expanded Field”. Many of those who have chapters in this book presented outstanding papers at this conference. From the moment that the first paper was presented by Professor Emeritus Anthea Callen, the auditorium was packed. While aware of the burgeoning interest in vitalism, the long and vigorous engagement shown by our audience surpassed all of our expectations, especially in a hot and stuffy lecture theatre where many had to stand for long periods. It was also at this conference that I was first able to discuss publication of these papers as chapters in a book with Isabella Vitti, the Art History and Visual Studies Editor of the Routledge Taylor & Francis Group. Given the strength of her support for this book, she is the first person that we wish to thank, followed by the excellent Series Editors for *Science and the Arts since 1750,* Barbara Larson and Ellen Levy.

For their generosity with archival resources, we also wish to thank ADAGP, Paris; Alamy Stock Photos, Archives & Special Collections, University of Manitoba; Artists Rights Society (ARS), New York; Arp Museum Bahnhof Rolandseck; Art Gallery of Ontario; Bridgeman Images; Countway Medical Library, Harvard University; DACS, London; Estorick Collection, London; Fondation Arp-Hagenbach, Locarno; Galleria d’Arte Moderna, Villa Reale, Milan; Harvard University; History and Special Collections for the Sciences at UCLA Library; Kunsthandel Achenbach, Dusseldorf; Kunsthaus Zurich, Dada-Sammlung; Luisa Ricciarini Collection; Medical Historical Library, Harvey Cushing/John Hay Whitney Medical Library, Yale University; McCord Museum; Metropolitan Museum of Art, New York; Musée d’Art Moderne de la Ville de Paris; Musée d’Orsay; Musée Picasso de Paris; Museum of Modern Art, New York; National Gallery of Art, Washington D.C.; National Library, Oslo; New York Academy of Medicine Library; RMN-Grand Palais; Sprengel Museum, Hanover; State Mayakovsky Museum, Moscow; St. John’s College Oxford; Tegner Museum, Sweden and the City of Copenhagen; State Darwin Museum, Moscow; The Gothenburg Museum of Art; The Menil Collection, Houston; The Museum of Fine Art, Boston; The Wellcome Library, London; University of Oslo Munch Museum; The Vatican Museum and Wellesley College Library. We wish also to thank Justin Fleming for his invaluable help with proofing each chapter. What follows are personal thanks.

For Chapter 1, Anthea Callen is greatly indebted to George Shackelford and Sylvie Brame for their generous help in sourcing the Caillebotte illustrations, and John Locke, via MagazineArt.org, Michael Ward at www.ephemeraforever.com and Erik Tweed at *The Magazine* in San Francisco, for allowing her to use the New York *Physical Culture* magazine illustrations. She would like to express her gratitude to The Leverhulme Trust for the Emeritus Fellowship that supported her new research on Impressionism, and to thank Fae Brauer for her unstinting work not only in organizing the *Vitalist Modernism* session at AAH 2019, but also for marshalling these and other contributions into the present volume. For the research and writing support she received for Chapter 4, Serena Keshavjee wishes to acknowledge the Social Science and Humanities Research Council. For their commentary and editorial input, she also wishes to acknowledge Fae Brauer, Oliver Botar, Tim Pearson, Paul Dutton, Geneviève Riou, Murray Leeder and Emma Dux. For Chapter 8, Donna Roberts wishes to acknowledge the research support she received from the Kone Foundation in Helsinki and the editorial responses from Fae Brauer, Sami Sjöberg and Kai Alhanen. For Chapter 10, Pat Simpson would like to express profound thanks to Anna Kliukina, Director of the State Darwin Museum, Moscow, for permitting her to reproduce illustrations from the Darwin Museum publications purchased on her research visit there, particularly works by Konstantin Flerov (Figures 10.3–10.5) and archival photographs (Figures 10.1–10.3). She also wishes to express her sincere gratitude to her son, the innovative graphic and UX designer, Edward Dillon, for expertly re-photographing and correctly formatting the illustrations provided for this volume. All of the contributors would also like to thank the Project Manager of Books at KnowledgeWorks Global Ltd., Cole Bowman, and his team of proofers, for so patiently and professionally transforming our manuscript into this book.

Creating this book during the pandemic has proven not to be a simple task. Unfortunately, I was not alone in suffering severe Covid throughout most of 2020. Three other contributors suffered Covid while an earlier contributor needed to withdraw because of it. In 2021, our contributors then experienced the Delta and Omicron variants while one also suffered malignant cancer. For continuing to contribute to this book in these extraordinary circumstances and for never losing sight of what we hoped to achieve with it, I wish to acknowledge the strength, tenacity and ingenuity of the contributors and thank them profoundly for their invaluable and generous support.

Fae Brauer

Contributors

Mark Antliff is the Mary Grace Wilson Distinguished Professor Emeritus of Art, Art History & Visual Studies. He is author of *Inventing Bergson: Cultural Politics and the Parisian Avant-Garde* (Princeton University Press, 1993), *Avant-Garde Fascism: The Mobilization of Myth, Art and Culture in France, 1909-1939* (Duke University Press, 2007; Les presses du réel, Paris, 2019) as well as co-author of *Fascist Visions: Art and Ideology in France and Italy* (with Matthew Affron, Princeton University Press, 1997), *Cubism and Culture* (with Patricia Leighten, Thames & Hudson, 2001; French edition 2001), and *A Cubism Reader: Documents and Criticism 1906-1914* (with Patricia Leighten, University of Chicago Press, 2008 and Les presses du réel, Paris, 2019). With Vivien Greene, he co-curated the exhibition *The Vorticists: Rebel Artists in London and New York, 1914-1918*, which opened at the Nasher Museum of Art and travelled to the Peggy Guggenheim Collection in Venice and Tate Britain in London (ex. cat. London: Tate Publishing, 2010–11). The conference associated with this exhibition resulted in *Vorticism: New Perspectives* (Oxford University Press, 2013). His most recent book is titled, *Sculptors Against the State: Anarchism and the Anglo-European Avant-Garde* (Penn State Press, 2021).

Patricia Berman holds the Feldberg Professorship in Art History at Wellesley College where she teaches modern and contemporary art, the history of photography and propaganda studies. From 2010 to 2015, she held a faculty position at the University of Oslo working on the research project, *Edvard Munch, Modernism, and Modernity.* Her books, curatorial work and articles include studies of art and visual cultures of the centuries, especially in Scandinavia, as exemplified by *Luminous Modernism; In Munch's Laboratory: The Path to the Aula; Edvard Munch and the Modern Life of the Soul;* and also, *Edvard Munch and Women: Image and Myth.* She is currently curating an exhibition on Edvard Munch's photography for the Art Museums in Bergen, Norway, and working on a study of the intersections of eugenics, solar medicine, and sunbathing cultures.

Fae (Fay) Brauer is Professor Emeritus of Art and Visual Culture at the University of East London Centre for Cultural Studies Research; Honorary Professor of Art History and Art Theory at The University of New South Wales, and a Commissioning Editor for the Rowman & Littlefield International Radical Cultural Studies Series. She is an elected Fellow of the Royal Society of Arts with MA and PhD from The Courtauld Institute of Art, University of London. Her interdisciplinary publications explore the Anthropocene and Ecoaesthetics; Darwinian and Neo-Lamarckian evolution and the visual cultures of eugenics; the body, sexualities and the fitness imperative; neurology, hysteria and mesmerism; vitalism, medical humanities and occultist sciences, as well as the cultural politics of art institutions. Her books include the award-winning *Art, Sex and Eugenics: Corpus Delecti; Picturing Evolution and Extinction: Degeneration and Regeneration in Modern Visual Cultures; The Art of Evolution: Darwin, Darwinisms and Visual Culture;* and *Rivals and Conspirators: The Paris Salons and the Modern Art Centre.*

Anthea Callen is Professor Emeritus of the University of Nottingham and also The Australian National University. An art historian, painter and art expert with a television and documentary profile, Callen is the author of many scholarly books including *Renoir* (Olympic Marketing Corp, 1978); *Women Artists of the Arts and Crafts Movement, 1870-1914* (Pantheon Books, 1979); *The Spectacular Body: Science, Method, and Meaning in the Work of Degas* (Yale University Press, 1995); *The Art of Impressionism: Painting Technique & the Making of Modernity* (Yale University Press, 2000); *Techniques of the Impressionists* (Chartwell Books, 2002), followed by her book on landscape painting, *The Work of Art: Plein Air Painting and Artistic Identity in Nineteenth-Century France* (London: Reaktion Books, 2015). Her most recent book examining the sexual politics of representing the modern male body, *Looking at Men: Art, Anatomy and the Modern Male Body,* was published by Yale University Press in 2018. She was a Leverhulme Emeritus Fellow from 2016 to 2018.

Serena Keshavjee is a Professor at the University of Winnipeg in the History Department, where she coordinates the Curatorial Practices stream of the Cultural Studies Masters program. She is a Co-Applicant on *The Space Between Us,* a Canadian Social Sciences Humanities Research Council (SSHRC) Partnership grant, exploring public space and inclusivity. Most of her publications focus on the intersection of art and science in visual culture. With Fae Brauer, Keshavjee co-edited *Picturing Evolution and Extinction: Regeneration and Degeneration in Modern Visual Culture* (Cambridge Scholars Press, 2015), and in 2009, edited a special issue of *Canadian Art Review (RACAR)* on "Science, Symbolism and Fin-de-Siècle Visual Culture" (no. 34, vol. 1, 2009). Her chapter on Emile Gallé's *Pasteur Vase* was published in *Symbolist Roots of Modern Art* (eds. Michelle Facos and Thor Mednick, Ashgate Press 2015), and her article outlining Camille Flammarion's relations with the French Symbolists appeared in a special issue of *Aries* (13 (1) (2013): 37–69). Keshavjee's current research projects, funded by a 2019 SSHRC grant, include the forthcoming book from the University of Manitoba Press, *Photographing Ghosts*, and the exhibition, *Undead Archive,* based on the scientific photographs of ectoplasm taken by Dr T. G. Hamilton and his wife Lillian Hamilton during the 1920s and 1930s in Winnipeg, Canada.

David S. Mather is an historian of late nineteenth-century and early twentieth-century European art, media, and science, centering on ideas of visuality. In 2013–14, he was the inaugural Andrew Mellon Postdoctoral Fellow in The MIT's Architecture Department (History, Theory, and Criticism Program) in conjunction with the Center for Art, Science, and Technology. He co-edited the volume, *Experience: Culture, Cognition, and the Common Sense* (The MIT Press, 2016). Prior to that, he completed a postdoc at the Getty Research Institute in Los Angeles. For several years, he has taught courses and mentored undergraduate and graduate students at Stony Brook University on Long Island, New York. His book-length study entitled *Futurist Conditions: Imagining Time in Italian Futurism* was published in 2020 by Bloomsbury Academic (London).

Donna Roberts received her PhD from the Department of Art History and Theory at the University of Essex on the *Grand Jeu* and the Czech painter Josef Šimá, after which she conducted researched as a post-doc at the Instituto de Investigaciones Estéticas, UNAM, Mexico City on Surrealism, nature, and politics. She has published on the *Grand Jeu,* Czech Surrealism, the writings of Roger Caillois, and Surrealist engagements with evolutionary theories and the natural world. Currently a researcher at the University of Helsinki, she is working on the interdisciplinary group project entitled *Science, Literature, and Research: Avant-Garde Encounters with Biology and Ecology.* Her forthcoming publications include a co-edited anthology on Surrealism and ecology and a monograph on the relations between Surrealism, natural history, evolutionary theories, and ecological thought.

Pascal Rousseau is Professor of Contemporary Art History at the University of Paris 1, the position he has occupied since 2011. Previously he held the positions of Professor at the University of Lausanne (UNIL) and Visiting Professor at the University of Geneva (UNIGE). A specialist in historical avant-gardes linking the arts, sciences, and technical cultures, he has curated numerous exhibitions in France including *Robert Delaunay. From Impressionism to Abstraction,* Centre Pompidou, Paris, 1999; *The Origins of Abstraction,* Musée d'Orsay, 2003; *Pasteur's Spirit*, Institut Pasteur, 2010; *Cosa Mentale. Art and Telepathy in the Twentieth Century*, Centre Pompidou, Metz, 2015, while also curating such exhibitions beyond France as *Robert and Sonia Delaunay,* Fundacion Thyssen Bornemisza, Madrid, 2003. His groundbreaking book, *Under the influence, Hypnosis as a New Medium: 100 Notes, 100 Thoughts,* was published in the 2012 Documenta Series 080, Hatje Cantz, followed by his award-winning *Hypnose: Art et hypnotism de Mesmer à nos jours* (Beaux-Arts de Paris et Musée d'arts de Nantes), which received the 2021 Pierre Daix Prix de Livre d'Art. Rousseau is also the author of over 100 articles in magazines, journals and exhibition catalogues, including numerous contributions on early Modernism and the beginnings of abstraction; *intermediality* (synesthesia and the total work of art) and the links between art, cognitivism, and experimental psychology.

Pat Simpson is Reader in Social History of Art and Research Tutor at the University of Hertfordshire School of Creative Arts. She specializes in Russian/Soviet and post-Soviet art and has published in a range of anthologies and scholarly journals relating to the fields of the histories of art and the histories of bioscience. Her interdisciplinary research project, "Art and Soviet Bioscience", has been funded by grants from the Arts and Humanities Research Board (AHRB) and the British Academy. Her most recent publications include "Revolutionary Evolution in Apes and Humans in the 1920s: Sculpture and Constructs of the New Man at the Darwin Museum Moscow", *The Art and Science of Making the New Man in Early Twentieth Century Russia,* eds. Nikolai Krementsov and Yvonne Howell (New York, Oxford, New Delhi, Sydney: Bloomsbury Academic, 2021); "Lysenko's 'Michurinism' and Art at the Darwin Museum, Moscow 1935-1964", *The Lysenko Controversy as a Global Phenomenon,* eds. William DeJong-Lambert and Nikolai Krementsov (Cham, Switzerland: Springer Nature, 2017) and “A Cold War Curiosity? The Soviet Collection at the Darwin Memorial Museum, Down House, Kent”, *The Journal of the History of Collections*, vol. 30 (November 2018).

Brandon Taylor is Professor Emeritus of History of Art at the University of Southampton, and a visiting D. Phil. Tutor at the Ruskin School of Art, University of Oxford. He has held research appointments at the Paul Mellon Center for British Art at Yale University, the Getty Research Institute, The Institute of Art History of the Czech Academy of Sciences in Prague, the Slovak Academy of Sciences in Bratislava, and at the Hans Arp and Sophie Taeuber-Arp Foundation in Berlin. His many books include his two-volume, *Art and Literature under the Bolsheviks (1991; 1992); The Art of Today (1995); Art for the Nation: Exhibitions and the London Public* (1999); *Collage: The Making of Modern Art* (Thames & Hudson, Ltd., 1994); *Art Today* (Laurence King, 2004), and *Sculpture and Psychoanalysis* (Taylor & Francis Ltd., 2006; Routledge, 2016). Among his most recent books are *After Constructivism* (Yale University Press, 2014), *St Ives and British Modernism* (Chichester: Pallant House, 2015), *The Life of Forms in Art: Modernism, Organism, Vitality* (Bloomsbury Publishing, 2020), and *Make It Modern: A History of Art in the Twentieth Century* (Yale University Press, 2022).

Introduction

Vitalizing Energies, Science, Creativity and Evolution

Fae Brauer

Today energy is not just a keyword: It's a science and a culture. The enhancement of individual energy is conceived as integral to physical fitness, well-being, efficient performativity and mental health to be achieved by healthy lifestyles, nutritious diets, vigorous exercise and modern sporting cultures alongside such movement cultures as Aerobics, Yoga and Pilates. While energy cultures in residential and working environments have become significant in businesses, organizations, schools and universities, they have also become a top priority for many individuals and those concerned with sustainability transitions. Vital to sustaining this energy culture is its production, consumption and preservation, particularly in the face of Climate Change and ecological disasters. Yet this concern with energy is not new.

It can be traced to the discovery of increasing degeneration and depopulation in the mid-nineteenth century amidst the discourses of devolution and ecological catastrophism. It can be correlated to the efflorescence of evolution theories to regenerate populations through populationist organizations, including eugenics. It is highlighted by the efflorescence of modern sporting cultures including gymnastics, boxing, cycling, swimming, diving, rowing, dancing, eurythmics, weight-lifting and the popularization of rugby and soccer. It is encapsulated by revival of the Olympic Games in 1896, particularly by the Baron Pierre de Coubertin, and his popularization of French energies acknowledged by the book, *An Artisan of French Energy: Pierre de Coubertin*.[1] It is also encapsulated by the science of *energeticism* developed by German physicist, Georg Helm, with physical chemist, Wilhelm Ostwald. Rejecting scientific materialism, these scientists demonstrated that energy is the substrate of all phenomena with all observable changes arising as transformations of one kind of energy into another.[2]

By the fin de siècle, as Helge Krage points out, mechanistic materialism and positivist empiricism came increasingly under attack by the new investigators into energetics, radioactivity and electromagnetism, as well as by Modernist artists.[3] In 1895, Ostwald gave a programmatic address in which he argued that energetics would overcome the inherent limitations of scientific materialism to become the scientific world view of the future. "The most promising scientific gift that the closing century can offer the rising century", he declared, "is the replacement of the materialistic world view by the energeticist world view".[4] Dynamically, he concluded, "Do not squander energy. Utilize it!"[5] Energy was also conceived as integral to evolution, as illuminated by Charles Darwin and other evolutionary theorists.

Elaborating his theory of "natural selection", Darwin had linked evolution to competition between tribes and races, the fittest being the ones that generally survived.

DOI: 10.4324/9781003045595-1

Their fitness did not just depend upon their fertility, numbers and what Darwin called "the grade of their civilization" but their good health, vigour and energy.[6] Following the legacies of Darwin and Georg Hegel after the humiliating defeat of France in the 1870 Franco-Prussian War, the French philosopher Hippolyte Taine had insisted that it was time for his generation to judge itself according to "race-milieu-moment": Its own time, its environment and most of all, its energy and fitness as a race.[7] Across America, Asia, Australia, Britain, Canada, Europe and Russia, there was an increasing concern with generating energies through fitness and healthy environments entailing hygiene and the embrace of natural elements, particularly exposure to unpolluted air, water and sunlight, as illuminated by Chapters 1 to 3 in this book. Regeneration of the body and its energies was also to be propelled through the practice of such "modern" sports as rowing, yachting, swimming, diving, body-building and weight-lifting, as revealed in Chapters 1 and 2, as well as by rugby and soccer, as illustrated by the Modernist artists examined in Chapters 3 and 6. For many artists, especially Modernists, these new energies did not just entail physiological invigoration.

Following research into neurology, psychology, psychiatry and psychoanalysis linking unconscious emotions with free-energy, these new energies also entailed psychological liberation.[8] In separating the study of the brain from the body around 1650, Thomas Willis is invariably identified as the founder of neurology, alongside the Swedish inventor, scientist, theologian and visionary, Emanuel Swedenborg. Due to the investigations conducted from the late eighteenth century by the French comparative anatomists, François-Xavier Bichat and Félix Vicq d'Azyr, plus the Scottish anatomist and surgeon, Sir Charles Bell, the study of neurological anatomy and pathology began to converge. After further explorations of neurology and pathology, particularly by Pierre Briquet who in 1859 published his epidemiologic study of 430 cases of hysteria, the École et Clinique des maladies du système nerveux was founded in Paris at the hôpital de la Salpêtrière at the beginning of the French Third Republic under the professorship of Jean Martin Charcot. Joined by Charcot's former students, Pierre Janet, Gilles de la Tourette, Paul Richer and Joseph Babinski, for the next 20 years, hysteria, neurological and dissociative disorders were psychoanalysed and documented at Salpêtrière in drawings, paintings, photographs and in its anatomo-pathological museum.[9] Amidst the prevalence of hypnosis, especially its medical demonstrations in Charcot's "theatres" of hysterical patients, both Henri Bergson and his close friend, Pierre Janet (with whom he had planned to study medicine) experimented with it. Janet focused upon the notorious magnetist and clairvoyant somnambulant hysteric, *Léonie*.[10] While Bergson taught psychology and neurology from Janet's *Traité élémentaire de philosophie,* Janet drew upon Bergson's *Matter and Memory* to develop his thesis on "psychic automatism" as "the involuntary exercise of memory and intelligence", and to explore how the concept of dissociation from psychological trauma and "traumatic memory" entailed the progressive loss of psychological energy.[11]

After studying neurology with Charcot at Salpêtrière in 1885, one year later Sigmund Freud opened his private practice in Vienna, publishing with Dr Josef Breuer *Studies on Hysteria* in 1895. Connecting hysteria to traumatic repression and unconscious energies, another year later Freud defined his techniques to identify its sources as psychoanalysis. Four years later, he published *The Interpretation of Dreams,* referenced in Chapter 8, and in 1910 founded the International Psychoanalytical Association. In 1874, Wilhelm Wundt had published the first textbook on experimental psychology and five years later, opened the first experimental psychology laboratory at the University of Leipzig to investigate unconscious energies.[12] In 1882, the Society for Psychical Research had been founded in London followed by the Neurological Society of London four years later. A spate of psychology publications followed including John Dewey's first American textbook on psychology, William James' *The Principles of Psychology* and Havelock Ellis' *Sexual Inversion,* to name but a few key texts. Hence the physiological regeneration of the body and its energies was conceived as inextricably intertwined with liberation of the unconscious through neurology, psychology and psychoanalysis. The psychological energies to be released by this process were conceived as entailing the emancipation of creative energies, particularly through the unleashing of raw emotions, feelings, instincts, intuition, memory, fertility and virility. This fusion became identified with the modern concepts of Vitalism that proved paramount to Vitalist Modernists ranging from Hugo Ball's Dadaist performances demonstrated in Zurich and Salvador Dalí's exploration of his paranoiac critical method to unleash his memories and unconscious in Paris, to Kazimir Malevich's Suprematism in Moscow, as is revealed by Chapters 7, 8 and 10 in this book.

While Vitalism has a long history, modern Vitalism became extensively theorized in America, Britain and Europe, particularly France and Germany during the late nineteenth and early twentieth century. Despite Bergson becoming one of the most popular philosophers of Vitalism, by no means was he alone in his theorization. This is demonstrated by the Vitalisms researched by Hans Driesch – the zoologist who had studied with Ernst Haeckel – alongside Friedrich Nietzsche and George Bernard Shaw, who used the same title of *Creative Evolution* as Bergson although with very different meanings. To elucidate their modern vitalist theories, particularly their similarities and divergences, the different conception of modern Vitalism by these theorists is charted in the next section of this *Introduction,* before the relationship of their Vitalisms to Modernism is explored in my survey of the chapters in this book.

Philosophizing Modern Vitalism: Bergson, Driesch, Nietzsche and Shaw

Bolstered by the philosophies of Vitalism by Bergson, Driesch, Nietzsche and Shaw, such concepts as life-force and the vital force, *l'élan vital* and *creative evolution* became widely and diversely engaged by Modernist artists and writers. This entailed the conception of life as a constant process of metamorphosis impelled by the free flow of energies, undisrupted by mechanistic materialism, to generate what Bergson aptly called *L'évolution créatrice,* the title of his fourth book published in 1907.[13] Some fourteen years later Shaw appropriated the term in English, *creative evolution,* although with distinctly different meanings. Imbricated within Transformist ecological evolutionary theories, Vitalism was embraced for being anti-mechanistic and materialistic, anti-positivist and anti-rationalist, particularly in its opposition to Thomas Huxley's conception of plants and animals as machines. Vitalism entailed a reconception of organic life as inspiring organisms within unspoiled nature, perpetually mutating into increasingly complex species and solidarist colonies following the Transformist concept of "life-force". While these philosophies of Vitalism were relatively new, they had long roots as revealed by Nietzsche's treatises, Driesch's lectures and his book, *The History and Theory of Vitalism,*[14] as well as the lectures and philosophies published by Bergson.

In *The Twilight of the Idols,* Nietzsche had argued passionately for the need to re-evaluate the idols of theology, spurn positivism, reject technologized modernity and ascend "high, free even terrible nature".[15] "Our whole modern world is caught in the net of Alexandrian culture and recognizes as its ideal the man of theory", lamented Nietzsche, "equipped with the highest cognitive powers, working in the service of science and whose archetype and progenitor is Socrates".[16] Nietzsche's reconception of the artist as Apollonian and Dionysian, a vessel of energy rather than neurasthenic, in contact with raw nature and their primal instincts and vitalized by the life-force, was welcomed by many Modernists. This is illuminated in Chapter 2 by Edvard Munch's murals for the newly built Aula, Festival Hall illustrated by Figures 2.1, 2.2a to 2.2e, and 2.7 to 2.9. Following Nietzsche's conception of the *Übermensch* in *Thus spake Zarathustra*, it is the energized "overman" able to create new values that emerges in these Vitalist Modernist murals. Reconceived as *Superman* in Shaw's 1903 play, *Man and Superman,* the *Übermensch* is referenced in Chapter 6 on Vitalist Futurism and features in Robert Delaunay's 1913 painting, *L'Équipe de Cardiff,* in Chapter 3, as the heroic French overman able to soar above all other players in the rugby lineout in order to catch the ball (Figure 3.1).

For the German biologist, Driesch, who had studied with Haeckel, Vitalism was conceived in relation to Entelechy as a vital force guiding the evolution of organisms. In his Gifford Lectures delivered in Aberdeen in 1907-8 and his London Lectures on Vitalism in 1913, Driesch traced what he called "the old Vitalism" to Aristotle's *Metaphysics* and *De Anima,* particularly his bio-philosophy of *entelecheia* entailing different stages of the soul.[17] "Aristotle's theory of life is pure vitalism", Driesch maintained, "and I may call it primitive or naïve vitalism for it arose from an entirely impartial contemplation of life's phenomena".[18] While Driesch highlighted the relationship of Vitalism to evolution and epigenesis through the theories of William Harvey and Georges-Louis Leclerc (Comte de Buffon), he did not explore its development in the eighteenth century at the Montpellier School of Medicine by François Sauvages de Lacroix, Théophile de Bordeu and Paul-Joseph Barthez into a *principium vitale.*[19]

Rejecting the body-machine concept, the Montpellier Animists postulated a distinction between living and other matter. It was, in fact, at Montpellier that the physiologist and vitalist, Marie-Jean-Pierre Flourens, had trained, as Barbara Larson points out, and whose theory of the "nœud vital" became Paul Gauguin's model for the soul.[20] By 1780, Montpellier vitalist therapies and those of physiologist, Marie François Xavier Bichat, regarding vital animal properties and the "milieu interieur", extended to Franz-Antoine Mesmer's theories and practices of animal magnetism.[21] The linkage that was forged between mesmerism and Vitalism then provided, according to Elizabeth A. Williams, a viable alternative to the Paris Clinical School,[22] as illuminated in Chapters 4 and 5. With the development of the microscope and observations of living tissue, medical vitalists and mesmerists considered that they were then able to furnish empirical evidence that the phenomena of life could no longer be explained by the laws of mechanics.

Despite these vitalist-mechanist debates prominent in medicine, Driesch confined his exploration to Immanuel Kant and what he called "the vitalism of the Nature-philosophers", including Arthur Schopenhauer whom he posited as bringing an end to "the old Vitalism".[23] "Modern Vitalism", as he called it, was then identified by Driesch as an Anti-Darwinian theory of descent, his scientificized theory of autonomy being correlated with biology and "universal teleology".[24] Extending this theory to the concept of *becoming*, which became so significant for Bergson, Driesch related it to "the so-called stream of consciousness", the term introduced by William James in his *Principles of Psychology*.[25] "What if *becoming* could be formulated as if an earlier phase of it were always the reason of a later phase", Driesch speculated, "and a later phase the consequence of an earlier one? If this were possible, then we might claim to understand *becoming*, to have rationalised it".[26] "Nature", he then concluded is "the proper field of a theory of *becoming*".[27]

To avoid falling into mechanistic traps that he detected in Darwinian and Lamarckian theories, Driesch posited Entelechy as the vital force within a non-material and non-spatial process of *becoming*.[28] From 1905, Driesch's theorisation of Entelechy extended to occult Vitalism, particularly psychic phenomena and parapsychology, which included hypnotism, levitations, phantoms and telekinesis – the movement of objects without human contact. As Serena Keshavjee reveals in Chapter 4, while Driesch was writing *The History and Theory of Vitalism*, he attended séances with the German physician, psychiatrist and psychical researcher, Albert von Schrenck-Notzing and considered that protoplasm could be expelled out of the body into an endless production of creative forms. When asked to lecture at Cambridge University, Driesch met Henry Sidgwick who intensified his interest in psychic phenomena and motivated him to become a member of the Society for Psychical Research of London in 1913 – the very year that Bergson was elected its President. So extensive and intensive did Driesch's psychical research become after the First World War that in 1926, Driesch was elected President of this Society. That year, he published *Psychical Research and Established Science*, as part of his Presidential address, followed by *Psychical Research and Philosophy*.[29] In 1931, he published a methodology of parapsychological research and two years later, *Psychical Research: The Science of the Super-normal*.[30] Closely associated with British physicist, Oliver Lodge, and Schrenck-Notzing, also examined in Chapter 4, as well as such renowned mediums as Rudi Schneider and "Margery" – Mrs. Osborne Leonard – Driesch concluded that paraphysical phenomena represented an "enlarged" Vitalism, which he called "superVitalism".[31]

The new sciences, including psychic phenomena and "psychic states", proved integral to the Vitalism explored by Driesch's French colleague, Bergson, as illuminated in Chapter 5 and its focus upon Pablo Picasso's creation of durational portraits. "Before Bergson forged his ideas", surmises Robert C. Grogin, "intellectuals were presented with an almost standard construction of their world, a materially enclosed world which was moved by purely mechanical and mathematical laws".[32] Reconceiving time from the perspective of memory, intuition and "psychic states", in Bergson's first book published in 1889, *Essai sur les données immédiates de la conscience (Time and Free Will)*, time was explored as durational, like an unfolding melody – not discrete instants of chronometric measurement within mechanistic materialism. In his next book, *Matière et mémoire*, published in 1896, Bergson closely engaged in the new sciences. Drawing upon the work of Faraday, Maxwell and Lord Kelvin, Bergson deduced: "Matter thus resolves itself into numberless vibrations, all linked together in uninterrupted continuity, all bound up with each other, and travelling in every direction like shivers through an immense body".[33] Given Bergson's own grounding in ether physics, as Linda Dalrymple Henderson has revealed, and his friendship with Gustave Le Bon, not surprisingly Bergson's theory dovetailed with the reconception of matter as penetrable, interacting with space and in turn conceived as a tangible vibrating atmosphere.[34]

Drawing upon the life of insects, as indicated in Chapter 8, and the evolution of intelligence from simians to homo-sapiens, in his fourth book, *L'Évolution créatrice,* Bergson theorized evolution as *l'élan vital.* A vital impulse and creative force, synonymous with new inventions, Bergson likened *l'élan vital* to the momentum of a continually surging wave. Energized and propelled by this vital impulse and creative energy, the human subject was then posited by Bergson to be in a constant state of *becoming,* as illuminated by Patricia Berman in Chapter 2 and illustrated by Figures 2.2a, b, d and e. Pitted against mechanist models of being that included those of Darwin and Lamarck, as well as Monism and Weismann's theory of germ plasm, in *Creative Evolution* Bergson stipulated that "life does not evolve mechanically and rationally". As he explained succinctly in French: "Elle ne procède pas par association et addition d'éléments", he maintained, "mais par dissociation et dédoublement".[35]

The Anarchist and Socialist implications of this potent statement are unravelled by Mark Antliff in Chapter 9. Some ten years after Bergson had published *Creative Evolution,* Shaw developed his own theory of this concept inspired more by Nietzsche's concept of the *übermensch,* which Shaw identified as *Superman.*[36] In his 1921 play, *Back to Methuselah,* which he called his "Bible of Creative Evolution", Shaw endeavoured to define this concept through his seven-hundred-year-old bisexual superhuman characters who needed no sleep and who were able to accelerate the evolutionary process by centuries. Hence *creative evolution* for Shaw entailed a life-force able to accelerate the process of evolution from which a race of omniscient and omnipotent supermen could evolve. This is illuminated by his final play in 1950, *Farfetched Fables* in which indestructible humans were able to achieve incorporeality subsisting on nothing but air. Despite using a title identical to that of Bergson, Shaw's concept was then distinctly different from Bergson's *Creative Evolution.* Rather than deterministic methodologies, mechanistic control systems and finalistic models of evolution from which Shaw's super race would supposedly evolve, Bergson's *Creative Evolution* draws upon co-evolutionary interrelationships entailing the symbiosis of humans, plants and animals, as well as reciprocal interpenetration to unleash *l'élan vital.*

Seizing this vital impulse of life, endemic in nature and redolent in intuition would, Bergson suggests, enable human subjects to become self-creative and invent the new. This would entail immediate, instinctive and intuitive responses able to reach the heart of another, especially through empathy and what Bergson called "psychic states" examined in Chapter 5. It would also entail unfurling memory, intuition and "psychic states" in realising the duration of being and the perpetual flux within it as a creative process of *becoming,* as is illuminated by the artworks in Chapters 2, 4, 5, 7 and 8. Hence Bergson, concludes Stephen Lehan, "gave weight to the modernist belief that art is the highest function of our activity, and helped establish the modernist belief that the universe is inseparable from mind and that the self is created out of memory".[37] In his final chapter of *Creative Evolution* entitled "The Cinematographical Mechanism of Thought and the Mechanistic Illusion", Bergson reviewed the history of philosophical thought to detect how it had failed to acknowledge the importance of *becoming,* thereby falsifying the nature of reality by imposing static concepts and what he calls "theoretical absurdities".[38] This is explored in Chapter 6, particularly Bergson's conclusion: "The mechanism of our ordinary knowledge is of a cinematographical kind".[39]

To pursue *becoming*, Bergson insisted upon the need to "escape from the cinematographical mechanism of thought" and to acknowledge the constant interpenetration of life.[40] "In reality", Bergson maintained, "life is of the psychological order, and it is of the essence of the psychical to enfold a confused plurality of interpenetrating terms. In space, and in space only, is distinct multiplicity possible. ... I am a unity that is multiple and a multiplicity that is one".[41] Given the idiosyncratic fusion of the new sciences, particularly neurology and psychology with occultist sciences in France at this time, as well as in England and Germany, as explored in Chapters 4 and 5, these multiplicities entailed energies which, for Bergson, could be spiritual and phantasmatic, as demonstrated in Chapter 5 and by his publications anthologized in Bergson's 1919 book, *L'Énergie spirituelle*.[42] This idiosyncratic interrelationship was explored simultaneously by such scientists as Charles Richet, Camille Flammarion, Gustave Geley and Karl van Reichenbach, examined in Chapter 4, and by such magnetist scientists as Hector Durville investigated in Chapter 5. Bergson's concept of spiritual energies in relation to his concept of *durée* as the continuous flow of interpenetrating moments forging the vital impetus then seem to lend themselves to the notion of souls, spirits, phantoms and auras, as well as etheric and astral bodies taking the form of phantoms, as illuminated in both Chapters 4 and 5.

From 1900, Bergson lectured twice a week on all of these concepts, invariably without notes, at the Collège de France on Fridays at 5 pm and on Saturdays at 4 pm.[43] Simultaneously his friend Pierre Janet lectured in Psychology at the Sorbonne. Two years after Janet became Chair of Experimental Psychology at the Collège de France in 1902, Bergson was appointed its Chair of Modern Philosophy. So popular did Bergson's lectures become that students queued for hours in the surrounding streets just to hear them.[44] With publication in 1907 of *L'Évolution créatrice*, Bergson's fame skyrocketed. So popular did it become that its French publisher, Félix Alcan, issued twenty-one editions by 1918. Mythologized as a great artist, Bergson was called "the Corot or the Vermeer of the interior universe".[45] Whenever Bergson lectured at the Collège de France thereafter, the press was quick to report attendance by the prestigious "Five o'clock Bergsonians" alongside women who flocked to its doors and the weekly riots that erupted.[46] Cartoons showed members of the public precariously perched on ladders atop windows of the main auditorium at the Sorbonne just to spy him. So fashionable did Bergson become that the celebrity portraitist, Jacques-Émile Blanche, eagerly painted his portrait in 1911 (Figure 5.3). So *à la mode* did it become to attend Bergson's lectures that by 1910, Bergson's Room VIII at the Collège de France Amphitheatre achieved notoriety as one of the most "elegant places" in Paris with the Collège de France becoming known as "the house of Bergson".[47] Even in New York in 1913, his lectures caused a traffic jam on Broadway. In anticipation of his lectures, Columbia University compiled *Contribution to a Bibliography of Henri Bergson*, with an introduction by their philosophy professor, John Dewey, which was presented to Bergson on his arrival in New York City on 2 February 1913.

Attracting major international awards, in 1911 Bergson was invited to give the Huxley Memorial Lecture, which he entitled in English, *Life and Consciousness*, two years later being also invited to give the Gifford Lectures at the Universities of Aberdeen and Edinburgh. Following the metaphysical dimensions of his philosophy, Bergson was also appointed President of the British Society for Psychical Research. In his presidential speech, "Phantasms of the Living" and "Psychical Research", Bergson surmized: "There is, present and invisible a certain metaphysic unconscious of itself – unconscious and therefore inconsistent, unconscious and therefore incapable of continually remodelling itself on observation and experience as every philosophy worthy of the name must do".[48] To emphasis the resonance and sustenance of "l'élan vital" within this "métapsychique" reality, Bergson succinctly concluded, "it's organisms that die, not life".[49] Subsequently lecturing on this subject in France under the title, *L'Âme et le Corps*, Bergson's publication of this series as essays, investigated in Chapter 5 of this book, was entitled *L'Énergie Spirituelle: essais et conferences.* Translated into English as *Mind-Energy,* these ideas were explored extensively at the Institut Métapsychique International in Paris from its inception in 1919, as indicated in Chapter 4, by both Bergson and his close friend, the Noble Prize-winning physiologist, Charles Richet.[50]

By no means secluded from the First World War, in December 1914 Bergson's Presidential address to the Académie des Sciences Morales et Politiques, elucidating how metaphysical philosophy and free will could overcome materialist oppression, was subsequently translated and published as *The Meaning of the War: Life and Matter in Conflict.* By January 1917, he also became involved in international diplomacy. Due to his fluency in English gleaned from his English mother and his formative years in England, Bergson was enlisted by the French Government to help broker a deal with President Woodrow Wilson. This entailed France and Britain supporting the creation of a League of Nations dedicated to maintaining world peace after the First World War if the United States entered the war on the side of the Allies. Successfully negotiated, Bergson was appointed President of the League's International Committee for Intellectual Cooperation, a position he retained until 1925 – much to the chagrin of André Breton and some other Surrealists, as is elaborated in Chapter 8 of this book.

Following Albert Einstein's visit to the French Society of Philosophy in 1922 and his response to a lecture by Bergson, Bergson published *Durée et simultaneité apropos de la théorie Einstein* in which he pointed out their differences, as was highlighted by Maurice Merleau-Ponty.[51] Despite "Bergsonmania" waning later during the 1920s, especially after Bergson's retirement from the Collège de France, *L'Évolution créatrice* was awarded the Nobel Prize in 1927. Sadly by this time Bergson was too crippled with rheumatoid arthritis to travel to Stockholm to accept it. Nevertheless he continued to be honoured and published, three years later being awarded France's highest honour, Grand-Croix de la Legion d'honneur, followed by publication of his book, *The Two Sources of Morality and Religion,* in which he explored the *mécanique/mystique* duality: The experiences of mysticism and mechanicism.[52]

On the Nazi Occupation of France, Bergson wrote on 1 July 1940: "I have seen this coming for several years now. We have touched the bottom of the abyss. At least now we know what is evil".[53] In December 1940 aged 81, Bergson declined the Vichy government's offer of exemption from Nazi regulations requiring the resignation of all Jewish employees from State positions. Offering solidarity with the suppression of the Jews, Bergson resigned his position from the Collège de France and registered at the Gendarmerie as "Academic. Philosopher. Nobel Prize Winner. Jew". Despite being seriously ill, he left his sick-bed to queue with other Jewish people to receive the Yellow Star required of all those with Jewish heritage. Contracting pneumonia, Bergson died on 3 January 1941. After his death in Nazi occupied Paris, Bergson's philosophies were explored most notably by Georges Canguilhem, Merleau-Ponty, Jean-Paul Sartre and Gilles Deleuze during the 1950s. Yet most of all it was Deleuze's *Bergsonism* published in 1966 in which Bergson's intuition and memory were re-examined as an interpenetrating concept of *becoming* in which the past co-existed with the present that gave, in the words of Craig Lundy, "new life to Bergson's thought".[54]

So overwhelming was its impact during Bergson's lifetime that many Modernists identified Vitalism with Bergsonism, as Brandon Taylor illuminates in his exploration of Dada and Neo-Dada in Chapter 7. This is endorsed by the quotation provided by Roberts in Chapter 8 extrapolated from Stephen Lehan: "If the moderns did not have Bergson, they would have had to invent him".[55] Yet while the dominance of Bergsonism and Bergson's philosophies are examined in Chapters 5, 7, 8 and 9, the Modernists who pursued the vitalist philosophies of Nietzsche are explored in Chapters 2, 3 and 6. How integral Driesch's vitalism became to the nexus between the "psychic force" and the "vital force" is illuminated in Chapter 4.

Despite the rapid growth of interest in the interrelationship of these philosophies, sciences and energies to Modernist artists, this is the first major exploration of their different engagements with Vitalism.[56] That there appear three main dimensions to *Vitalist Modernism,* arising from these philosophies, sciences and energies in relation to Modernist art is signified by the organization of the chapters of this book into three main parts. Posited against positivist empiricism, mechanism and rationalism, the first part of this book reveals how Vitalism at the fin de siècle became correlated with the pursuit of raw nature, wholesome environments, aquatic therapies, heliotherapy, modern sports, water sports, physical culture and eurythmics to energize the human body and, in so doing, vitalize its life force. The second part of this book explores how Vitalism became simultaneously aligned with anthroposophy, esotericism, magnetism, mesmerism, mysticism, occultism, parapsychology, spiritism, theosophy and what Bergson called "psychic states", alongside such new sciences as electromagnetism and radiology. During and after the devastation of the First World War, the third part examines how Vitalism, particularly Bergson's theory of *becoming,* became associated with Dadaist, Neo-Dadaist and Surrealist notions of amorality, atemporality, dysfunctionality, entropy, irrationality, inversion, negation and the nonsensical, alongside John Cage's concept of *Nothing.* After investigating the widespread engagement with Bergson's philosophies amongst Anarchists, Marxists and Communists concerned with Vitalism and art during and after the First World War, the concluding chapter reveals the official rejection of Bergson and any form of Vitalism in the Soviet Union under Stalin. Yet it also reveals how Vitalism was smuggled back into Soviet culture during the Great Patriotic War through art. How this happened is elucidated in the following summary of the three parts, ten chapters and Modernist artists in their relationship to these multiple dimensions of Vitalism.

BioVitalism: Corporeal Regeneration, Environmental Purification and National Evolution

Faced with increasing depopulation, corporeal degeneration and rampant diseases in *The Sick City* and "a queasy sickening feeling that all was not right", many Modernists expanded the field of art into raw nature, ethnic communities, modern sport and physical cultures.[57] Disturbed by toxic industrial environments with rapidly inclining suicide rates, accelerating alcoholism, rampant syphilis, degenerating bodies and the prospect of national devolution, such Modernists as diverse as Frédéric Bazille, Paul Cézanne, Gustave Caillebotte, Henry Scott Tuke, Thomas

Eakins, Edvard Munch, Robert Delaunay and Albert Gleizes sought to picture the vitalization of energies through regeneration of the body. They did so through their imaging of such modern sports as swimming, diving, rowing, yachting, rugby, soccer and the new physical culture. This is explored in the first part of this book in which *BioVitalism* is the focus of its first three chapters by Anthea Callen, Patricia Berman and Pascal Rousseau.

In Chapter 1, Callen illuminates how manly vitality, virility and hygiene were perceived to arise through French male bodies becoming immersed in water sports. The concern with water throughout the nineteenth century was, Callen points out, not just as a source of hygiene and cleanliness. Following the new hygiene regimes, public baths were revalued for cleansing the body. Rivers, lakes and seas became highly valued as a locus for not just cleansing the body but invigorating it in raw nature, particularly through sailing, rowing, swimming and diving. Designed to counter degeneration in modern urban society and the prospect of devolution, Callen explores how water sports became integral to the generation of a body culture able to achieve the "vital life force". More specifically, Callen focuses upon how water-sports and hygiene became instrumental in virilizing, vitalizing and homoeroticizing the modern male body, as illustrated by her examples of Gustave Caillebotte's and Georges Leroux's naked male divers (Figures 1.6 and 1.7), Caillebotte's muscular naked man drying himself after his bath (Figure 1.8), Frederic Bazille's naked "wild bathers" (Figure 1.9), and Thomas Eakins and Caillebotte's rowers (Figures 1.10 and 1.11).

In the following chapter, Berman reveals how and why Munch's Nordic female and male bodies, which appear just as muscular, healthy and hygienic as those of Bazille, Caillebotte and Leroux, were meant to appear virilized and vitalized through sunshine, sea-water and exercise in order to regenerate the Nordic race. This is epitomized by the murals that Munch created between 1909 and 1916 for the newly built Aula (Festival Hall) in Norway's Royal Frederiks University, now the University of Oslo. Designed to capture, in Munch's words, "the great forces that govern humankind", they illuminate the University's research into electromagnetism, heliotherapy, open-air athleticism and sky science.[58] In Munch's paintings of the naked Norwegian men and women flanking the central mural of the sun, the solar rays are pictured as so energizing that their bodies appear to be brought to life by them (Figures 2.2a, b and e) while embryos seem to germinate within them (Figure 2.2d). Drawing upon Gilles Deleuze's elaboration of Bergson's theory of creative evolution, Berman perceives that the multiple contours with which these bodies are rendered make them appear as if they are in a radial state of *becoming*.[59] At the central core of these murals, Munch captured *The Sun* as the dominant life-giving force and ultimate signifier of Vitalism – the antithesis to fin-de-siècle decadence, pollution and decay. This is why *Munch's mural of The Sun* was chosen for the cover of this book.

For such artists as Caillebotte and Munch, *Biovitalism* could be explored beyond the fringes of such modern art centres as Paris, Oslo and Berlin. It could also be explored amongst seemingly untouched and uncontaminated environments that better represented an organicist concept of Vitalism. This is illustrated by the energizing air, light and water of Caillebotte's summer home on the Yerres River south of Paris and the estate where Munch lived and worked for some 28 years, Ekely at Skøyen. For such early twentieth century German Expressionists as Albert Bloch, Fritz Bleyl, Erich Heckel, Ernst Kirchner and Karl Schmidt-Rottluff who formed *die Brücke* and who were propelled by Nietzsche's Dionysian Vitalism, unspoilt nature became a source of biovitalism, particularly the countryside by the Moritzburg lakes and the Island of Fehman in which they could immerse their mostly naked bodies over summer from 1907 to 1911.[60] Yet for a new generation of young Modernists such as Robert and Sonia Delaunay, Picasso, Umberto Boccioni, Tristan Tzara and Kurt Schwitters, the source of new vitalizing energies they felt pulsating through their veins was the modern metropolis. From 1900, Paris seemed to have become the vitalizing centre, as illuminated by Chapters 3 to 5. Nevertheless the French capital was not without problems of degeneracy and devolution.

Despite the widespread pursuit of modern energizing sports in France since the Franco-Prussian War and the Commune, Paris still appeared to be so plagued by depopulation and degeneration by 1898 that renowned physiologist, Charles Richet, questioned whether France would perish.[61] For Baron Pierre de Coubertin, rugby was the panacea. Despite his revival of the Olympic Games, no other sport could revitalize the French body and its energies moreso than rugby, according to Coubertin, as exemplified by its inspirational practice at the English School of Rugby.[62] By 1892, the first international rugby match was played in Paris. Nearly twenty years later, after beating the strongest rugby international team, the Scots, the French Rugby team achieved their first major victory. When due to play their next match against the Scots in Paris in 1913, unsurprisingly it received sensationalist newspaper and billboard coverage which inspired Robert Delaunay's series of paintings, *L'Équipe de Cardiff* (Figure 3.1). Boosted by this new cult of action and Coubertin's Internationalist Vitalism, plus Bergson's and Nietzsche's philosophies, in Chapter 3 Pascal Rousseau reveals how Delaunay simulated the chromatic sensation of strident billboards and colourful rugby guernseys at football stadiums in his paintings to generate the intense experience at rugby matches of vitalist sensations.[63] In his comparison of Delaunay's painting with *Les Joueurs de football* by Albert Gleizes (Figure 3.2), Rousseau also reveals why Delaunay abandoned Gleizes' vitalist group dynamic for the Vitalism of heroic individualism epitomized by Nietzsche's "overman".[64] Yet for Delaunay's Paris-based Spanish contemporary, Picasso, Vitalism took a very different course (Figure 5.1), as is revealed in the second part of this book.

Occultist Vitalism: Magnetism, Parapsychology, Spiritism and Theosophy

Some "Metrovitalists" pursued Vitalism through Occultism, particularly its relation to anthroposophy, magnetism, parapsychology, spiritism and theosophy. At the same time many of those pursuing Occultism and the esoteric facets of Vitalism embraced the new sciences, particularly electromagnetism, Roentgen's X-rays and radiology, wireless telegraphy and theories of the luminiferous ether through which electromagnetic waves were perceived to travel. Given the flow of knowledge between the new sciences and occult sciences, both were regarded as vitalisers of an invisible energy that could be physically, emotionally and spiritually transporting, as well as creatively liberating. "Rather than occultism being on the fringe of culture in the late 19th and early 20th century", as Linda Dalrymple Henderson succinctly surmises, "the occult was closely connected to the newest developments in science in a period when the two fields were not seen as so clearly demarcated as later in the 20th century".[65] This is the focus of the second part of this book.

Throughout the nineteenth century, the new research into science and medicine did not necessarily exclude magnetism, mesmerism, hauntings, phantoms and ghosts in what became known as the psychic sciences. In 1855, Fellows at Trinity College, Cambridge, began to discuss psychic sciences and paranormal phenomena, including ghosts. Seven years later, The Ghost Club was formed with such members as Charles Dickens. Dissolved in 1870 following the death of Dickens, The Ghost Club was revived on All Saints Day in 1882. This was the very year that the Society for Psychical Research (SPR) was incepted. Four years later, one of the earliest exhibitions dedicated to ghost imagery opened at St. James Hall in London. Entitled *The Phenomena of Materialization,* 28 images of ghostly materializations were displayed in this exhibition.[66] These included, as Keshavjee illustrates in Chapter Four, James Tissot's celebrated mezzotint, *Apparition Mediunimique* [sic], *Dark séances d'Eglinton du 20 May 1885, Londres* (Figure 4.3) showing the materialized ghost of Tissot's deceased lover, Kathleen Newton, with a "Spirit Guide". Simultaneously Spiritualism grew rapidly with visualization of apparitions by spirit photographers including John Gerard Keulemans' "miniature face forms" emerging from excreted ectoplasm as illiustrated by Figures 4.4 to 4.6. As Florence Raulin-Cerceau surmises, it became widely assumed that life could begin without the input of an external agent.[67]

With comparable research and visual cultures emerging in Paris, a parallel may be drawn between these two centres. However, it was in Paris at the fin de siècle, as Keshavjee reveals, that "the vital force" and "the psychic force" were being extensively investigated in relation to the unconscious. This was happening particularly through hysterical trances and psychic phenomena, as demonstrated by spiritualists, mediums, séances and the excretion of ectoplasm. Hence, while Ostwald was exploring "energetism", the "idea that energy is the substrate of all phenomena", Dr Charles Richet and Camille Flammarion were conducting extensive research into the psychic sciences of Vitalism that long preceded Bergson's publication of *Creative Evolution.* Seven years before Bergson theorized Vitalism in *Creative Evolution*, Keshavjee points out, "Flammarion had widely disseminated his teleological Vitalist theory of evolution whereby the law of progress ... regulates all life".[68] It was Flammarion's concept of a directing dynamism moreso than Bergson's "l'élan vital" that Keshavjee considers was appropriated in 1919 by Dr. Gustave Geley when he became Director of the Institut Métapsychique International of Paris. Seminal to their research were visualizations of these concepts of the vital force that shifted, following Keshavjee's examples, from glowing luminous energy to mechanical forces. This is illuminated by extensive photographs of the performative medium, Eva Carrière, excreting whitish, jelly-like material alongside Juliette Bisson's and Schrenck-Notzing's magnesium flash photographs of ectoplasm seemingly directed by a vital force illustrated in Figures 4.9 and 4.10.

What is sometimes overlooked are these extensive and intensive artistic and culturo-scientific explorations of occultisms being conducted in *avant-guerre* Paris. Yet from the time that Picasso moved in with the poet, Max Jacob in 1901, Jacob acted as a conduit or point of mediation for Picasso elucidating alchemy, astrology, chiromancy, magic, palmistry and the Tarot. Picasso was also exposed to magnetism, mediumism, mesmerism, mysticism, spiritism and Bergsonism, the creative outcome of which is explored in Chapter 5. In fact, the very year that Picasso began exploring what is now called Cubism, Bergson's *L'Évolution créatrice* was published. At this time, the renown Parisian magnetist Hector Durville also began experimenting with phantasmatic photography, Baraduc produced *psychicones,* while Albert de Rochas was documenting and photographing magnetized performers and endeavouring to capture "the vital force".[69] The ramifications of these uncanny conjunctions are illuminated by my examination of Picasso's Hermetic Cubism which I have called Occultist Cubism in its relation to phantasmatic luminescence and Bergson's concept of "psychic states". Since it is possible to identify Picasso as perpetually playing with concepts of Vitalism, I have called this chapter, *Vitalist Picasso*.

In scrutinizing Picasso's fusion of the new sciences and Bergsonian philosophies with Occultisms, what I have endeavoured to illuminate is how Picasso's Vitalism differed substantially from his earlier explorations of the experience that I have called "becoming simian" and his encounter with Iberian culture and African tribal masks.[70] To unravel its relationship to Bergson's "psychic states", durational being and *l'élan vital*, as well as to phantasmatic luminescence, his Vitalist Cubism is then explored from three multidimensional perspectives.[71] Initially Picasso's rupture with positivist empiricism and scientistic realism is conveyed by comparing Picasso's durational portrait of Daniel-Henry Kahnweiler with Léon Bonnat's empirical *Portrait of the President of France: Armand Fallières* (Figure 5.2) and Jacques-Émile Blanche's *Portrait of Henri Bergson* (Figure 5.3). Unlike either Bonnat or Blanche, Picasso did not render Kahnweiler's portrait from life, in the same stationary position with a single light source, let alone from direct observation of his art dealer sitting in the same position every time they met. Instead I consider how Picasso's portrait was rendered as a felt experience from sensory memory to capture Kahnweiler's "psychic states" in a non-sequential time and "tactile" space as conceived by Bergson's duration and, following Deleuze's extrapolation from Bergson, a body without organs.[72] In the second part of the chapter, Kahnweiler's portrait is rescrutinized through the lens of spiritist photography captured by Baraduc, as well as the magnetist photography of phantoms with uncanny luminescence captured by Durville and Rochas (Figures 5.4 and 5.5). In the final part, the fusion of *l'élan vital* and *la force vitale* is explored through the conjunctions of Bergsonism with hermeticism, mysticism and occultism in Picasso's illustrations for Jacob's semi-autobiographical prose-poems, *Saint Matorel* and *Le Siège de Jérusalum* (Figures 5.9–5.11). This entailed a very different exploration of Bergsonian Vitalism to that pursued by the Futurists from 1910, particularly by Umberto Boccioni, as illuminated by Chapter 6.

From 1910 to 1911, an affinity between Boccioni's stated explorations of Futurism and Picasso's Occultist Cubism may be discerned. In the 1910 *Technical Manifesto of Futurist Painting*, Boccioni's claim to penetrate "the opacity of bodies" seemed to endorse the transparency of Picasso's Cubist bodies.[73] Their relative affinity appeared to be reinforced by the occultism of Boccioni's Vitalism alongside his relationship to Bergsonism. Yet as David Mather reveals in Chapter 6, Boccioni denounced Picasso's Cubism as "the result of an impassive scientific calibration" drained of life.[74] Mather then considers how Boccioni's deployment of a full chromatic scale may be regarded as an outright rejection of what he called Picasso's "ghostly emanations" in his dark Cubism. In response to the question, "What Divides us from Cubism?", Boccioni did not hesitate to identify Cubism as "a dead stop" supposedly "lacking all vitality".[75] In contrast, Boccioni argued passionately in favour of Futurism's chromatic intensity as the prime signifier of its Vitalism, particularly in its engagement with the vitalist dimensions of Bergson's philosophy. Following Bergson's metaphor for duration as a kaleidoscope blazing with a full range of vitalistic colours and intensities, Mather explores how Boccioni then developed this into Futurism's chromatic strategy to vitalize intuitive perception of all energizing phenomena, transform the Italian beholder and inspire new social configurations.[76] Nevertheless, the life-enhancing and socially regenerative dimensions of Futurist chromatic Vitalism were counter-balanced, as Mather reveals, by the incendiary dynamism of vitalist destruction and Italy's interventionist strategies in the First World War. Ending the second section of this book with the First World War then provides an historical segue to the third part, *Bergsonian Neo-Vitalism* in which *becoming* is explored in relation to absurdity, dysfunctionality, inversion and negation in Dada, Neo-Dada and Surrealism during and after this War.

Neo-Vitalism: Absurdity, Dysfunctionality, Inversion and Socialism

In Chapter 7 provocatively entitled *Was Dada Vitalist?*, Brandon Taylor stresses the hostility of Hugo Ball to utilitarianism and mechanization, particularly given their linkage to the mass destruction of war. So illogical and inhumane was this carnage that this was why, as Taylor points out, Theo van Doesburg's "neo-Vitalism" entailed such inversions as amorality, atemporality, discord and indifference alongside tactics of absurdity, contradiction, detachment, illogicality, irony, parody, randomness, negation and the nonsensical. Underlying these tactics, as Taylor reveals, was the dialectical interplay of absence and presence, assertion and denial, construction and destruction. While Dada Neo-Vitalists were identified as Albert Einstein and Charlie Chaplin, due to Bergson's phenomenology of negation and "The Idea of Immutability and the Idea of Nothing", Taylor singles him out, particularly the dialectic of extremes Bergson theorises in *L'Évolution créatrice* "to escape from the cinematographic mechanism of thought".[77] In performing these dialectical strategies in protest against what Hugo Ball called "the mechanical world", Taylor reveals how these Dadaists illuminated the absurdity of mass devastation while affirming intuitive rhythm, instinctive performativity and *becoming* as enhancing energies that could be unleashed in unpredictable interdisciplinary art. How this was pursued is demonstrated by Taylor's incisive readings of the *antipoetry* of Ball, *Bruitism* of Richard Huelsenbeck, *Ready-mades* of Marcel Duchamp, "non-zoological vitality" of Picabia, *Counter-Compositions* of Theo van Doesberg alongside Kurt Schwitters' vitalist intuition in his *Merz*. These readings encompass Bergsonian paradoxes in which nothing energizes something, culminating in Taylor's brilliant decoding of Cage's *Lecture on Nothing*.[78]

In the following Chapter 8, Donna Roberts examines Bergson in relation to Surrealism, "the vital impetus" and what she calls, after Salvador Dalí's painting, "The Persistence of Memory". Due to André Breton's resistance to Bergsonian Vitalism, the connection between surrealism's and Bergson's philosophies has received relatively little scrutiny. Yet paradoxically, as Roberts points out, both Bergson and Breton outrightly rejected what Breton calls "chop logic" for open-ended flux. Both explored the flow of time, the organic nature of memory, the self-creative impetus, experiential environmental relationships, co-evolution and interspeciality through what Bergson called "reciprocal interpenetration".[79] Deploying a post-Deleuzian methodology, Roberts then examines how Bergson's melding of memory with organicism in an intuitive and "an *eco-logical* vision of the contemporary world",[80] may be correlated to Surrealism. From this perspective, Roberts considers that Surrealism may have engaged, moreso than any other Modernist movement, "the radically transformative possibilities of an open-ended, spontaneously creative principle of life and art", as illuminated by Julien Gracq's identification of the concept of "automatism" articulated in the First Manifesto of Surrealism as "highly Bergsonian" and by the correlations between Roger Caillois' explorations of instinctual sympathy and Bergson's concepts of co-evolution. This is also illuminated by her exploration of such contemporary philosophers of vitalism as Monica Greco and Elizabeth Grosz, as well as Robert Mitchell's concept of "experimental Vitalism". Having navigated their theories in the first part of her Chapter 8, Roberts then deploys them as tools in her second and third parts to dissect the Bergsonian vitalism in Dalí's *The Persistence of Memory* (Figure 8.1) and Yves Tanguy's two paintings, *And there it is! (The Evening Before)* (Figure 8.2) and *Heredity of Acquired Characteristics* (Figure 8.3).

That Dalí was able to portray the incongruity of measured time in relation to the organic flow of time and memory is illustrated by Robert's scrutiny of *The Persistence of Memory*. Bergson's juxtapositions of artificially measured clock time with the fluid psychological time of memory in which the past flows into the present and anticipations of the future is the context for her decoding of Dalí's dream painting. Memories of the golden Cap de Creus cliffs of the Costa Brava, its aquamarine seawater by the clay-filled Catalan earth of Dalí's childhood seem to form his unconscious mind-place on which Roberts discerns Dalí's somnambulistic phantom in dream states. Despite the conundrum between physical time and psychological time that Paul Davies calls a "glaring mismatch", Dalí's dream states seem to be echoed by Dalí's notorious pocket watches melting, which Roberts relates to Bergson's experiences of the sensation of clock chimes melting into one another in unbroken continuity – each moment permeating one another through intuition. Roberts then explores how Tanguy captured the coming-into-being of primitive and aquatic life forms and their relationship to the process of painting as an open-ended inner duration in his paintings, *And there it is! (The Evening Before)* and *Heredity of Acquired Characteristics*.

To identify the character of deep-sea life in the planet's oceans and determine the penetration of light, British oceanographers within The Royal Society of London had launched the Challenger Expedition in 1875. Some 4,700 unknown species and oceanic trenches were discovered and measured, the depth of the Mariana Trench being found to well exceed the height of Mount Everest. Already the zoologist Ernst Haeckel at the University of Jena had embarked upon explorations of marine biology in the Mediterranean and the Canary Islands from 1866, naming 150 new species of radiolarians while demonstrating the affinities between the embryos of fish, salamanders, chickens, cows, pigs, rabbits and humans in his 1874 tome, *Anthropogenie*.[81] From 1899 to 1904, radiolarians, cnidarias, sea anemones and medusas were also copiously illustrated in Haeckel's book, *Artforms in Nature*.[82] These explorations were well-known to the French Neo-Lamarckian naturalists, Alphonse Milne-Edwards and Edmond Perrier, at the Muséum National de l'histoire naturelle in Paris. From 1880 to 1883, scientific expeditions funded by the French Education Ministry had been conducted by French marine biologists abord the *Travailleur* and *Talisman*, which led to Perrier's reports on *Annélides, brachiopods, crustacions, mollusques, Cirrhipèdes* and *Echinodermes* and Perrier's 1891 book, *Les Explorations sous-marines*.[83] By that year, France hosted the largest number of marine laboratories.[84] Hence scientific research into marine biology and its relationship to evolution was well underway in France by the time that Bergson was drafting *Creative Evolution* and Tanguy travelled to Africa and America as a merchant marine officer cadet in 1918, returning to Africa in 1930.

Given Tanguy's time in the merchant navy and his family origins in Brittany, in Chapter 8 Roberts points out that his artwork has been invariably correlated with the submarine. Yet it was the experience of being deeply underwater, far removed from sunlight, amidst billowing gases, star-like phosphorescence and hybrid humanoid forms, according to Roberts, that was captured in Tanguy's painting with a title inverting the progressive chronological imperative, *And there it is! (The Evening Before)*. Imaging what Roberts calls "sprouting flora, spermatazoic organisms and cloudy gaseous emissions", she perceives that this painting appears "to ripple with vital energy and a poetic pleasure in ambiguity". At the same time, she also explores how it invokes "a vivid image of involution, harking back to deep ancestral states of primordial indistinction" that Bergson theorizes in *Creative Evolution* as evolutionary anteriority. Conceiving of Tanguy's artwork as less Freudian than Bergsonian, more concerned with capturing concepts of memory and *déjà vu* than unravelling the unconscious, Roberts then observes how the solitary spermatozoon in Tanguy's painting seem to "imply a picture of life as pre- or non-human anteriority in the process of *becoming*, with the darkness making of space an infinite temporality". While acknowledging the Lamarckian evolutionary derivation of Tanguy's title for Figure 8.3, *Heredity of Acquired Characteristics*, Roberts deduces that the amorphous nature of its life-forms in an indeterminate space signify deep time. This deduction seems to be affirmed by the correlation of the Lamarckian title and images in Tanguy's painting to the deep-time geological research conducted by Jean-Baptiste Lamarck into the earliest and simplest of microscopic organisms for his *Hydrogéologie* (1802), *Recherches sur l'organisation des corps vivans* (1802) and *Philosophie zoologique* (1809).[85] Roberts then concludes that Bergson's explorations of the organic vitalism imbricated within different forms of duration appear to come "most vividly into play" in the artwork of Tanguy with the coming-into-being of Tanguy's indefinite life forms also configuring the gestational process of painting as one of inner duration precipitating matter and consciousness over deep-time. In the following chapter, the relationship of vitalism to cultural and political ideologies at this time is explored.

In the *Surrealist Map of the World*, published in 1929, all of Europe, except for Paris, plus the United States of America have disappeared along with other countries identified with colonial imperialism.[86] Inverting the Eurocentric world view on this map, the margins seem to have become the centre with the Pacific Ocean appearing at its focal point while Russia looms so large that it appears to dominate. Although Tanguy's political ideologies rarely appeared explicit, unlike those of Dalí, he endorsed Socialism. A political activist at that time aligned with the Vitalist Left, in 1931 Dalí became involved with the Communist Group in Spain, *The Workers and Peasants Bloc*. Yet from the time that *La Révolution Surréaliste* was launched on 1 December 1924, the Surrealist revolution had become identified as inextricably intertwined with Marxist theories and the Communist revolution, Breton and four other Surrealists joining the Parti communiste française in January 1927, although not for long. That the Vitalist Left and its artists became inextricably intertwined with Communism, Bergsonism and creativity is examined in Chapter 9 by Mark Antliff.

Even though Bergson was a major influence upon Surrealism, Dada and Futurism, as this book demonstrates, the three other important sites of vitalist thinking were, as Antliff identifies in Chapter 9, the Paris-based journal *Action d'Art* edited by the poet, philosopher and theatrical performer, André Colomer; *Creative Revolution:*

A Study of Communist Ergotocracy, by Eden and Cedar Paul, and the Bergsonian Vitalism of Georges Sorel. So closely involved with *Action d'Art* did the Paris-based Futurist, Gino Severini, become that he planned to launch a Théâtre d'Action Art.[87] Not only was Bergson deployed by Colomer to promote Max Stirner's radical nominalism but also to release intuition, achieve the anarchist creative self and turn their lives into works of art.[88] Outraged, Pieter Kropotkin attacked Bergson's *Creative Evolution* as "elegantly fantastical assertions" with no basis in scientific facts.[89] This elicited what Antliff calls Colomer's "stinging rebuke" entailing a Bergsonian critique of communist concepts of collectivity and a defence of intuitive knowledge for overcoming habitual thinking, releasing willed empathy and grasping the creative life force permeating the cosmos.[90]

While the syndicalist and "shop-steward" movements were identified by the Pauls as manifestations of the vital impulse, Antliff points out that they considered the First World War activated "a highly creative break" enabling Bergsonian Vitalism to transform history. More specifically Bergsonian Vitalism corresponded to, as Antliff reveals, Lenin's "soviet dictatorship", able to nurture "individual freedom". For the Pauls, the Soviet state born of the revolution constituted a vital order, "flexible" and "mobile" like duration itself, with Lenin as the "Great artist" producing by means of creative revolution "other artists". In his focus, Antliff then portrays a very different Bergsonian conception of the Soviet Union to that conveyed in the following chapter by Patricia Simpson.

From the time that *L'Évolution créatrice* was published in 1907, Moscow University students avidly read and discussed Bergson's philosophy.[91] By 1914, all of Bergson's writings had been translated into Russian, as had Dreisch's *Vitalism: Its History and System*. While the Russian Formalists, Viktor Shklovsky and Yuri Tynianov, engaged with Bergsonism, such Russian Modernists as Wassily Kandinsky drew upon Bergson's reconceptions of time and space.[92] In turn Kazimir Malevich correlated non-objectivity in his art to creative intuition, while defining his Suprematism as Bergsonian unbounded creativity able to encompass all aspects of life.[93] However, as Simpson points out in Chapter 10, after the Stalinist shut-down of avant-garde cultures, by 1930 "vitalist" and "Vitalism" constituted terms of abuse signifying enemies of the state.

Nevertheless ironically by the Great Patriotic War against Nazi Germany, Simpson reveals that Vitalism was seemingly smuggled into the Darwin Museum in Moscow. As "evolutionary psychotherapy", it was designed to revitalize wounded soldiers suffering "shell-shock". Artworks were commissioned by the Directors of the State Darwin Museum, Dr Aleksandr Kots and his wife, Nadezhda Ladygina-Kots, from the zoologist and "animalist" sculptor and painter, Konstantin Flerov, featuring "animals as a means and weapon of war" alongside models of superhuman physiological strength identifiable with the Soviet aspirational model of the "New Man".[94] Designed to illustrate the stories that Aleksandr Kots told the Soviet soldiers about the roles played by animals and humans, these paintings would have been also used to identify the heroes or villains with whom the soldiers should feel empathy or *sobornost'* – the covert vitalistic elements of Soviet "Creative Darwinism". By no means was this to be the last encounter with vitalism. After the Second World War, it would resurface with Dada Neo-Vitalists, particularly through John Cage, as indicated in Chapter 7, and would become reinvigorated some 40 years later by, amongst others, Deleuze in new philosophies of *becoming*.

Hence in these ten chapters exploring the conception of life as a constant process of metamorphosis impelled by the free flow and momentum of physiological and psychological energies to generate *creative evolution,* this book endeavours to reveal how Vitalism became widely and diversely engaged by Modernists. It endeavours to chart how their engagement entailed the rejection of positivist empiricism, mechanical materialism, "chop logic" and chronometric imperatives for such new vitalist concepts as those philosophized by Bergson, Driesch and Nietzsche, amongst others, amidst the new corporeal and psychological energies being unleashed by modern sports, physical cultures, the new psychology, spiritisms and occultisms. It reveals how those Vitalist Modernists who did so included Hans Arp, Sophie Taeuber-Arp, Juliette Bisson, Hugo Ball, Giacomo Balla, Umberto Boccioni, Georges Braque, John Cage, Gustave Caillebotte, Eva Carrière, Carlo Carrà, Charlie Chaplin, Bruno Corra, Salvador Dalí, Robert Delaunay, Marcel Duchamp, Albert Gleizes, Richard Huelsenbeck, Louis Lumière, Kazimir Malevich, Edvard Munch, Francis Picabia, Picasso, Enrico Prampolini, Ivan Puni, Barbara Reichel, Luigi Russolo, Kurt Schwitters, Gino Severini, Yves Tanguy, Vladimir Tatlin, Tristan Tzara and Theo Van Doesburg. It examines how these Vitalist Modernists explored intuition, instinct, spontaneity, chance, empathy, intense emotion, memory, hypnosis, mesmerism, occultisms, "psychic states", unconscious states, uncanny vibrations, socialism, communism, deep time and a continuous flow of time through duration fusing the present, past and future within the concept of *becoming.* Nevertheless despite the immense significance of Vitalism to these Modernist artists from Caillebotte in *Impressionism* France to Cage's *Lecture on Nothing,* first performed in Manhattan in 1949, this book ends with its decline during the Second World War and a paradox: That even when outlawed by Stalin, it was smuggled back in through art during the Great Patriotic War in order to revitalize traumatized Soviet soldiers.

Notes

1 Ernest Seillière, *Un Artisan d'Énergie Française : Pierre de Coubertin* (Paris: Henri Didier, Librairie-Éditeur, 1917).
2 Wilhelm Ostwald, "The Modern Theory of Energetics", *The Monist,* 17/4 (1907): 481–515.
3 Helge Krage, *An Introduction to the Historiography of Science* (New York: Cambridge University Press, 1989).
4 Paul Walden, *Wilhelm Ostwald* (Kessinger Publishing: 1904) 231.
5 Wilhelm Ostwald, *Der energetische Imperativ* (Leipzig: Akademische Verlagsgesellschaft m.b.H., 1912).
6 Charles Darwin, *The Descent of Man, and Selection in Relation to Sex* (London: John Murray, 1871) 212.
7 Hippolyte Taine, *Les Origines de la France contemporaine,* vol. 1 (Paris: Hachette, 1896) 70–71.
8 Michael T. Michael, "Unconscious Emotion and Free-Energy: A Philosophical and Neuroscientific Exploration", *Frontiers in Psychology* (21 May 2020) https://doi.org/10.3389/fpsyg.2020.00984
9 Fae Brauer, "Capturing Unconsciousness: The New Psychology, Hypnosis and the Culture of Hysteria", *A Companion to Nineteenth Century Art,* ed. Michelle Facos (London et al: John Wiley & Sons, Inc., 2019) 243–262.
10 Henri Bergson, "De la simulation inconsciente dans l'état d'hypnotisme", *Mélanges* (1882): 331–341; refer Giuseppe Bianco, "Henri Bergson's Failed Career as a Psychologist: Dealing "seriously" with Philosophy?", *Academia.edu.* (2018): 14.

11 Pierre Janet, *L'automatisme psychologique, essai de psychologie expérimentale sur les forms inférieures de l'activité humaine,* Thèse d'État, Faculté des lettres (Paris: Félix Alcan, 1889); Olivier Walusinski and Julien Bogousslavsky, "Charcot, Janet and the Models of Psychopathology", *European Neurology* (17 June 2020): 333–340.
12 Wilhelm Wundt, *Grundzüge der Physiologischen Psychologie* (Leipzig: W. Engelman, 1874).
13 Henri Bergson, *L'Évolution créatrice* (Paris: Félix Alcan, Éditeur, 1907); *Creative Evolution,* trans. Arthur Mitchell (Mineola, New York: Dover Publications, Inc., 1998); hereinafter referred to as *Creative Evolution.*
14 Hans Driesch, *The History and Theory of Vitalism,* trans. C. K. Ogden (London: Macmillan and Co., Limited, 1914).
15 Frederick Nietzsche, *Twilight of the Idols, or, How to punish with a Hammer,* trans. Duncan Large (Liepzig: 1889; Oxford: Oxford University Press, 2008) ix.
16 Frederick Nietzsche, *The Birth of Tragedy, Or: Hellenism and Pessimism,* trans. Douglas Smith (1872; Oxford: Oxford University Press, 2008) 109.
17 Driesch, *The History and Theory of Vitalism,* 11–22: *The Old Vitalism.*
18 Driesch, *The History and Theory of Vitalism,* 19.
19 Georg Ernst Stahl, *Theora medica vera,* 1712; Elizabeth A. Williams, *A Cultural History of Medical Vitalism in Enlightenment Montpellier* (London and New York: Routledge, 2017) 80–101.
20 Barbara Larson, "Gauguin: Vitalist, Hypnotist", *Gauguin's Challenge: New Perspectives after Postmodernism,* ed. Norma Broude (London and New York: Bloomsbury Visual Arts, 2018) 180.
21 Larson, "Gauguin: Vitalist, Hypnotist", 180.
22 Williams, *A Cultural History of Medical Vitalism,* 147.
23 Driesch, *The History and Theory of Vitalism,* 93–123.
24 Driesch, *The History and Theory of Vitalism,* 187–188; also Horst Heinz Freyhofer, *The Vitalism of Hans Driesch: The Success and Decline of a Scientific Theory* (European University Studies, 1982).
25 Driesch, *The History and Theory of Vitalism,* 191; William James, *The Principles of Psychology* (New York: Henry Holt and Company, 1890): Chapter 27 is entitled *Hypnotism.* Modes of Operating and Susceptibility. James introduces this chapter with the statement: "The 'hypnotic,' 'mesmeric,' or 'magnetic' trance can be induced in various ways."
26 Driesch, *The History and Theory of Vitalism,* 192.
27 Driesch, *The History and Theory of Vitalism,* 192.
28 Driesch, *The History and Theory of Vitalism,* 204–205.
29 Hans Driesch, "Psychical Research and Philosophy", *The Case For and Against Psychical Belief,* ed. Carl Murchison (Worcester: Clark University, 1927).
30 Hans Driesch, *The Science of the Super-normal* (London: Bell, 1933).
31 Germana Pareti, "Hans Driesch's Interest in the Psychical Research. A Historical Study", *Medical Historica,* Vol. 1, No. 3 (2017): 152–162.
32 Robert C. Grogin, "Henri Bergson and the University Community, 1900-1914", *Historical Reflections,* 2/2 (1976): 209.
33 Henri Bergson, *Matière et Mémoire : Essai sur la relation du corps à l'esprit* (Paris: Félix Alcan, Éditeur, 1896); *Matter and Memory,* trans. Nancy Margaret Paul, W. Scott Palmer (New York: Zone Books, 1988) 200–201.
34 Linda Dalrymple Henderson, "Writing Modern Art and Science – An Overview. Cubism, Futurism and Ether Physics in the Early Twentieth Century". *Science in Context,* 17/4 (December, 2004): 423–466.
35 Bergson, *Creative Evolution,* 89.
36 In *The Perfect Wagnerite* published in 1895, Shaw mentions an impulse in nature he calls the "life-force" able to change matter and generate new forms.
37 Stephen Lehan, "Bergson and the Discourse of the Moderns", *The Crisis in Modernity: Vitalist Controversy,* eds. Frederick Burwick, Paul Douglass (Cambridge: The Cambridge University Press, 2010) 311.
38 Bergson, *Creative Evolution,* 313.

39 Bergson, *Creative Evolution,* 306.
40 Bergson, *Creative Evolution,* 313.
41 Bergson, *Creative Evolution,* 257–258.
42 Henri Bergson, *L'énergie spirituelle: essais et conférences* (Paris: Librairie Félix Alcan, 1919): Its English translation as *Mind Energy,* appears inaccurate given the contents addressing *L'au-dela,* 29, *L'Âme et le corps,* Chapître II, and *Fantômes et vivants,* Chapître III, in which Bergson explores *la "Recherche psychique";* his philosophy was referred to as "a new spiritualism".
43 Grogin, *Henri Bergson and the University Community,* 213.
44 From the time that Bergson began teaching at the Collège de France, he proved so popular that he was blamed for students rejection of other academics, particularly Charles Péguy. By 1906, such strident student criticism of most of the other Sorbonne professors was made that a senatorial scrutiny of the complaints was conducted.
45 Gabriel Marcel, *The Bergsonnian Heritage,* ed. Thomas Hanna (N.Y. 1962) 126.
46 *La Vie Parisienne,* 1912, 509; quoted by Grogin, 215.
47 Jacques Chastenet, *La France de M. Fallières* (Paris: Fayad, 1949) 213.
48 Henri Bergson, "Phantasms of the Living" and "Psychical Research": Presidential Address to the Society for Psychical Research, London, 23 May 1913, *Mind-Energy,* trans. H. Wildon Carr, ed. Keith Ansell Pearson and Michael Kolkman (Basingstoke, 2007) 77.
49 J. Saussman, ""It's organisms that die, not life"; Henri Bergson, Psychical Research and the Contemporary uses of Vitalism": Chapter 1 in *The Machine and the Ghost: Technology and spiritualism in nineteenth- to twenty-first-century art and culture,* S. Mays and N. Matheson (eds) (Manchester: 2013): 16–36.
50 Jesse Hong Xiang, *The Outline of Parapsychology* (London: 2009) 42.
51 Maurice Merleau-Ponty, *Signs,* trans. Richard C. McCleary (Northwestern University Press, 1964).
52 Henri Bergson, *Les deux sources de la morale et de la religion* (Paris: Félix Alcan, 1932).
53 *Archives of Literary Life under the Occupation* (Paris: Hôtel de Ville Exhibition, 2011).
54 Craig Lundy, *Deleuze's Bergsonism* (Edinburgh University Press, 2018) 111.
55 Stephen Lehan, "Bergson and the Discourse of the Moderns", *The Crisis in Modernity: Vitalist Controversy,* eds. Frederick Burwick, Paul Douglass (Cambridge: The Cambridge University Press, 2010) 311.
56 *The Spirit of Vitalism: Health, Beauty and Strength in Danish Art, 1890-1940,* eds. Hvidberg-Hansen and Gertrude Oelsner, provides invaluable explorations of the interrelationship of nature, health and sport in Danish Vitalism but does not venture beyond Denmark. The excellent anthology, *Bergson and the Art of Immanence: Painting, Photography, Film,* eds. John Ó Maoilearca and Charlotte de Mille (Edinburgh University Press, 2013; paperback 2015) explores Bergsonism in relation to Duncan Grant, Roger Fry, Alfred Manessier and Jean Fautrier.
57 Sharon L. Hirsch, "The Sick City", Chapter 4, *Symbolism and Modern Urban Society* (Cambridge: Cambridge University Press, 2004) 121.
58 Edvard Munch, Konkurransen om den Kunstneriske Utsmykning av Universitetets Nye Festsal (Competition for the Artistic Decoration of the University's New Festival Hall), (Helsinki: Pamphlet, 1911) x.
59 Gilles Deleuze, *Bergsonism,* trans. Hugh Tomlinson and Barbara Habberjam (New York: Zone Books, 1988); Eric Alliez, "On Deleuze's Bergsonism", *Discourse,* Vol. 20, No. 3 (Fall 1998) 226–246; Craig Lundy, *Deleuze's Bergsonism* (Edinburgh: Edinburgh University Press, 2018).
60 Reinhold Heller, *Brücke: The Birth of Expressionism in Dresden and Berlin, 1905-1913* (New York: Neue Gallery; Ostfildern: Hatjie Cantz, 2009); Chapter Two: *Bridge of Art,* Nietzsche's Dionysian Vitalism, his concept of the "tightrope walker" and Albert Bloch's *Slack Wire (Seiltänzer)* 1913: Amanda du Preez, *Art, The Sublime, and Movement. Spaced Out* (New York: Routledge, 2022).
61 C. Richet, "Faut-il laisser la France périr?", *Revue bleue* (1896): 620–622.

62 Pierre de Coubertin, *L'Éducation en Angleterre : Collèges et Universités* (Paris: Librairie Hachette et Cie, 1888).
63 Also refer Pascal Rousseau, *Robert Delaunay: L'invention du pop* (Paris: Hazan, 2019).
64 Also refer Fae Brauer "Vitalist Cubisms: The Biocultures of Virility, Militarism and *La Vie Sportive*", *Sport and the European Avant-Garde (1900-1945)*, eds. Andreas Kramer and Przemyslaw Strozek (Leiden/Boston: Brill Avant-Garde Critical Studies, 2022) 19–56.
65 Linda Dalrymple Henderson, "Rethinking Modern Art, Science and Occultism in Light of the Ether of Space: Wassily Kandinsky, Umberto Boccioni, and Kazimir Malevich", *The History of Art and 'Rejected Knowledge': From the Hermetic Tradition to the 21st Century,* ed. Anna Korndorf (Moscow: The State Institute of Art Studies, 2018) 218-37.
66 *The Phenomena of Materialisation*; described in *Light,* vol. 6 (January 16, 1886): 34.
67 Florence Raulin-Cerceau, *À la recherche d'intelligences extraterrestes* (Paris: Upper Editions, 2017); *Les origines de la vie. Histoire des idées* (Paris: Editions Ellipses, 2009).
68 Flammarion, *The Unknown*, x.
69 Fae Brauer, "Scientistic Magnetism and Hauntological Metarealism: The Phantasmatic Doubles of Duchamp and Durville", *Realisms of the Avant-Garde,* eds. David Ayers, Moritz Baßler, Sascha Bru, Ursula Frohne and Benedikt Hjartarson (Berlin & Boston: Walter de Gruyter, 2020) 42-75; "Magnetic Modernism: František Kupka's Mesmeric Abstraction and Anarcho-Cosmic Utopia", *Utopia: The Avant-Garde, Modernism and (Im)possible Life,* eds. David Ayers, Benedikt Hjartarson, Tomi Huttunen, Harri Veivo (Berlin & Boston: Walter de Gruyter, 2015) 123-153.
70 Fae Brauer, "Becoming Simian: Devolution as Evolution in Transformist Modernism", *Picturing Evolution and Extinction: Degeneration and Regeneration in Modern Visual Culture,* eds. Fae Brauer and Serena Keshavjee (Lady Stephenson Library, Newcastle upon Tyne: Cambridge Scholars Publishing, 2015) 127–156. The state of "becoming simian" arose during Picasso's midnight forays into the *Jardin des Plantes.* This was followed by his exploration of Iberian culture and "becoming tribal" after Picasso's notorious epiphany experienced amidst the Musée de l'Homme collection of African tribal masks and sculptures. Also refer, Fae Brauer, "Exposing "The Venereal Peril": Fournier's Sphilography, Munch's Heredo-Syphilitic, "La Syphilis Arabe" and Picasso's Prostitutes", forthcoming, *Contagion, Hygiene and the Avant-Garde* (Routledge, 2023).
71 With Occultist Cubism, as is revealed in Chapter 5, Picasso appears to have abandoned any confinement of the human body to a contour with a wholesome surface and a clear distinction between its interior and exterior, let alone a distinct separation from the world it inhabits.
72 Gilles Deleuze and Félix Guattari, *Anti-Oedipus,* trans. Robert Hurley, Mark Seem and Helen R. Lane (London and New York: Continuum, 2004).
73 Umberto Boccioni et al, "Futurist Painting: Technical Manifesto", April 1910; *Futurist Manifestos,* ed. Umbro Apollonio, trans. Robert Brain et al (New York: Viking Press, 1970) 28.
74 Umberto Boccioni, "What Divides Us from Cubism"; Ester Coen, *Umberto Boccioni: A Retrospective* (New York: Harry Abrams and the Museum of Modern Art, 1988) 248.
75 Boccioni, "What Divides Us from Cubism"; Coen, *Umberto Boccioni,* 243–244.
76 Bergson, *Matter and Memory,* 268–269.
77 Bergson, *Creative Evolution,* 272–298; 313.
78 John Cage, *Lecture on Nothing* (1949; Incontri Musicali, August 1959) 109–126.
79 Bergson, *Creative Evolution,* 178.
80 Richard A. Cohen, "Philo, Spinoza, Bergson: The rise of an Ecological Age", *The New Bergson*, 1999, 25.
81 Ernst Haeckel, *Anthropogenie, oder, Entwikelungsgeschichte des Menschen* (Leipzig: W. Engelmann, 1874).
82 Prof. Dr. Ernst Haeckel, *Kunstformen der Natur* (Leipzig und Wien: Bibliographisches Institut, 1904).
83 Alphonse Milne Edwards and Edmond Perrier, *Expéditions scientifiques du Travailleur et Talisman pendant les années 1880, 1881, 1882, 1883* (Paris: G. Masson, 1888-1906); Edmond Perrier, *Les Explorations sous-marines* (Paris: Librairie Hatchette, 1891).

84 Benoît Dayrat, "Henri de Lacaze-Duthiers and the creation of the Laboratoire de zoologie expérimentale, Roscoff, France", *Revue d'Histoire des Sciences,* 2/69 (2016) 335–368.

85 For more on the relationship of Lamarck and Neo-Lamarckism to Modernism, refer Fae Brauer, "Becoming Simian: Devolution as Evolution in Transformist Modernism", *Picturing Evolution and Extinction: Degeneration and Regeneration in Modern Visual Culture,* eds. Fae Brauer and Serena Keshavjee (Lady Stephenson Library, Newcastle upon Tyne: Cambridge Scholars Publishing, 2015) 127–156.

86 "Le Monde au Temps des Surrealists", *Variétés,* "Le Surréalisme en 1929", Belgium, 1929: Anonymous, this countermap has been attributed to Paul Éluard.

87 Mark Antliff, "Cubism, Futurism, Anarchism: The 'Aestheticism' of the Action d'Art Group, 1906-1920", (Oxford: Oxford Art Journal, 21/2 (1998): 99–120.

88 Also refer Mark Antliff, "Revolutionary Immanence: Bergson Among the Anarchists", *Bergson and The art of Immanence, 96.*

89 Jesse S. Cohn, *Anarchism and the Crisis of Representation: Hermeneutics, Aesthetics, Politics* (Selisgrove: Susquehanna University Press, 2006).

90 Antliff, *Revolutionary Immanence,* 103.

91 Hilary L. Fink, *Bergson and Russian Modernism 1900-1930* (Evanston: Northwestern University Press, 2012).

92 Paul Atkinson, *Henri Bergson and Visual Culture: A Philosophy for a New Aesthetic* (London, etc.: Bloomsbury Academic, 2021) 215.

93 Isabel Wünsche, "Creative Intuition: The Russian Interpretation of Henri Bergson's Metaphysics", *Experiment,* 23/1 (October 2017): 215–228.

94 Patricia Simpson, "Imag[in]ing Post-Revolutionary Evolution: The Taylorized Proletarian, "Conditioning", and Soviet Darwinism in the 1920", *The Art of Evolution: Darwin, Darwinisms and Visual Culture,* eds. Barbara Larson and Fae Brauer (Hanover: University Press of New England, 2009) 226–261.

Part I BIOVITALISM

Corporeal Regeneration, Environmental Purification and National Evolution

1 The Manly Water Arts

Hygiene, Vitality and Virility at the fin-de-siècle

Anthea Callen

Dedicated to my wonderful *sportif* brother Henry Vane Young

My theme here is water, both in hygiene and in sport, primarily in France but touching too on Northern Europe, Britain and America. Water is an apt vehicle for examining the wider meanings of the fin-de-siècle corporeal health and hygiene movements given credence by both Darwinist and Neo-Lamarckian theories of evolutionary regeneration.[1] These movements would evolve hand in hand with the rise of a Bergsonian Vitalism to offer adherents the desirable new *élan vital*: A 'vital life force' to counter and transform the degenerative forces of modern urban society.[2] In this chapter, I want to demonstrate the links between the visual arts, water-sports and hygiene in the construction of ideals of the modern male body – virilized and vitalized. Although finding renewed credence within the modern Vitalist movements encouraging sport and physical culture, this perfected male body nevertheless had its origins in the ideal physiques of classical sculpture. And as is evident from the outsize 'fig-leaves' deployed in Napoléon Sarony's New York studio portraits and George Steckel's 1894 Los Angeles studio portraits of the naked physical culturist Eugen Sandow (1867–1925), classical proportions were not the only preoccupation: Sexual potency was high on the masculine agenda.[3]

Nineteenth-century European constructions of the muscular male body were both classed and racialized, whether associated with the white labouring body (the gentleman's 'bit of rough') or, in the colonial context, with the powerful but disempowered Black 'other'. Among middling and upper classes, the popular rise of the healthy athletic body, with its roots in the military and the widely emulated British Public-school sporting ethos, was legitimized by health and fitness imperatives at all national levels. Given that sex between men was recognized as central to Ancient Greek culture, the emergence of this aesthetic at the fin-de-siècle – especially with the advent of sexology – raised further complex and often conflicting questions of virile male sexuality, race and class. As Michel Foucault noted: '[A] medical category of homosexuality was constituted from the moment it was characterised [by Westphal in 1870] ... The sodomite had been a temporary aberration; the homosexual was now a species'.[4] Whether in pose and attitude, or in the form of hyper-muscular Herculean strongman, athletic Apollonian warrior-God or nubile adolescent *ephebe*, Greco-Roman sculpture offered exemplary ideal models of 'natural' physical beauty: These were the aspirational *beaux-idéals* for modern man. In the fin-de-siècle context therefore, this virile athletic ideal arguably provided the perfect guise for 'deviant' masculinity: A heroic muscular body inside which subversive sexualities could discreetly reside, attracting admiration

DOI: 10.4324/9781003045595-3

rather than persecution. Equally, the 'respectable' admiring gaze on this athletic body could as well be male as female, as in the case of bi-sexual body-builder Sandow.[5]

Regardless of whether any of the individual artists discussed here came consciously to align themselves with particular anti-rationalist Vitalist theories or theorists, their fascination with water and aquatics mirrored broader social concerns over population degeneration in this period, and the prevailing imperative towards national regeneration. Indeed, the work of many of these artists predates the cultural consolidation of Vitalist theories, and as such anticipates, or was actively formative in, their emergence in the European *zeitgeist*. I shall argue that these artists' representations of water and water-sports affirm an ongoing continuity in concerns over national manly health and hygiene that predates the expansion, in France at least, of industrial mechanization as such – rather it coincides with rapid urbanization in the period. I shall begin with a study of water and its meanings, with hygiene and bathing, and then focus on water-sports – especially swimming and rowing. The obvious advantage of course, to these sports – especially swimming – was the naked or near-nude display of the athletic body.

Hygienic Ideals

The extensive volume of visual material on my theme demonstrates that manly athletic ideals were not limited to the new cult of body building but can be found throughout 'physical culture' – including aquatics. The ubiquity of the athletic *beau idéal* is evident in the wide circulation of photographs, cast models, postcards and prints of Greco-Roman sculptures. Alongside these highbrow exemplars were innumerable widely disseminated popular variants, from cartoons, caricatures and advertisements to photographs both of artists' naked models in 'classical' poses, of gymnasts, wrestlers and of muscular performance artists – like those circulated from around 1885 onwards by French medical physical culturist Edmond Desbonnet and, from the early 1890s, the highly commercial collectable 'series' starring the incomparable Sandow.[6] Dedicated body-building courses, available from the mid-1880s, were not the only route to muscular fitness and virile masculinity: Team sports increasingly provided opportunities for all-male *camaraderie* and healthy exercise, often outdoors. In water sports, this ethos dates back at least to *c.*1800 and the first public swimming pools that already offered opportunities for the display and performance of the male athletic body. In Europe and the Anglophone nations, illustrated sport, health and physical culture books and magazines addressed primarily at men proliferated from the 1890s.[7] Indeed, ideals of hygienic sporting health and manly virility coincided to such a degree in this period that they became virtually synonymous, as Olympic revivalist Pierre Baron de Coubertin (1863–1937) himself observed in 1913: 'Sport appears as the very symbol of virility'.[8]

Here the classical *Apoxyomenos* provides my example (Figure 1.1).

Around 330 BCE, Lysippos used eight-head proportions to give his sculpted male figures small heads, elongated limbs and characteristically slim body proportions. This made his figures appear tall, elegant and statuesque: They were typically both *mouvementé* and balanced in pose. His *Apoxyomenos* is exemplary: It shows an ideal classical athlete cleansing his body after sport with a *strigil*, a curved instrument used to scrape off oil, dirt and sweat. The *strigil* also removed body hair. This sculpture thus gave its modern audiences a classical role-model of virile athleticism combined with attentive male hygiene (Figure 1.2). It was not enough to be tall.

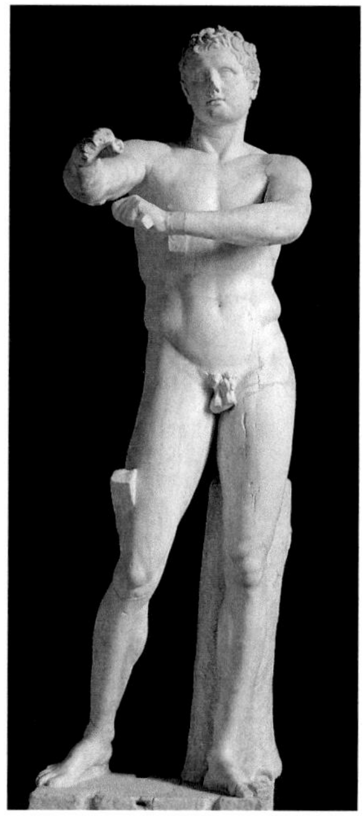

Figure 1.1 Roman copy after Lysippos, Apoxyomenos (The Scraper), first century CE, marble, h. 205 cm. Vatican Museum. (Photograph courtesy: Marie-Lan Nguyen. Open Source, https://commons.wikimedia.org/wiki/File:Apoxyomenos_Pio-Clementino_Inv1185.jpg.)

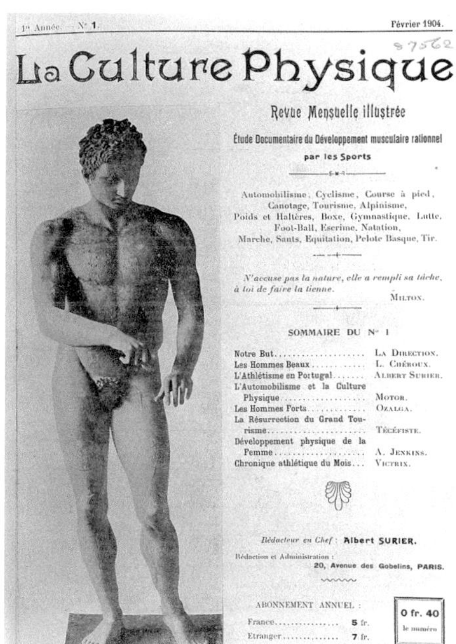

1re Année. — N° 1. Février 1904.

La Culture Physique

Revue Mensuelle illustrée

Étude Documentaire du Développement musculaire rationnel

par les Sports

Automobilisme, Cyclisme, Course à pied, Canotage, Tourisme, Alpinisme, Poids et Haltères, Boxe, Gymnastique, Lutte, Foot-Ball, Escrime, Natation, Marche, Sauts, Equitation, Pelote Basque, Tir.

N'accuse pas la nature, elle a rempli sa tâche, à toi de faire la tienne.

MILTON.

SOMMAIRE DU N° 1

Notre But	LA DIRECTION.
Les Hommes Beaux	L. CHÉROUX.
L'Athlétisme en Portugal	ALBERT SURIER.
L'Automobilisme et la Culture Physique	MOTOR.
Les Hommes Forts	OZALGA.
La Résurrection du Grand Tourisme	TÉCÉFISTE.
Développement physique de la Femme	A. JENKINS.
Chronique athlétique du Mois	VICTRIX.

Rédacteur en Chef : **Albert SURIER.**

Rédaction et Administration :
20, Avenue des Gobelins, PARIS.

ABONNEMENT ANNUEL :

France .... 5 fr.
Etranger .... 7 fr.

0 fr. 40 le numéro

Wellcome Images

Figure 1.2 "Ephesian Apoxyomenos", front cover, *La Culture Physique*, Revue Mensuelle Illustrée, February 1904. Paris: Albert Surier (Editor-in-chief). After Greek bronze, AD 1-90, original in Kunsthistorisches Museum, Vienna. (Photo courtesy: The Wellcome Library, London.)

The ideal modern male physique needed to appear youthful and strong: Lightly muscular, graceful, well-proportioned, clean-shaven, smooth, and above all clean.

Water

Water at the start of the nineteenth century was often mistrusted and little understood, as indeed were its possibly negative effects on the skin or the internal organs: Indeed, most people drank wine or 'small' (weak) beer rather than water for fear of disease. Despite the emergence of public hygiene in 1802 and its foundation as a branch of public medicine in Paris in the 1820s, water was not automatically associated with bodily cleansing in France. During the nineteenth century 'hygiene', aptly derived from the Greek *hygienòs* meaning 'that which is healthy' spawned a new set of knowledges that reimagined water as a protective medium, with public measures in place to promote healthy maintenance of the social body – and also the private body.[9] Water was increasingly appropriated by the medical profession as a *hygienic* medium which, depending on the temperature of the bath, had different therapeutic effects, not all of which were deemed beneficial for male physical or moral health. A moral as well as a medical imperative meant the hygienic injunction to cleanse the body became a key sign of male sexual virility. Baths and bathing thus required regulation.[10]

From the time of the ancient Greeks seawater was thought restorative and invigorating and, in Thalassotherapy, it retained this classical authority and associations into the modern era. Outdoor, or what is now popularly called 'wild' bathing, was a widely popular pastime, whether in the sea, rivers, or lakes, while for less-privileged men such naked river bathing was a hygienic necessity as well as a pleasure. As much as it was healthy exercise and recreation, 'wild' swimming also constituted an important space for all-male homosociability.

Like outdoor water bathing, water sports offered men a heteronormative space for male bonding, camaraderie, manly competition, and virile physical display. Yet crucially, the opportunities that aquatics afforded men to see and be seen naked or semi-nude opened up legitimate spaces for the potentially desiring male gaze – a potentially libidinous gaze scoping homo-virility that was effectively shielded from accusations of effeminacy or 'inversion'.[11]

Water, Aquatics and Hygiene

The discourses of water and health, health and hygiene, hygiene and sport were all intimately linked since vital *force* and virility were key factors for men. The call in the later nineteenth century for revitalization – for the regeneration of an apparently ailing French manhood and a depleting population – came from all political sides and from the medical professionals too. Following the disastrous war with Prussia in 1870, the state realized not only that French leadership in the doomed campaign was appalling but also that the French army was insufficiently fit, disciplined and militaristic.[12] Yet this fear was not new.

In the aftermath of Napoleon I's Empire, similar appeals to sporting fitness and healthy manhood had identified swimming as one effective means to renew the nation's military strength and readiness. At least by the 1840s, swimming lessons to build fitness were already the norm in military colleges, as too for school-boys in Paris.[13]

Anglophilia is key here, in the French admiration of England's team-sporting prowess due, it was thought, to the education of their ruling classes in rigorous public schools. The English public school system was widely considered as a model in Restoration France, while in the 1890s this Anglophile discourse was again invoked by physical culture educationalist and Olympic Games revivalist, Pierre Baron de Coubertin. By promoting what he saw as the liberal democratic and character-building properties of English school sport, Coubertin sought to instil a new physical vigour and discipline in the men of France. Eight years after the introduction of compulsory gymnastics into French education in 1880, Coubertin made physical education compulsory in the French school curriculum.[14] The 'father' of modern Olympics, Coubertin, was instrumental in the revival of the ancient Greek games in the mid-1890s. Water sporting events included from the first Olympics held in Athens in 1896 were sailing, swimming and rowing. This was followed in 1900 by the Paris Olympics where water polo also featured while in the 1904 St. Louis Olympic Games diving was introduced to the official program. In these early years, all the water sports were exclusively for *male* athletes.[15]

Yet water is a slippery phenomenon still linked even then in the popular imaginary with disease and death, rather than health and 'vitality'. Cholera had reached London in Spring 1832 and Paris in March 1834. From a population of 650,000 Parisians, 20,000 lost their lives within the first six months, and in total 100,000 people throughout France died from this dread disease. Initially attributed to 'miasmas' or fetid air, in 1854 Englishman John Snow traced the cholera outbreak in London's overcrowded Soho area to a faecal-tainted public well.[16] Regular cholera epidemics throughout the nineteenth century saw water increasingly feared as dangerous, indeed life-threatening. By the 1880s, the link between cholera and contaminated water had seeped into the popular imagination, paradoxically at a time when personal hygiene was becoming more customary – among the bourgeoisie at least.

Regulation by the Conseil de salubrité meant a growing surveillance of public bathing on moral grounds in country and city alike.[17] Certainly by the late 1830s, the Municipal police in Paris were pursuing young men engaged in 'wild bathing', except for them it was also personal hygiene. Several of Daumier's significantly large series of 30-odd lithographic caricatures, *Les Baigneurs* (published in *Le Charivari* between 1839 and 1842), show this pursuit – in both senses – as does Bertall's illustration in Eugène Briffault's description of the practice in *Paris dans l'eau* of 1844.

Briffault's text emphasizes that of course this bathing was *naked*. This should equally inform our reading of Daumier's caricatures of bourgeois shocked by sudden exposure to wild swimmers on the outskirts of Paris.[18] Nakedness for men was the norm, too, in the early Parisian bathing barges. Yet what river scenes depicted by the likes of Daumier (Figure 1.3) and Bertall reveal of this period is not only the widespread moral surveillance of public use of the Seine (which constrained laundresses too) but also the popularity at this early date of sporting and recreational exploitation of the river.[19] Organized rowing teams and leisure sailing boats were already visible on the Seine in the 1840s.

When in 1853, Baron Haussmann first began improving Paris water, air and sewerage, the old city's sanitation was primitive with most effluent spilling directly into the Seine. In this same river-water, laundry was washed and Parisians bathed – and drowned. The Paris morgue was located until 1864 on the Quai de l'Archevêché on the south-eastern tip of the Île de la Cité, behind Notre-Dame Cathedral.

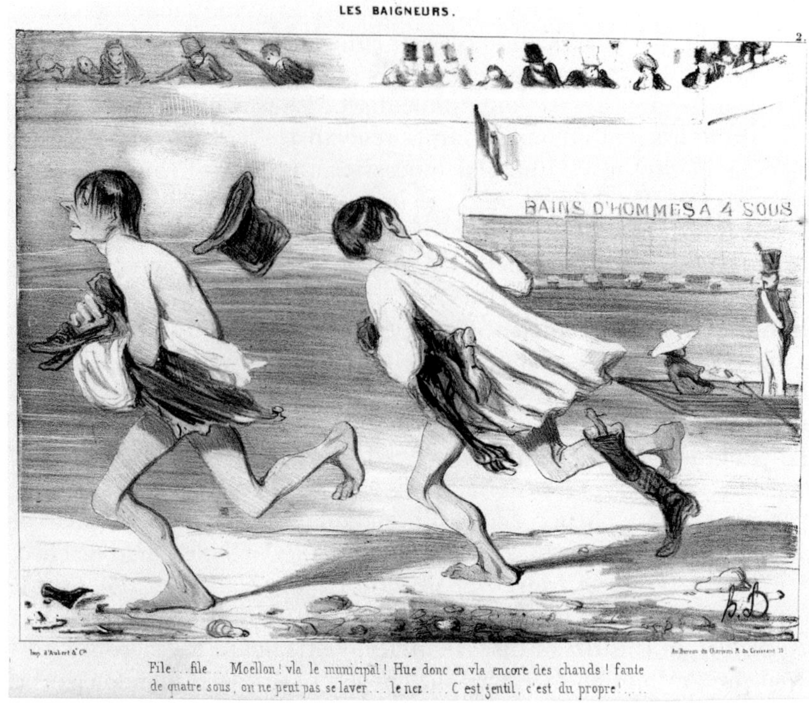

Figure 1.3 Honoré Victorin Daumier (1808–1879), "File ... file... Moellon ! vlà le municipal ! Hue donc en vlà encore des chauds ! Faute de quatre sous, on ne peut pas se laver... le nez... C'est gentil, c'est du propre ! ..." ("Run... Run... Moellon! here comes the municipal! So giddy-up, its getting hotter [pricier]! Lacking four sous, you can't wash your ... nose [penis]... How nice, that's clean [dirty]! ... ", [nineteenth-century French slang alternatives in square brackets]), 1839. From series "Les Baigneurs" ("The Bathers"). Lithograph, image: 21.5 × 27.2 cm. Paris: *Le Charivari*. (Alamy Stock photos.)

With drowned corpses frozen and displayed there for identification, it became a freak-show spectacle for gawpers.[20] Bathing establishments located downriver from the morgue and the city, at the Quai d'Orsay – like that of the first entrepreneur Poitevin in 1762 – were plagued by urban detritus and effluent both human and animal.[21] In contrast, Petit's *école de natation*, located at the Quai de Béthune on the eastern tip of the Île St Louis, and thus upstream from the morgue, had the acknowledged virtue of its clear pure water. The cold 'bains à quat'sous' – the four 'sous' of Ancien Régime currency dates their origin – were sited 'all over the Seine' by the 1840s, when their use cost 20 centimes in the Napoleonic decimal currency. For locals, who kept the habitual name 'bains à quat'sous', these gender-segregated mass baths were the mainstay of Parisian hygienic bathing.[22] While hot baths were also available, they were out of the reach of most Parisians, as Briffault pointed out:

> The hot baths in the town centre boasted a luxury approaching oriental magnificence, and two steps away from this animation the cold baths remained immobile in their native roughness.[23]

Clearly marked by class difference, as evidenced in Georges Seurat's monumental *Bathers at Asnières* (1883–84, National Gallery, London), water was also a gendered medium. In the male imaginary, water was stained by its association with the feminine: Venus-Aphrodite being the classical reference here. The upper classes aside, warm bathing was deemed suitable only for women – and even then, with caution. It was especially problematic for the vigorous male since warm baths induced a softness and lassitude redolent of sensual pleasures beyond the pale: They literally undid his manhood, it was thought. This was particularly the case with youth: The warmth and privacy of bathing were considered to encourage masturbation, and required close supervision.[24]

Popular in English public schools, cold bathing however rapidly gained traction. By the 1840s, it already had medical advocates in France who recommended ice cold showers as both health-giving and invigorating (Figure 1.4). Sea-bathing too was sanctioned by this date and widely practiced on health grounds for 'cures' – for those who could afford it. Indeed, whether cold or hot springs, or the sea, Spas, Bains and Baden rapidly became all the rage throughout Europe for the ailing rich elite.

At first bathing, or aquatics as sport and as (aptly named) 'recreation', was not strictly distinguished from 'bathing' to cleanse the body. Indeed, bathing for personal

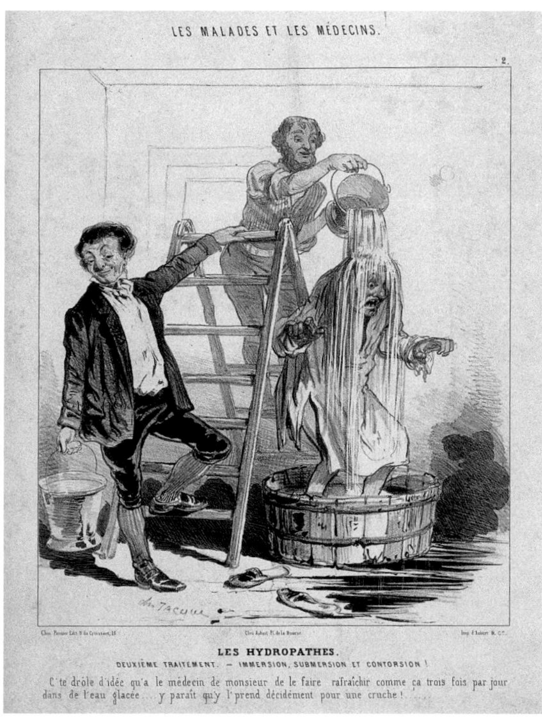

Figure 1.4 Charles-Émile Jacque, "Les Hydropathes: Deuxième Traitement: Immersion, Submersion, Contorsion". ("The Hydropaths: Second Treatment: Immersion, Submersion, Contortion"). A man is treated to a cascade of iced water in the name of hydrotherapy. From series "Les Malades et les Médecins" ("The Sick and The Doctors"), 1843. Lithograph, image 23.1 × 19.7 cm. Paris: Imp. Aubert & Cie. (Photo courtesy: The Wellcome Library, London.)

hygiene, and more particularly the entrepreneurial provision for such effete practices among the Parisian upper classes at least, dates back to the pre-Revolutionary eighteenth century. The first river-barge bathhouse of M. Poitevin was, tellingly, built with all the outward appearance of a classical villa; Diderot and d'Alembert's *Encyclopédie* documented his grand bathhouse in the year of its establishment, 1762.[25] Poitevin's vast, elegant, floating emporium included not only private bath cubicles but also cafés, restaurants, smoking rooms and body-care facilities offering pedicures and a barber. Still in place in the 1840s under different management, Briffault wittily noted its orientalist attractions to the sedentary *parvenue*:

> At his bath, the Parisian bourgeois dreams of the Orient, its delights, its voluptuousness, its perfumes and its odalisques, opium and its ecstasies, and partakes in a pot snack.[26]

It was from such luxury bathing establishments that public bathhouses and swimming schools evolved. These latter became the more disciplined health-promotion side of the emerging passion for water. For the masses, river bathing was at first the only option for bodily cleansing which is why, in the early nineteenth-century, bathing and swimming for the general public remained broadly undifferentiated. By 1847, the Paris *Almanac de Commerce* distinguished six 'bains de natation' as against the generality of 'bains publiques'.[27] The earliest dedicated *bain* or *école de natation* was in fact opened by Turquin at the Quai Bethune on the Île St Louis in 1786 – the first in Europe to be constructed and advertised as such.[28] With swimming recognized by the medical profession as a positive exercise for training the military and enhancing the physical development of youth – and subsequently for improving the manly vigour and vitality of an increasingly enfeebled urban population – entrepreneurs were quick to capitalize on swimming lessons. For schools and the 'Hautes Écoles', swimming lessons became compulsory, and students who failed to take at least weekly swimming classes during the summer months were required instead to bath every 15 days.[29]

The architecture of these establishments was of course distinct: Swimming pools required large expanses of water for forward motion rather than the individual cubicles found in posh bathhouses. Yet for the masses in the 'bains à quatre sous' these functions overlapped in a bizarre melange of recreation and bodily hygiene. For individual bourgeois (hygienic) bathing and organized swimming, there are significant political, architectural and commercial histories dating back well into the eighteenth century. And since, as noted, there are important distinctions of class and gender at play here, these histories are also inflected by social, medical and sexual histories. Indeed, the police ruling of 3 June 1783 prohibiting 'wild' swimming (generally male) in the Seine within the City limits reveals the desire to 'civilize' bathing practices, as well as to control naked public exposure on moral grounds (see Figure 1.3). Under this new regime, ordinary citizens were obliged to pay fees to wash in the (river-water) baths that began to proliferate along the Parisian *quais* during the Restoration and the July Monarchy.[30]

Bathing, especially these populist bathing rituals, constituted key themes exploited by Daumier as early as the 1830s (Figure 1.5). The 'bains à quatre sous' as they continued to be known, were designed for mass bathing, either male or female, segregated, and not for swimming.

In his lithograph c.1840 (see Figure 1.5), Daumier caricatured the 'cleanliness' of these public baths: A youth, left, drinks the water, while another adjacent to him

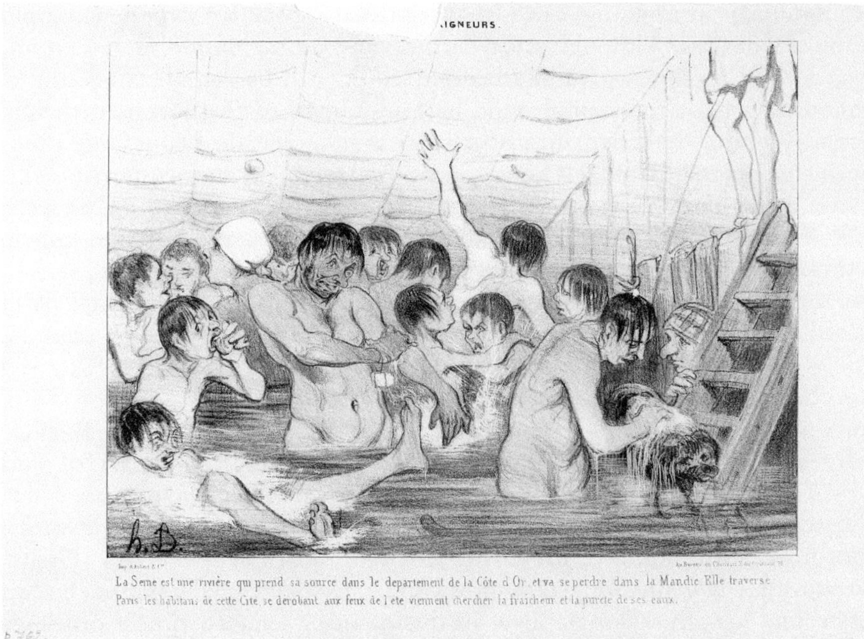

Figure 1.5 Honoré Victorin Daumier (1808–1879), "Le bain à quatre sous". (The bath at four *sous*'). "La Seine est une rivière qui prend sa source dans le départment de la Côte d'Or et va se perdre dans la Manche. Elle traverse Paris : les habitans de cette Cité, se dérobant aux feux de l'été viennent chercher la fraicheur et la pureté de ses eaux". ("The Seine is a river with its source in the Côte d'Or and its estuary in the Channel. It crosses Paris whose inhabitants, hiding from the fires of summer, come seeking the freshness and purity of its waters".), 1839. Plate 3 from series "Les Baigneurs", ("The Bathers"). Lithograph in black on ivory wove paper, 204 × 270 mm (image), 253 × 342 mm (sheet). (Alamy Stock photos.)

scrubs himself and, far right, a third man lathers his dog. Meanwhile, prominent in the foreground, faeces float freely among the bathers. Such unhygienic conditions were equally rife, however, in the summer overcrowding of higher-class establishments like the École Deligny, caricatured in Daumier's *'Croquis d'été'*, c.1858. Thanks to their unsanitary conditions and their democratizing undress, the 'conceit of social distinction' was exposed as just that: A conceit.[31] The numerous lithographs notably in Daumier's series of *Baigneurs* and 'Croquis d'été' (c.1840 and c.1858, respectively), confirm the popularity of Parisian water-based recreation. They also highlight two key aquatic truths – not just the democratizing effects of disrobing, but the stark reality of 'actual' bodies. Bathing made bluntly clear that distinctions of class depended on **dress**, and that not all bodies were 'sportif'. While clearly caricatures such as Daumier's heightened truth to create their acid humour, they nonetheless reminded their audience that few men attained the Apollonian physique of ideal manhood. Indeed, most were sagging, scrawny, or severely obese, their often swaggering body-language expressly designed by Daumier to mock his characters' inflated self-importance and ignorance of their own physical degeneration.

In Paris, the increasing popularity from the 1830s of public baths for hygiene[32] and recreational bathing afforded not just a theatre for competitive male display and 'healthy' homosocial bonding, but also a legitimate space for experiencing pleasurable same-sex desires. Although river bathing and its regulation remained an issue in Paris, beyond its limits greater freedoms could still be found. Given the dearth of bathhouses outside the capital, wild bathing remained the norm. At the close of the Second Empire, the emergence of regular Sunday leisure haunts like that at La Grenouilliere, on the Seine at Croissy, near Bougival, on the western outskirts of Paris, gave a questionable legitimacy to clothed heterosexual bathing, as well as to dining and boating for the middling and working classes and Parisian bohemians. It was this convivial abandon that Monet and Renoir famously painted, side-by side *en plein air,* in the summer of 1869. How different the narrative in their paintings, dominated by the landscape setting, to that figured in the river bathing scenes I shall examine of Gustave Caillebotte (1848–1894), Frédéric Bazille (1841–1870) and, referencing the younger generation, Georges Leroux (1877–1957).

Leroux was openly homosexual, while the older Caillebotte and Bazille were reputedly so. At his family summer residence at Yerres, south east of Paris, in 1877–1878 Caillebotte painted numerous *plein air* scenes of boating and bathing. In 1878, the painter probably combined studio-posed figures with a *plein air* riverscape for his painting, *Bather preparing to Dive: Banks of the Yerres* (Figure 1.6). The extraordinarily mauve-tinged shadows and blueish reflected light that he captures on and around his athletic male diver in this painting lend an almost other-worldly air of seclusion and intimacy to the scene. It is only the addition of a more prosaic

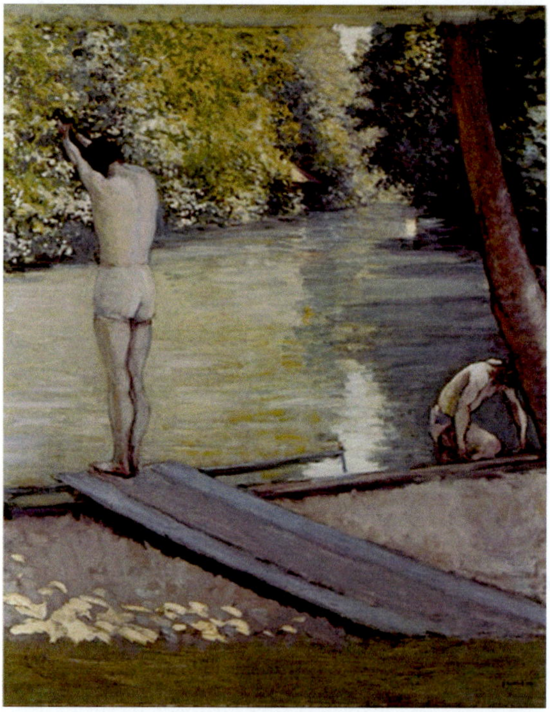

Figure 1.6 Gustave Caillebotte, *Baigneur s'apprêtant à plonger, bords de l'Yerres.* (*Bather Preparing to Dive, Banks of Yerres*), 1878. Oil on canvas, 117 × 79 cm. (Private collection, USA. By kind permission of the owner and the Comité Caillebotte, Paris.)

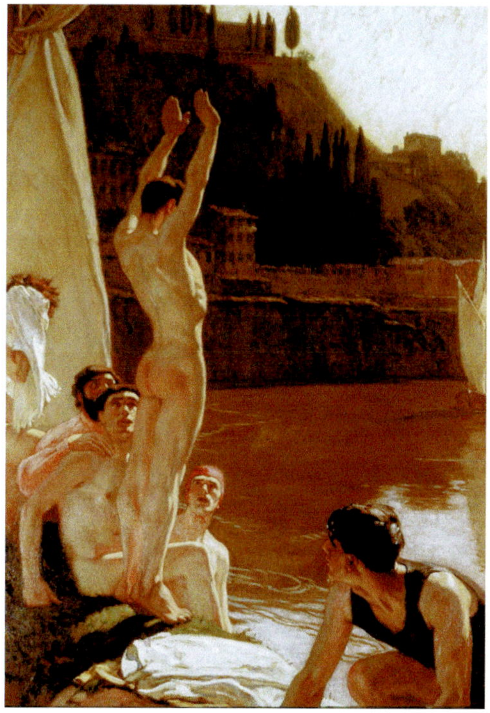

Figure 1.7 Georges Leroux, *Les Baigneurs du Tibre.* (*Bathers in the Tiber*), 1909. Envoi de Rome. Oil on canvas, 2.57 × 1.72 cm. Musée de l'Oise, Beauvais. (Photo courtesy: Fae Brauer.)

second swimmer emerging from the river into sunlight, far right, that grounds the depiction in normalcy. Caillebotte's Impressionist palette and touch make this painting – with its brilliantly unbalanced composition – shimmer with heat and light. In stark contrast is the academic panache of Leroux who, while based in Rome, submitted his *Bords de Tibre* (Figure 1.7) to the 1909 Paris Salon des Artistes Français.[33] Here, immediately signalling a Roman delight in the athletic body on the banks of the Tiber, Leroux distances his subject from the insistent modernity seen in Caillebotte: The Rome that Leroux references might as well be Ancient as modern. Equally shown from behind, but naked, Leroux employs a similar device of shadow and reflected light on his figure poised to dive. However, he uses it to demonstrate his own prowess in flesh modelling, emphasizing his diver's slim waist, taut ruddy buttocks and flexed muscular legs. Leroux's group of young bathers all turn to admire their fellow sportsman's exquisite body stretched up to perform his dive – sight-lines that of course both invite and authorize our own admiring gaze. In Caillebotte's painting, there is no such explicit invitation: Indeed we are voyeuristic intruders on this private moment.

'Wild Swimming', Hygiene and Virility

Throughout continental Europe, the rejuvenating powers newly associated with open water, sunshine and fresh air gave legitimacy to aquatic practices within the discourse of the new vitalism: They represented a vigorous and pleasurable route to the restoration of urban man's flagging *élan vital.* 'Wild' bathing in isolated country rivers offered men of all classes sustained opportunities for male homosociality, naked display and desire. Meanwhile, within the seclusion of the bourgeois domestic setting, increasing approval was accorded to the act of private personal hygiene despite, as we have seen, male bathing's troubling association with masturbation and voluptuousness. Indeed, the social *difference* personal hygiene conferred was crucial in a bourgeois society keen to mark its distinction from the 'great unwashed'. Thus, modern medical discourses of hygiene and cleanliness permeated not only the personal feminine *toilette* and the brothel but also the world of the virile male.[34]

Cleverly combining themes of sporting manhood and personal hygiene, Caillebotte's radical *Man at his Bath*, painted in 1884 (Figure 1.8) reveals the toned physique of a naked male apparently unaware of his spectator. His back turned towards us, his towelled hair attractively mussed up, this bather's endearing disarray echoes the youthful contrivance of Bazille's *Fisherman with Net,* painted 15 years earlier (Figure 1.9). Subverting the stock image of the female bather, Caillebotte's startling male nude (Figure 1.8) assures us that 'real men' take baths. At the same time, it offers the viewer a delectable homoerotic spectacle of an athletic body and firm manly buttocks plus a tantalizing glimpse of balls, atop legs tautly flexed against his energetic rubbing. Indeed, the positioning and contrasting texture of his bath towel serves to emphasize his two beautifully developed gluteus maximus, further drawing them to our attention. The crusted weight of paint, made thickly opaque with flake (lead) white, gives extraordinary materiality to the man's solid muscular form, almost sculpting him

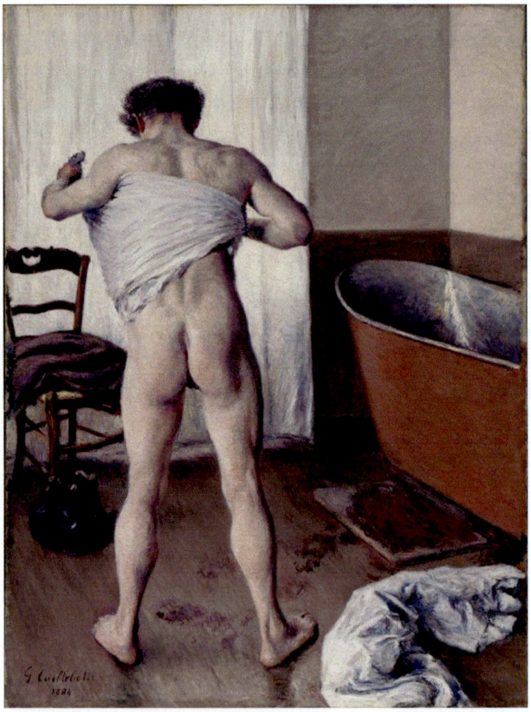

Figure 1.8 Gustave Caillebotte (1848–1894), *Homme au bain.* (*Man at his Bath*), 1884. Oil on canvas, 144.8 × 114.3 cm. (Courtesy: The Museum of Fine Art, Boston, MA. Public domain: open access.)

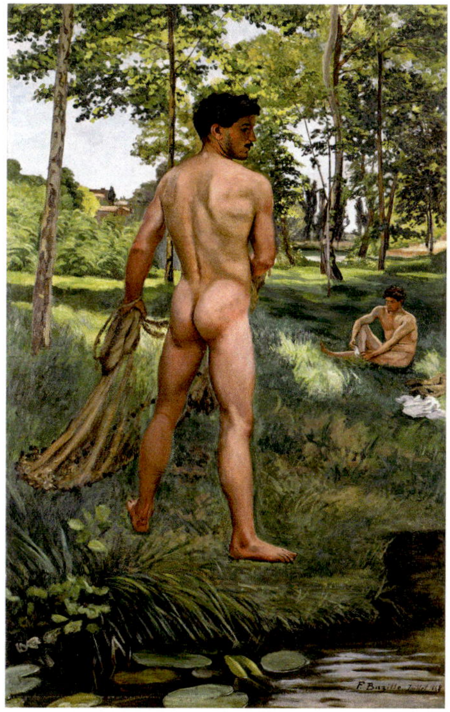

Figure 1.9 Jean-Frédéric Bazille (1841–1870), *Pêcheur à l'épervier*. (*Fisherman with Net*), 1868. Oil on canvas, 137.8 × 86.8 cm. © Remagen, Arp Museum Bahnhof Rolandseck/ Collection Rau for UNICEF, Inv. Nr. GR 1.653. (Photo: Mick Vincenz, Essen.)

in paint in a manner that pre-figures Lucien Freud's mature nudes. His visible hand is tanned, as is his neck, suggesting exposure to sunshine at a time and in a society where most gentlemen wore gloves. The pallor of his flesh elsewhere indicates he was not exposed to outdoor labour, suggesting the tanned parts of his anatomy may have resulted from rowing, which would account too for his well-built physique and notably muscular legs. Certainly, the elegantly polished black boots, neatly set, bespeak a man of Caillebotte's own class.

This is not effete *sensuality*, we are told, but virile personal *hygiene*. The man's athletic form and thorough ablutions combine to extol the sanitary value for healthy manhood of a designated bathroom (at least for the wealthy classes), in a modern Parisian luxury apartment – in this case, Caillebotte's own.[35] His tin-lined copper bath would have been vastly expensive at the time. It would have held the heat well but would not have been readily portable. Indeed, the dado décor here suggests a permanent bathroom with the bath plumbed in. Yet Caillebotte does not overtly reveal this to be the case. Many bourgeois 'bathrooms' were still only temporary affairs, with the majority of those later nineteenth century Parisians who bathed at home renting tin tubs plus hot-water supplied as required (generally not often).[36] It was radical indeed to represent a male *toilette* – let alone one as heroic and virile as Caillebotte's bather. In a culture that still associated the practice at best with feminine lassitude, at worst with disease, full body immersion in warm water conjured the spectre of illicit sexual thoughts and practices.[37] Posed indoors as a hygienic bather,

Caillebotte's male nude is legitimated by its bathroom context. In contrast, Bazille's outdoor naked male has no equivalent rationale (see Figure 1.9).

Like all his bather compositions,[38] Bazille's *Fisherman* is staged in the Southern milieu of his family estate at Méric near Montpellier. This naked figure, awkwardly superimposed on the local landscape setting, reveals all the signs of a studio composition designed to emulate *plein air*, and thus appear authentically modern. In fact, a less authentic modern 'fisherman' is hard to imagine. His muscular back, taut buttocks and firm upright stance form a visual metonym for the erect penis – that organ hidden from us but exposed to the naked male companion seated facing him. With only the flimsy disguise of a fishing net, there is little here to legitimize the viewer's pleasurable enjoyment of the naked male rear. Far from promoting the national procreation imperative, this scene is surely Bazille's celebration of the homoerotic delights of virile athletic manhood.

Common from youth onward, rural bathing was considered a vigorously healthy practice. Yet it was also a potentially subversive one. The normative constraints of civic behaviour in a public space could readily be cast off at the riverbank or beach, along with the clothes of modern civilization. Thus unconstrained, the male bather could experience a vitalist open-air abandon, unself-conscious in the pleasurable company of his peers. By the close of the Second Empire, mixed river-bathing in Parisian resorts along the Seine had (as we have seen) become a popular Sunday pastime with a heterosexual *frisson*. Renoir's later *Luncheon of the Boating Party* depicted the boating camaraderie among his friends at the Restaurant Fournaise on the Île de la Grande Jatte at Chatou, which was also west of the city.[39] Seurat's *Bathers at Asnières* was set in the same Parisian *banlieue* facing the Île de la Grande Jatte, and focused instead on urban working men on a bathing break from their labours.

Beyond the immediate environs of the capital, in regions more remote from Paris like the Midi and the Languedoc, wild naked bathing for health and pleasure grew alongside the rise of nudism and the popularization of revitalizing outdoor sports. The bodily exposure to sunshine entailed in nude bathing came to be seen as powerfully restorative in its own right.[40] Whether as 'passive' recreation or, at first, purely as healing, from the early 1890s onward, medical professionals opened specialist sanatoria along the French Midi coast offering dedicated heliotherapy in the treatment of such conditions as tuberculosis, psoriasis, lupus and rickets.[41] Significantly too, in art 'white' skin colour was gendered: Male nudes were typically depicted with darker 'tanned' (therefore more manly) skin than female nudes. Gradually, and especially from the 1920s, the class and racial stigma attached to tanned skin gave way to its aesthetic appreciation, and as a sign of bodily 'health'.

As in Bazille's depiction of bathing in the Languedoc, so in the Midi male same-sex bathing was celebrated by Cézanne, notably in paintings after 1880. From the late 1880s, sea-bathing at Åsgårdstrand in Norway first occurs in Edvard Munch's *plein air* paintings of powerful nude men at the beach, as indicated by Pat Berman in Chapter Two of this book (see Figure 2.7). Sea-bathing was often recommended on medical grounds, as in Munch's own case it was prescribed to restore his mental health. In 1907, Munch was photographed on the beach working on his large-scale *Male Bathers* (Ateneum, Helsinki), and there, in his all but naked performance, the heroic masculine ideal fuses with the radical modern artist at work *en plein air*. This Nordic enthusiasm was doubtless encouraged too by the growing German *Freikörperkultur* [Free Body Culture] movement from c.1900, which advocated 'nudism, regular bathing, and soaking up the sun as a path to physical and spiritual renewal'.[42] From the 1880s in England, Henry Scott Tuke likewise focused on idyllic scenes of male *plein air* nudity, his nubile ephebic youths being portrayed bathing and fishing naked on the Cornish coast.[43] The naked male bathers of both Tuke and Munch were modelled by actual contemporary individuals. However, in Cézanne's painting, male bathing was reimagined retrospectively, the artist gazing back nostalgically to his youth with Émile Zola bathing *en plein air* in the countryside around Aix-en-Provence. While Tuke's ephebes were overtly homoerotic and 'Uranian', Munch and Cézanne both heroicized their muscular bathers as the embodiment of strength, virility and morality. Shielded from the urban gaze as a practice, yet openly consumed in art exhibitions – at the Salons and picture dealers' galleries – naked male bathers were a potent theme for fin-de-siècle male painters, providing the opportunity for heroic modern nudes. Far from being displaced by the female nude, the male nude was in fact equally ubiquitous. Yet given the heteronormative sexual imperative of the period, with its fears of degeneracy and persecution of 'inversion', such depictions of males displaying virile nudity inevitably draw our attention to their homoerotic undertow.

Sculling, Rowing and Yachting

In parallel to ideas of outdoor physical revitalization, rowing grew rapidly in popularity from the 1860s onward. Like bathing, rowing afforded men not only the opportunity for healthy stripped-down homosociability but also an energetic and invigorating competitive sport: Rowing clubs and teams proliferated. Bathing and boating often co-habited, as on the rivers around Paris, and in Philadelphia where Thomas Eakins painted river sports. Caillebotte's yachting paintings date from the 1880s after his move to Gennevilliers, on the Seine riverbank opposite Argenteuil west of Paris. Yet his earlier series at Yerres in 1877–1878 focused not just on swimmers and divers, but on fishermen, rowers and canoers. Eakins trained in Philadelphia at Pennsylvania Fine Art Academy and, in 1860s, at the Paris École des Beaux-Arts under academicians Jean-Léon Gérôme and Léon Bonnat. After his return to Philadelphia, Eakins painted sculling scenes on the Schuylkill River in 1871–1874, while his bathing pictures date from 1884 to 1885. Eakins's portrait of high school friend Max Schmitt celebrated him as single-scull champion, a painting whose contemporary sporting theme shocked conventional Philadelphian sensibilities (Figure 1.10).[44]

Identifying Eakins's rowing paintings as emblematic of his masculine insecurities as well as a bid to assuage them, Martin Berger sees the artist's inclusion of his own self-portrait as a vigorous rower in the central middle distance behind Schmitt in *The Champion Single Sculls* (1871), as him 'metaphorically strengthen[ing] his links to the male world'.[45] Berger suggests Eakins's fusion of science and art in this painting echoed that found in the design and manufacture of modern racing craft, a fusion found too in Caillebotte. However, Eakins's work was not recognized as vigorously muscular by his critics until the early twentieth century, when it was identified as embodying a modern American 'virile' masculinity.[46] In his later bather paintings like *The Swimming Hole*, Eakins echoed an Emersonian Transcendentalism that located male vitality in the embrace of 'raw nature', a specifically American sensibility Eakins shared with his friend Walt Whitman.

In Caillebotte's 'rowing' paintings, he depicted not only single sculls and canoes but rowing 'à deux'. His extraordinary 1878 painting, *A Boating Party* or *Rower in a Top Hat* (Figure 1.11), has the artist-spectator seated facing a well-endowed Sunday

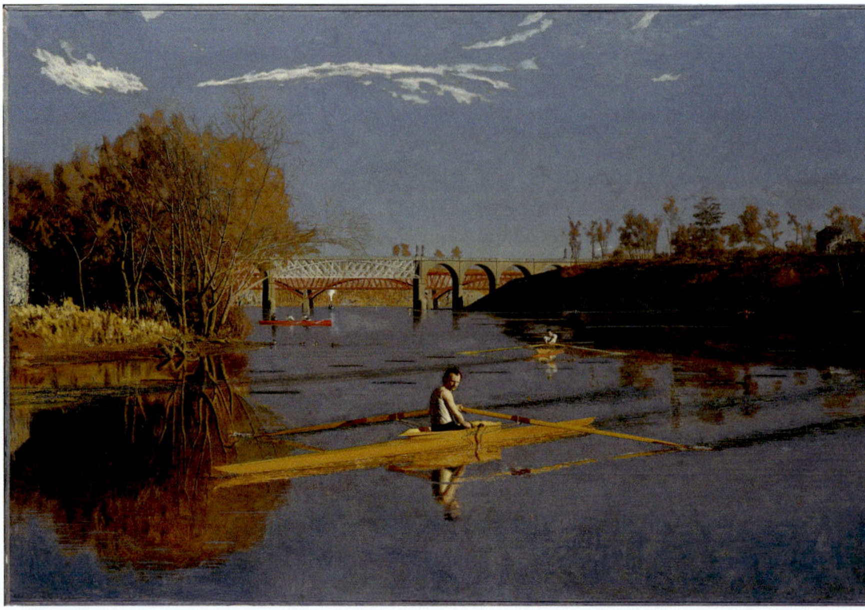

Figure 1.10 Thomas Eakins (1844–1916), *Champion Single Sculls*, 1871. Oil on canvas, 81.9 × 117.5 cm. (Courtesy: The Metropolitan Museum of Art, New York. Public domain: open access.)

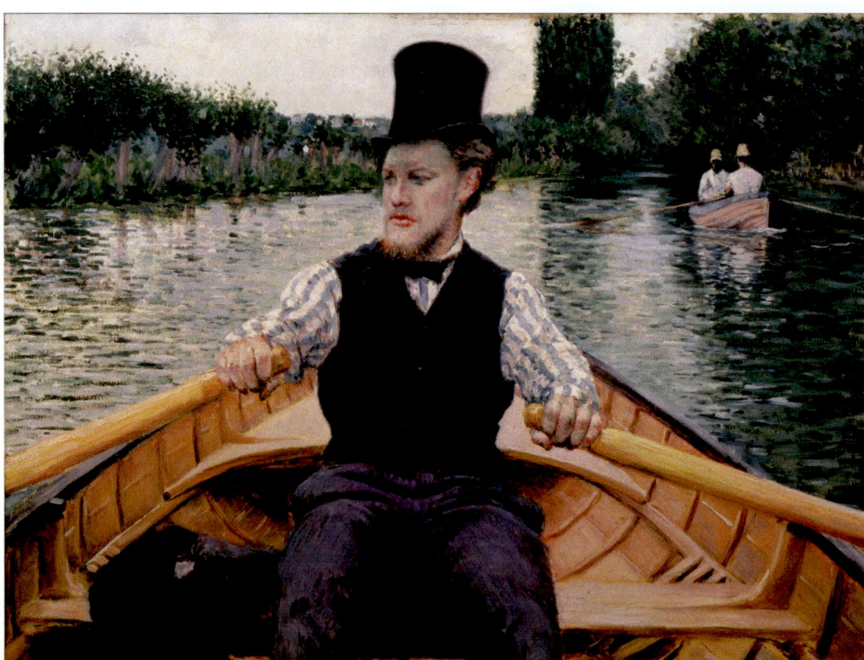

Figure 1.11 Gustave Caillebotte (1848–1894), *Canotier au chapeau haute de forme.* (*The Boating Party [Rower in a Top Hat]*), 1878. Oil on canvas, 90 × 117 cm. (Private collection. Courtesy: The Comité Caillebotte, Paris.)

rower: As viewer, we partner him in the boat. We appear within touching distance of this vigorously powerful male stripped down to his shirtsleeves for an unexpected jaunt – not dressed for the part as are the two men in the boat beyond. Like theirs, this is a two-man affair. Caillebotte's handsome rowing partner has a well-built physique, his muscular arms and torso straining against the limits of his city clothes. Clearly a sportsman, he has massive bulging thighs and a resplendently pronounced crotch positioned centrally before us. His powerful large hands suggestively grip the oars, their slight a-rhythm suggesting him deftly manoeuvring rather than simply rowing the craft. However, with his formal bowtie and top hat still in place, this is surely a companionable exercise in restorative pleasure after the pressures of city life, rather than serious *sportif* rowing. As Mary Morton argues, Caillebotte in this work, as in his male bather paintings, subverts the genre by substituting a delectable male specimen for the normative female object of desire, as found in paintings like Manet's *Boating* (1874) or Renoir's various *canotiers* from the 1870s.[47] Citing Anne Distel, Morton notes the words of Maupassant (a friend and fellow *canotier*), on 'the strange feeling of pleasurable self-abandonment that often overcomes a person gliding over the water in a boat', the writer describing the boisterous groups of 'half naked' men playing and joking, with their bare arms and trunks clad in tight-fitting cotton shirts.[48] Lacking even a singlet, the 'built' Olympian body of the sculler aggressively confronting us on the September 1903 cover of American *Physical Culture* magazine (Figure 1.12) aptly represents both the proximity and distance of these two oarsmen: 25 years, an Olympic revival and a continent apart.

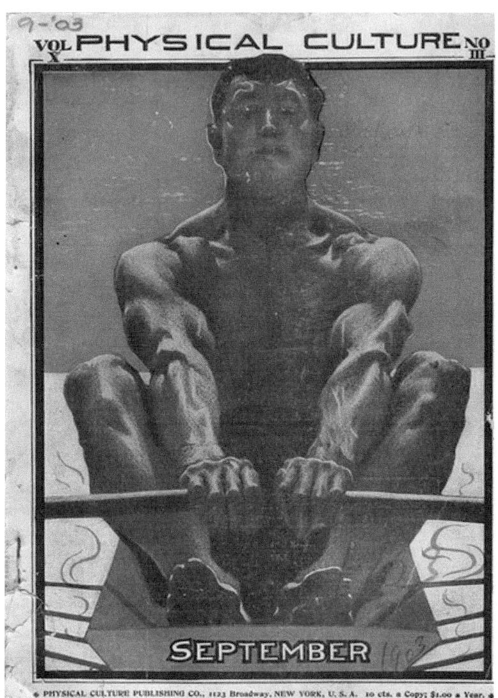

Figure 1.12 "Scullsman", cover picture, *Physical Culture* magazine, Vol. III. New York, Physical Culture Publishing Co., September 1903. (Courtesy: "The Magazine", San Francisco.)

Given the fears of urban 'degeneration' voiced in Third Republican Paris, where one's virile manly heterosexuality needed constant affirmation, *Regatta at Argenteuil* served Caillebotte in similarly multi-layered ways to his *Male Bather.* Such affirmation was especially urgent for unmarried men without offspring or those associated with the effete decadence of the Paris art world. Like fellow artist-sailor, Paul Signac in Saint Tropez, Caillebotte by the 1880s had moved into yachting, a vigorously manly sport, racing his own boats near his home on the Seine at Petit-Gennevilliers and on the Normandy coast. Akin in resonance to Eakins' *Single Scull*, Caillebotte's *Regatta at Argenteuil* (1893, private collection) includes a self-portrait of the artist at the helm of his own racing yacht, '*Roastbeef*', which had the added male kudos of having been designed by him and built in his own boatyard.[49] A personal manifesto, *Regatta at Argenteuil* thus celebrated all of Caillebotte's vitalist manly attributes. These included not merely his personal wealth for leisure and boat-building, and his achievements in yacht racing as the French champion. Caillebotte was also France's foremost yacht designer and a successful artist able to portray himself demonstrating both his sailing skills on '*Roastbeef*', and its fingertip-sensitive rudder design.

These attributes served not only to secure Caillebotte's 'manliness' but also his virility, which was identified with the 'man of action' who turned his art and ingenuity to heroic achievements. During this Darwinian and Neo-Lamarckian period of depopulation and urban degeneration, when Republican masculinity was under renewed scrutiny, male heterosexual virility required constant reaffirmation. The visual trope of the upright muscular male body served as an apt metonym for the erect phallus and its virile performance – traits embodied both in Caillebotte's male *Bather* and Bazille's *Fisherman.* Arguably the male bodies of Bazille and Caillebotte may be regarded as prescient signs of a nascent Vitalism: What Baron Pierre de Coubertin would theorize as the role of sport in revitalizing the body, as elaborated in Chapter Four by Pascal Rousseau. For Coubertin, the body energized by sport and exercise could cure 'universal neuroses' and provide a 'vital form of adaptation to modern life'. Paradoxically, however, in the face of Republican advocacy of heteronormativity and population regeneration, the vigorously muscular, athletically 'vitalized' body could equally be a virile homosexual body. Given the same-sex (or bi-sexual) orientation associated with Eakins, Caillebotte, Bazille, Tuke and equally Leroux, the celebration of the virile male nude in painting – and indeed in Vitalist philosophy – takes on meanings at once more complex and more overtly homoerotic.

Aquatics and the visual arts worked closely in the construction of a healthy athletic ideal for the modern male body: *Élan vital* was figured in the well-scrubbed and 'naturalized' *beau idéal.* This anatomical ideal became more 'scientific' during the early twentieth century through the work of anatomist-sculptor, Dr Paul Richer and physiologist, Étienne-Jules Marey, both of whom were concerned too with the moving body in action, whether athletic, military, or laboring. Concurrently a more 'holistic' approach involving diet and lifestyle as well as exercise, was adopted by early 'bodybuilders' and physical culturists like Desbonnet, Sandow and American fitness pioneer Bernarr MacFadden (1868–1955).[50] The modern morphological ideal and the research supporting it were themselves closely imbricated in the visual arts, whether popular, fine art, or photography, with their enduring debt to classical body prototypes. During the twentieth century, the sporting ideal become more Herculean, more 'perfected', more artificial, more drug-driven. However as distinct from the body built for theatrical performance, athletic masculinity in competitive sports like aquatics entailed exhaustive training as well as physical fitness. For participants and spectators alike, such sports offered a license to enjoy the display of semi-naked male bodies deemed at once virile and heroic.

While raising national *élan vital*, sport may have reinforced potentially rigid ideals of male honour, valour and fitness, but simultaneously it created a space of sanctuary for a newly 'virilized' homosexuality.[51] This occurred at a time when – in the wake of the 1895 Oscar Wilde trials and the rise of sexology and sexual surveillance – the 'effete degenerate' faced growing exclusion and persecution.[52] Conterminously, the persistent ideal of virile white 'healthy' (heterosexual) manhood – of heroism, militarism, courage, stoicism – worked through the suppression of its multivalent inverse, of what was seen as normative masculinity's shadow: The 'feminine' abject, Blackness and homosexuality. This process of denial gave credence to a more sinister vitalism emerging at the level of populations after 1900: Not just sexology, but nationalist racial 'science' and eugenics.

Notes

1 See Fae Brauer and Serena Keshavjee, *Picturing Evolution and Extinction: Regeneration and Degeneration in Modern Visual Culture* (Newcastle upon Tyne: Cambridge Scholars Publishing, 2015); Fae Brauer and Barbara Larson, *The Art of Evolution: Darwin, Darwinisms and Visual Culture* (Lebanon, NH: The University Press of New England, 2009). See also https://www.ncbi.nlm.nih.gov/pmc/articles/PMC3730912/ accessed 14 March 2020.

2 Henri Bergson's *Creative Evolution* was published in Paris in 1907; English translation 1911; see also F. Burwick and P. Douglass (eds.), *The Crisis in Modernism: Bergson and the Vitalist Controversy* (New York: Cambridge University Press, 1992). For more information and a bibliography, see https://plato.stanford.edu/entries/bergson/ accessed 14 March 2020.

3 Sarony 'cultivated' his own extraordinarily large tin fig-leaves for use in his photographs of the naked Sandow, refer Fay Brauer, 'Virilizing and Valorizing Homoeroticism: Eugen Sandow's Queering of Body Culture before and after the Wilde Trials', *Exploring Queer Cultures and Lifestyles, 1885-1967*, Reina Lewis and Andrew Stephenson (eds.), *Visual Culture in Britain* (18, no. 1, May 2017): 35–67.

4 Michel Foucault, *The History of Sexuality (L'Histoire de la Sexualité: La Volonté de savoir,* 1976); vol. I, trans. Robert Hurley (New York: Pantheon Books, 1978) 43.

5 Whilst Sandow was performing in the US (1893–1894), his promoter Florenz Ziegfeld introduced paying backstage visitors who were permitted to feel Sandow's muscles; see R. Brandon Kershner, 'The World's Strongest Man: Joyce or Sandow', "Joyce and Advertising": *James Joyce Quarterly*, 30–31: 30/4–31/1 (Summer-Fall, 1993), 667–693; 669, Figure 7.

6 Edmond Desbonnet (1868–1953), Doctor of Philosophy and Doctor of Sciences, is considered the father of physical culture in France. In 1901, his book *La Force physique: culture rationnelle* included his own methods and those of Attila and Sandow, and in 1902 he founded his monthly magazine *La Culture physique*, which continued until 1967. Desbonnet's *La Méthode Desbonnet: Comment on devient athlète*, preface by Pierre Loti (Paris: Berger-Levrault, 1909 [6th edition]), described the techniques taught at his gyms in Lille, Roubaix, Paris and across Europe in Genève and Bruxelles, etc. Eugen Sandow (1867–1925), Prussian bodybuilder, showman and physical culturist, performed and advocated his methods throughout the Anglophone world. Between 1897-1907 he published a *Magazine of Physical Culture* (with title changes); *The Construction and Reconstruction of the Human Body* (1907) included a foreword by his admirer, Arthur Conan Doyle. See David Waller, *The Perfect Man: The Muscular Life and Times of Eugen Sandow, Victorian Strongman* (Brighton: Victorian Secrets, 2011).

7 Many health and physical culture magazines contained articles addressed to women too, and there were also dedicated feminine hygiene and exercise books.
8 Pierre de Coubertin, *Essais de psychologie sportive* (1913; Grenoble: Jerome Millon, 1992), 151.
9 These ideas are found in Georges Vigarello, *Le Propre et le Sale: L'hygiène du corps depuis le Moyen Âge* (Paris: Seuil, 1985), 182–184.
10 Vigarello, *Le Propre et le Sale* (1985), 182.
11 See Michael Hatt, 'The Male Body in Another Frame: Thomas Eakins' *The Swimming Hole* as a Homoerotic Image', *The Body*, special edition, Andrew Benjamin (ed.), *Journal of Philosophy and the Visual Arts* (London: Academy Group, 1993), 8–21; see also Fay Brauer, 'Virilizing and Valorizing Homoeroticism' (2017); Anthea Callen, 'Introduction', and Chapter 2, 'La Lutte', in *Looking at Men: Art, Anatomy and the Modern Male Body* (London and New Haven, CT: Yale University Press, 2018).
12 The key texts are: Robert A. Nye, *Masculinity and Male Codes of Honour in Modern France* (Berkeley and Los Angeles: UCLA Press, 1992); Christopher E. Forth and Bertrand Taithe, *French Masculinities: History, Culture and Politics* (Basingstoke and New York: Palgrave MacMillan, 2007); Christopher E. Forth and Peter Cryle (eds.), *Sexuality at the Fin de Siècle: The Making of a 'Central Problem'* (Cranbury, NJ: Associated University Presses, 2008); Anne-Marie Sohn, *'Sois un Homme!': La construction de la masculinité au XIXe siècle* (Paris: Seuil, 2009). See also Christopher E. Forth, 'Gender, Empire, and the Politics of Regeneration', *French Politics, Culture & Society* (36, no. 2, 2018): 149–156.
13 See Eugène Briffault, illus. Bertall, *Paris dans l'eau* (Paris: J. Hetzell), 1844) and Sun-Young Park's excellent, *Ideals of the Body: Architecture, Urbanism, and Hygiene in Postrevolutionary Paris* (Pittsburgh, PA: University of Pittsburgh Press, 2018).
14 In 1888, Coubertin founded the Comité pour la propagation des exercises physiques dans l'éducation (called Comité Jules Simon); refer Coubertin, Pierre de, *L'Éducation en Angleterre* (Paris: Hachette, 1888) and *Essais de psychologie sportive* (Lausanne and Paris: Payot, 1913).
15 Women competed for the first time at the 1900 Olympic Games. Of 997 athletes, 22 women competed in sailing, tennis, croquet, golf, and equestrianism.
16 The cholera bacterium was also identified that same year by the Italian microbiologist, Filippo Pacini, but was not widely recognized until the 1880s.
17 The Conseil de salubrité was created in June 1802, and a century later became the Conseil supérieur d'hygiène publique de France.
18 There are at least two variants of this theme: Daumier, '"Une facétie [a joke]", "Les Baigneurs"', 1839–1842, lithograph, no. 398, estate of George Longstreet; and Daumier, 'Les Baigneurs' no. 25 published 1842: 'Excuse me, M. le Maire – would you do me the pleasure of telling me if one is allowed to bathe here?', Coll. christchurchartgallery.org.nz
19 See Jaimee Grüring, 'Dirty Laundry …', 2011, PhD dissertation, Arizona State University.
20 For images and texts, see https://wellcomecollection.org/articles/W-RTBBEAAO5m-fQ3M consulted 16 March 2019.
21 Poitevin's successor was Vigier. For detailed information, see Park, *Ideals of the Body*, 255–257.
22 On the 'Bains à quat'sous', see Briffault, *Paris dans l'eau*, 34ff.
23 Briffault, *Paris dans l'eau*, 34.
24 Park, *Ideals of the Body*, quotes Dr Pavet de Courteille's 1827 treatise on boy's schools, *Hygiène des collèges*, 84, 184 and 324 n.106.
25 Briffault, *Paris dans l'eau*, 38, 39, gave the date as 1765; the correct date of 1762 is confirmed by the *Encyclopédie* entry.
26 Briffault, *Paris dans l'eau*, 38.
27 Park, *Ideals of the Body*, 258, and 330, n.60.
28 Park, *Ideals of the Body*, 257; she notes a legal battle with Poitevin forced Turquin to close this swimming pool, which he reopened in 1796 on the Quai d'Orsay. It continued through the mid-century under his son-in-law, Deligny.
29 Park, *Ideals of the Body*, 88.

30 See Park, *Ideals of the Body*, on the evolution in Paris of bathing establishments and their architecture, especially 255–265.
31 Park, *Ideals of the Body*, 267.
32 On the rise of public and private bathing in nineteenth-century France, see Vigarello, *Le Propre et le Sale,* 1985, esp. 199ff. Access to piped domestic water, drainage and sanitation were key problems in Paris, and systematic installation of plumbed water did not begin before Haussmann in 1853; '*tout-à-l'égout*' (sewerage) was not made law until 1894, and cess-pools were still in use in parts of Paris in the early twentieth century. Flush toilets were first introduced in France in the 1890s.
33 For more on Leroux and this painting, refer Fae Brauer, *Eroticizing Lamarckian Eugenics: The Body Stripped Bare during French Neo-Regulation,* Chapter Three, *Art, Sex and Eugenics: Corpus Delecti,* eds. Fae Brauer and Anthea Callen (Hampshire, UK; Burlington, USA: Ashgate, 2008) 96–138. A postcard version of the painting, indicating that it circulated widely, is now in The Wellcome Collection, London.
34 See Georges Vigarello, *Histoire des pratiques de santé: La sain et le malsain depuis le Moyen-Âge* (Paris: Éditions Seuil, 1987; 2015) and 'Hygiène du corps et travail des apparences' in Alain Corbin (dir.), *Histoire du corps. 2. De la Révolution à la Grand Guerre* (Paris: Éditions Seuil, 2005) part 2, 307–320; and Alain Corbin, *The Foul and the Fragrant: Odour and the French Social Imagination* (Leamington Spa, Hamburg and New York: Berg Press, 1986). On feminine hygiene and disease see Anthea Callen, *The Spectacular Body: Science, Method and Meaning in the Work of Degas* (London and New Haven: Yale University Press, 1995), esp. Chapters 2 and 5.
35 The setting is Caillebotte's apartment at 31 Boulevard Haussmann. On Caillebotte's *Man Bathing*, see Callen, *The Spectacular Body*, 145–147, 150–151; Tamar Garb, 'Muscularity, Masculinity and Modernity in Caillebotte's Male Figures', in *Visible Touch: Modernism and Masculinity,* Terry Smith (ed.) (Chicago, IL: University of Chicago Press, 1997), 53–74; Mary Morton and George T. Shackelford, *Gustave Caillebotte: The Painter's Eye*, exhibition catalogue, National Gallery of Art, Washington and Kimbell Art Museum, Fort Worth (Washington DC, National Gallery of Art, 2015) 180–185; they note that *Man at his Bath* was shown at *Les XX* in Brussels in 1888 (albeit hidden in the depths), but was then unseen in public until 1994.
36 Vigarello, *Le Propre et le Sale*, 1985, 203–204, 'Les quartiers les plus fortunés créent un espace privé de bain sans en construire ni vraiment en aménager la place' (204). See also Vigarello, 'Hygiène du corps et travail des apparences ', 3e Partie, 2, *Histoire du corps. 2. De la Révolution à la Grande Guerre,* Alain Corbin (dir.) (Paris : Librairie Decitre, 2005), 307–320.
37 On personal hygiene, water, and sensuality, see Vigarello, *Le Propre et le Sale*, 1985, especially 187–190; see also Park, *Ideals of the Body* (2018): 184–185 (discussing boys' and girls' schools) and Callen, *The Spectacular Body* (1995): 144–145.
38 On Bazille's large male bather painting, *Summer Scene* (1869–1870, Harvard University, Fogg Art Museum, Cambridge, MA), see Callen, *Looking at Men*, 2018, 95. Bazille mentioned in a letter to his father, 2 May 1869, that he planned his bathers and wrestlers in this painting to be nude, as shown too in his early studies: see Michel Hilaire and Paul Perrin (Dir.), *Frédéric Bazille: La Jeunesse de l'impressonnisme*, exhibition catalogue (Montpellier: Musée Fabre, 2016) cat. No. 82, letter quoted 138; studies illustrated 140–141.
39 See the Phillips Collection on-line catalogue for detailed information on the people in Renoir's *Luncheon of the Boating Party.* Renoir's earlier variant with just three diners, *Lunch at the Restaurant Fournaise* or *The Rowers' Lunch* (1879), is now in The Art Institute of Chicago.
40 For women especially, a pale skin denoted a leisured-class refinement, as distinct from the weathered skin of women laborers.
41 See Tania Woloshyn, 'Regenerative Tanning: pigmentation, Neo-Lamarckian eugenics and the visual culture of the cure de soleil', Chapter 10, *Picturing Evolution and Extinction: Regeneration and Degeneration in Modern Visual Culture,* eds. Fae Brauer and Serena Keshavjee (Newcastle-on-Tyne: Cambridge Scholars Publishing, 2016) 193–216; see also Tania Woloshyn, *Soaking Up the Rays. Light Therapy and Visual Culture in Britain, c. 1890–1940* (Manchester: Manchester University Press, 2017).

42 Quoted from The Art Institute of Chicago's entry for Munch's *Boys Bathing* (1899). https://www.artic.edu/artworks/17267/boys-bathing accessed 12 March 2019.

43 On Munch see Amanda O'Neill, *The Life and Works of Munch* (Bristol: Parragon Book Service, 1996); Sue Prideaux: *Edvard Munch: Behind the Scream* (New Haven, CT: Yale University Press, 2005). On Henry Tuke, see Emmanuel Cooper, *The Life and Work of Henry Scott Tuke 1858–1929* (London: GMP, 1988); Catherine Wallace, *Catching the Light: the art and life of Henry Scott Tuke 1858–1929* (Edinburgh: Atelier Books, 2008); and Cicely Robinson, *Henry Scott Tuke*, exhibition catalogue, Watts Gallery (Compton, Surrey) 2021.

44 Schmitt (a lawyer and champion rower) and Eakins were both members of the Philadelphia (or Pennsylvania) Barge Club, one of nine men's clubs in the Schuylkill Navy, and the 12 that rowed on the river. The Schuylkill Navy was organized in 1858, with approximately 300 members, and began hosting annual regattas in 1859. The new racing scull, a long thin craft designed for speed, was then gaining in popularity. See Helen A. Cooper (ed.), *Thomas Eakins: The Rowing Pictures*, exhibition catalogue (New Haven, CT: Yale University Art Gallery, 1996), especially Cooper's chapter, 'Rowing in Art and Life', 24–78, and Martin A. Berger's chapter, 'Painting Victorian Manhood', 102–123. On the reception of Eakins' rowing paintings in Philadelphia, see Marc Simpson, *Thomas Eakins* (Philadelphia, PA: Museum of Fine Art, 2001) 28.

45 Berger, 'Painting Victorian Manhood', *Thomas Eakins: The Rowing Pictures*, Cooper (ed.), 102–123.

46 Evident too in Eakins' portrayal of himself swimming naked with his male students in *The Swimming Hole* (1884–1885, Amon Carter Museum). For changing perceptions of masculinity and the artist during this period in America, see Michael Hatt, 'Muscles, Morals, Mind: The Male Body in Thomas Eakins' *Salutat'*, in Kathleen Adler and Marcia Pointon (eds), *The Body Images: The Human Form and Visual Culture Since the Renaissance* (Cambridge and New York: Cambridge University Press, 1993) 57–69. See also Jonathan Katz, 'Hide/Seek: Difference and Desire in American Portraiture', introductory essay, 'Before Representation', 17–21, and David C. Ward, 'Before Difference: 1870-1917', 63–67, in Katz and Ward, *Hide/Seek: Difference and Desire in American Portraiture*, exhibition catalogue (Washington DC: National Portrait Gallery and Smithsonian Institution, 2010).

47 Mary Morton, with Camille Mathieu, Galina Olmsted and George T. Shackelford, 'Suburban Views', in Morton and Shackelford, *Caillebotte*, 2015: 207–208. Morton notes that Manet showed his *Boating* in the Paris Salon of 1879, the same year that Caillebotte's *The Boating Party* appeared in the Fourth Impressionist exhibition: Hardly coincidental, she justly contends.

48 Morton and Shackelford, *Caillebotte*, 2015, 207; No. 3, citing Anne Distel, 'Petit-Gennevillier and Argenteuil', *Gustave Caillebotte, Urban Impressionist*, exhibition catalogue, The Art Institute of Chicago (1995): 268–269, and Peter Burger in Anne-Birgitte Fonsmark et al, *Gustave Caillebotte*, exhibition catalogue (Bremen: Kunsthalle, 2008), 23–24. The Maupassant short story is 'Une Partie de campagne', which first appeared in the literary revue published by Renoir's friend Georges Charpentier, *La Vie moderne*, on 2 and 9 April 1881, before being integrated into the Maupassant collection, *La Maison Tellier*.

49 On the importance of Caillebotte's boat design, see especially Morton and Shackelford *Caillebotte*, 2015.

50 On Paul Richer see, Michel Poivert, 'Variété et vérité du corps humain: l'esthétique de Paul Richer', *L'Art du nu au XIXe siècle : le photographe et son modèle*, Sylvie Aubenas (ed.), exhibition catalogue (Paris: Bibliothèque nationale de France, 1997); Natasha Ruiz-Gomez, 'The 'Scientific Artworks' of Doctor Paul Richer', *Medical Humanities* (39, no. 1, June 2013): 4–10; available online, doi: http://dx.doi.org/ 10.1136/medhum-2012-010279; Anthea Callen, 'Muscularity and Masculinity: Dr Paul Richer and Modern Manhood', *Paragraph* (26, no. 1–2, March & July 2003), special issue: 'Men's Bodies', Judith Still (ed.), 17–41; on Marey, see Marta Braun, *Picturing Time: The Work of Étienne-Jules Marey (1830-1904)* (Chicago, IL: University of Chicago Press, 1994); Eugen Sandow, *The Construction and Reconstruction of the Human Body* (London:

John Bale, Sons and Danielsson Ltd., 1907), with a foreword by Sir Arthur Conan Doyle; David L. Chapman, *Sandow the Magnificent: Eugen Sandow and the Beginnings of Bodybuilding* (Urbana, IL: University of Illinois Press, 2006), David Waller, *The Perfect Man* (Brighton: Victorian Secrets, 2011); Edmond Desbonnet, *La Méthode Desbonnet* (Paris: Berger-Levrault, 1909 [6th edition]).

51 Brauer, 'Virilizing and Valorizing Homoeroticism' (2017): 35–67.

52 See Régis Revenin, 'Homosexualité et virilité', Alain Corbin (dir.), *Histoire de la virilité. 2. Le tromphe de la virilité. Le XIXe siècle* (Paris: Éditions Seuil, 2011), 375–407. On the visual culture of the virile homosexual, see Brauer, 'Virilizing and Valorizing Homoeroticism' (2017): 35–67; Fae Brauer, 'Flaunting Manliness: Republican Masculinity, Virilized Homosexuality and the Desirable Male Body', *Masculinities, Australian and New Zealand Journal of Art* (6, no. 1, 2005): 23–42. On the rise of sexual surveillance, see Foucault, *The History of Sexuality*, vol. 1, 1981.

2 Edvard Munch and the Vitalized Bodies of *National Science*

Patricia G. Berman

From the 1890s until late into his career, the Norwegian artist Edvard Munch pictured bathers posing on the shore or immersed in water, paintings that in part define the visual culture of Nordic Vitalism.[1] Between 1909 and 1916, Munch created murals for the newly built Aula (Festival Hall) in Norway's Royal Frederik's University (now the University of Oslo) in which he translated his beach motifs to monumental scale as part of an allegory of the university. In the murals, the body and the beach are imbricated in a larger project of picturing the sciences. In a published text, Munch described his murals as "embodying the great forces that govern humankind."[2] In order to suggest this enormity, Munch enlisted images resonating with turn-of-the-century scientific research. Over the previous decades, the Royal Frederik's University had staked a new identity in the empirical sciences, in the rhetoric of material progress. Munch's paintings tease out the extent to which the academy's aspirations toward rationalism were inflected by metaphysical speculations of a life force generating and transforming matter. They articulate anti-mechanistic, Vitalist dimensions of the rational, technologized new sciences, especially the university's highly regarded research into electromagnetism and sky science.[3]

Cultural historian, Sven Halse has identified Vitalism in the Nordic countries as consisting of three epiphenomena: the philosophical, based in part on the writings of Friedrich Nietzsche and Ernst Haeckel; the visual and artistic; and, what he terms "pragmatic" Vitalism, popular leisure activities such as bodybuilding and swimming.[4] In Munch's paintings for the university, these three strands meet in a Neo-Lamarckian allegory as the artist pictured the philosophical or mystical as imbricated within the pragmatic and scientific.

When the Royal Frederik's University first announced its painting competition for the murals in its Festival Hall, Munch worked on the eleven monumental paintings between 1909 and 1916, when the murals were finally accepted.[5] The murals conjoin National romantic imagery on either side of the long, narrow hall with references to electromagnetism, open-air athleticism, heliotherapy, and the occult (Figure 2.1). The centerpiece is a five-canvas group anchored by an enormous sun whose rays alternately appear as protoplasmic expulsions as revealed by the image on the cover of this book, and as strings jumping through the architectural frames and radiating into the flanking canvases (Figure 2.2). The nude figures in two canvases to the right of the sun, and two canvases to its left, are dematerialized and rematerialized by the sun's multicolored rays, picturing the "new man," or what Pascal Rousseau calls a "new species."[6]

Seen from a distance, Munch's figures seem bounded. From close-up, however, they are spectral, smeared, dripping, and composed of contours that seem to unwind and intertwine. Comparing them to nudes appearing in other Nordic public art at

DOI: 10.4324/9781003045595-4

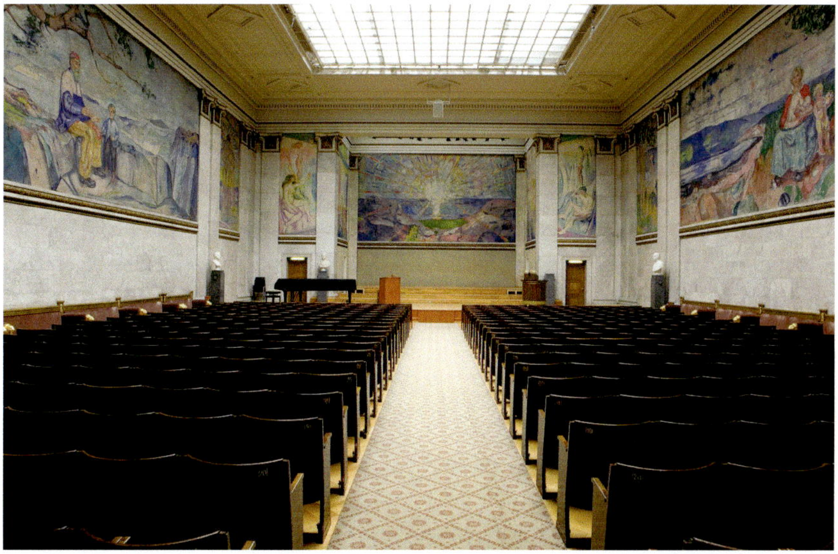

Figure 2.1 University of Oslo Festival Hall. (Courtesy: University of Oslo/Munch Museum.)

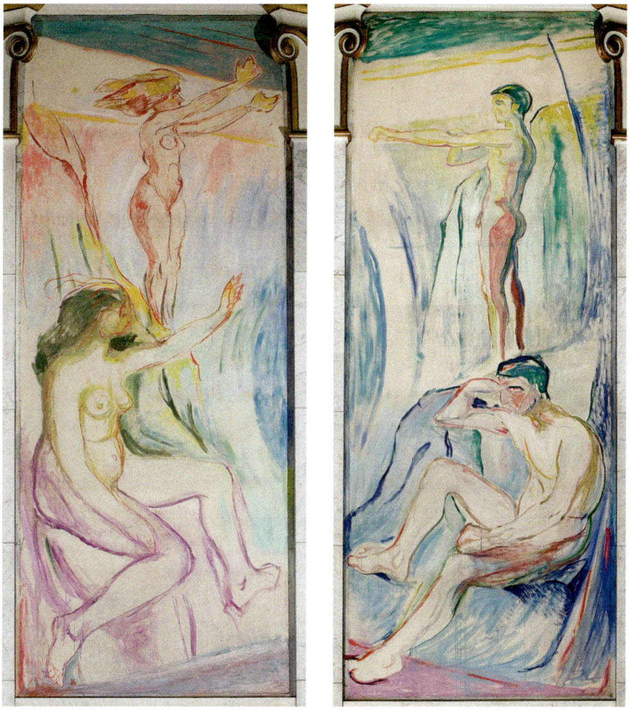

Figure 2.2a to 2.2e Edvard Munch's murals at the front of the University of Oslo Festival Hall. From left to right they are: Figure 2.2a *Women Reaching toward the Light*, 1914–1916; Figure 2.2b *Awakening Men in a Flood of Light*, 1914–1916; (*Continued*)

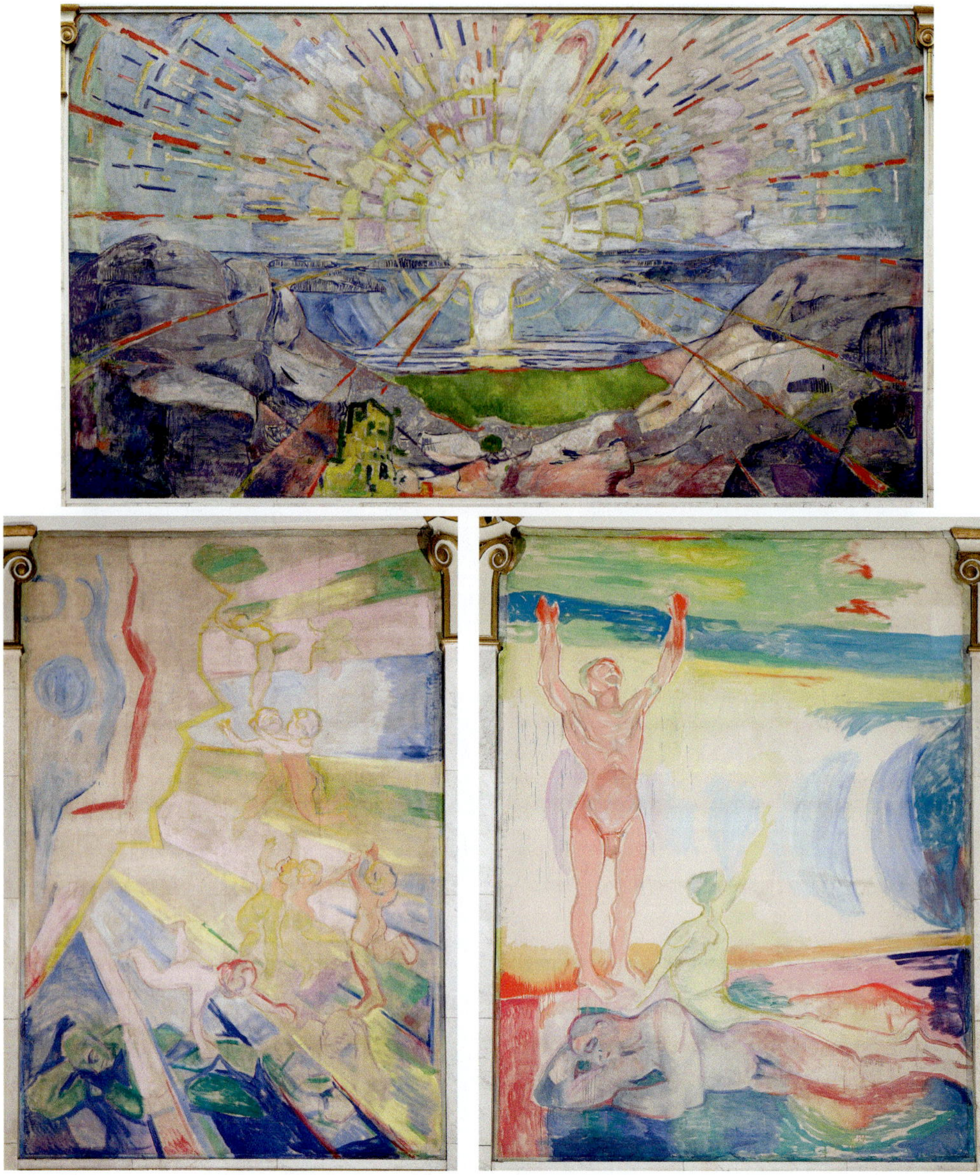

Edvard Munch, University of Oslo Festival Hall murals, 1914-1916 continued: Figure 2.2c *The Sun*, 1911; Figure 2.2d *The Genii of Light*, 1914 -1916; Figure 2.2e *Men Reaching toward the Light*, 1914–1916. (Courtesy: University of Oslo/Munch Museum.)

the time, such as the Swedish artist Georg Pauli's mural *Mens sana in corpore sano*, installed in the Jönköping High School in 1912 (Figure 2.3), Munch's bodies are boundaryless, seemingly in the process of becoming. Their slippery contours move in and out of the surrounding space, defying the conventions of public art in that period in that culture.[7] Even when viewed in concert with works by Munch's Nordic contemporaries in which muscular bodies seem hewn by nature and "transluminated" by spectral colors – a stylistic hallmark of Nordic Vitalist painting (Figure 2.4) – his

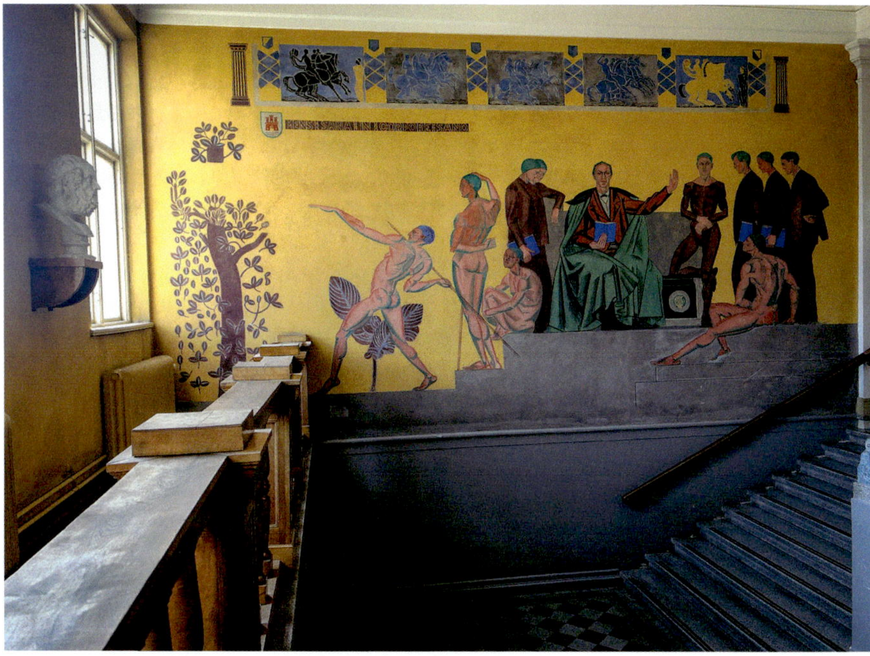

Figure 2.3 Georg Pauli, *Mens sana in Corpore Sano*, 1912, wall painting in the staircase of the Jönköping High School (now Per Brahe High School). (Courtesy: Per Brahe High School.)

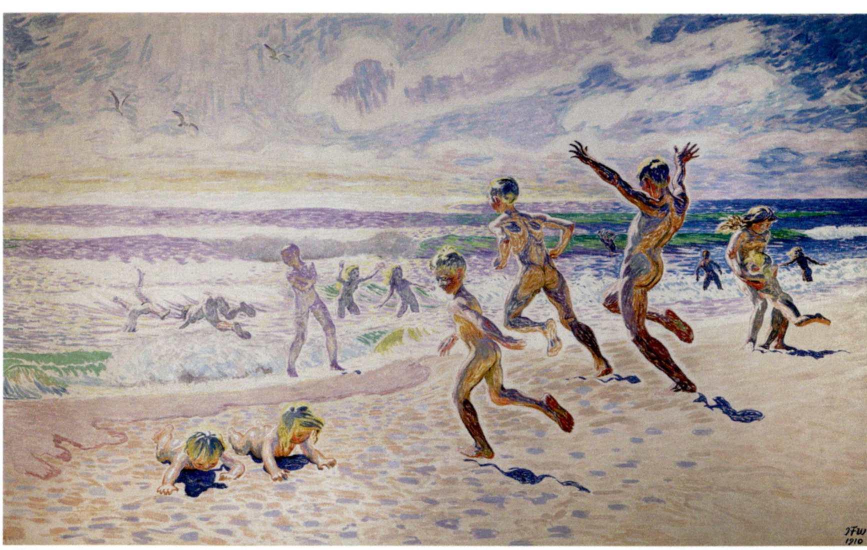

Figure 2.4 Jens Ferdinand Willumsen, *Sun and Youth,* 1910, Oil on canvas, 266 × 427 cm. (Photograph, Hossein Sehatlou; Courtesy of Göteborgs konstmuseum, Sweden, 2016, WL38.)

Aula nudes appear to be intermediary forms.[8] They are not merely acted upon by the light but seem, before our eyes, to come into being through its actions. These motifs picture a fluid connectedness between humans and the physical and cosmic environment that echo the writings of the University of Jena-based zoologist and philosopher, Haeckel, the most important interpreter of Darwin in the German and Scandinavian contexts. Writing of "Monist ethics" in his popularizing *Riddle of the Universe* (1899), Haeckel offered aspirations and perhaps inspiration for such institutional paintings:

> In the school of the future nature will be the chief object of the study; a man shall learn a correct view of the world he lives in; he will not be made to stand outside of and opposed to nature but be represented as its highest and noblest product.[9]

Public universities were special sites for public art, in which mural painting, illustrating and inspiring scholarship, was an exercise in epistemology as much as decoration.[10] The Royal Frederik's University commissioned its new Aula and its murals as a permanent monument to mark the university's centennial in 1911. Norway only became an independent nation in 1905, following over four centuries as a Danish colony and then nearly a century as a Swedish protectorate. As the Royal Frederik's University was the nation's only institution of higher learning, it had played a critical role in nation-building efforts. Like many universities throughout Europe, the Royal Frederik's University transformed from a school for the elites (associated with a Danish official class) to an educational institution for all social classes. The university had recently undergone a series of "crises,"[11] radical revisions to modernize and democratize its curriculum. A later university official identified this endeavor as the purview of "men of the future," training researchers from all social classes practicing a new experimentalism.[12] The work of Charles Darwin, adapted by younger scientists, in part fomented the institutional change, as caricatures in the Norwegian popular press suggest.[13] One from an 1879 satirical journal pictures an orangutan, a predigested racializing symbol of Darwinian evolutionary theory, attempting to divert a young scholar – an "unprofessor" as in the caption – from the university into the parodic "natural environment" for a Darwinian.[14] (Figure 2.5)

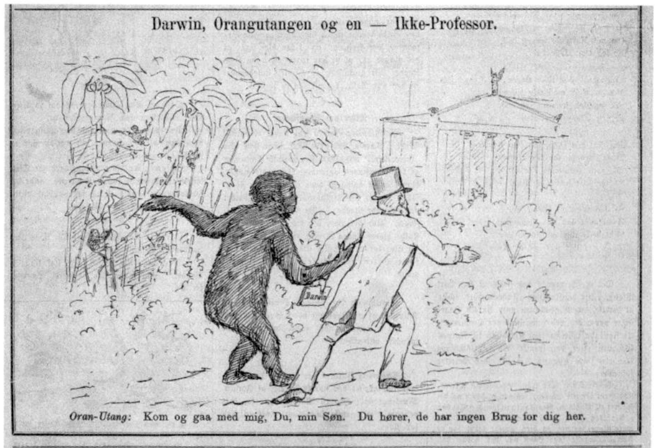

Figure 2.5 "Darwin, the Orangutan, and an – 'Un-Professor,'" *Vikingen*, 24 May 1879. (Courtesy: National Library, Oslo.)

One of the university's curricular reforms in the later nineteenth century was the implementation of a professionalized medical faculty and new therapeutic regimes, promising renewed emphases on health as a condition of modernity.[15] It was in these sites that new epistemological models were developed to advance sanitation and hygiene as public health initiatives, as well as to improve the health and even the perfectibility of the body politic. As elsewhere in Europe, a concern with degeneration had given impetus to the study of decreasing vitality among children, and therefore the future prospects of the Norwegian nation.[16] Concerns about illness and reproductive capacity within the population were amplified by the mass emigration of Norwegians to the United States. This demographic crisis in which over 1% of Norway's population emigrated each year between 1825 and the mid-1880s,[17] gave rise to the popular notion that the best were leaving, replaced by a lesser population.[18] A growing anxiety about Norway's future was a breeding ground for racializing thought at the turn of the century, and shortly thereafter the rise of eugenics movements.[19] The combination of population decline and public health concerns about substandard living conditions in the cities held social and scientific dimensions, converging in a multipronged strategy of hygienism. These extended from medical regimes to branches of the reform movement.[20]

A survey of international medical schools published in 1887 suggested that education in practical hygiene (today identified as "Public health") was one of the fastest growing areas in European medicine, acknowledging "the important role that hygiene was beginning to play in science as in practical life."[21] Beginning with the 1874 Congress on Hygiene and Demographics in Vienna, international medical and popular attention was brought to the problems of urban crowding and public sanitation through pedagogy and product display.[22] Such expositions instrumentalized the rigorous "science of hygiene" in place of "practical knowledge."[23] In May of 1911, a few months before the Royal Frederik's University's centennial celebration, and while Munch worked on his murals, the 1911 International Hygiene Exhibition in Dresden opened, attracting five million visitors during its run. The exhibition's popular educational section, "Der Mensch als Kunstwerk" (Man as a Work of Art), suggested that the body, as a form of architecture, could be perfected through study and discipline.[24] The image of the muscular naked male body, arms raised as if in an ecstatic embrace of nature, was deployed as a symbol of such instructive disciplined collective athleticism. Souvenir medals from the 1907 International Congress for Hygiene and Demography in Berlin, for example, were inscribed with this image (Figure 2.6). This kind of representation, reverberating in the works of Munch and in paintings throughout the north, asserted the perfected male body as an agent of cultural renewal.

In the 1880s, the British social and hygienic reformer Edwin Chadwick recommended that physical education could operate as a "preventative" measure against social ills. Referring to the Swedish system of educational gymnastics, developed in the early nineteenth century by Pehr Henrik Ling, that was adapted worldwide, including Norway,[25] Chadwick noted in the 1880s that Sweden was a world leader "in this branch of preventative science."[26] Offering, according to Hartvig Nissen, "exercise based on scientific principles and founded on physiological rules and on esthetical and psychological truth," the objective of the Swedish system was to "educate the mind and body," helping the student as well as the soldier "to master his own will, and to subordinate himself" as "part of a great totality."[27] Throughout Europe and North America, physical culture was folded into emerging mandatory educational systems because of its recognized support from both military

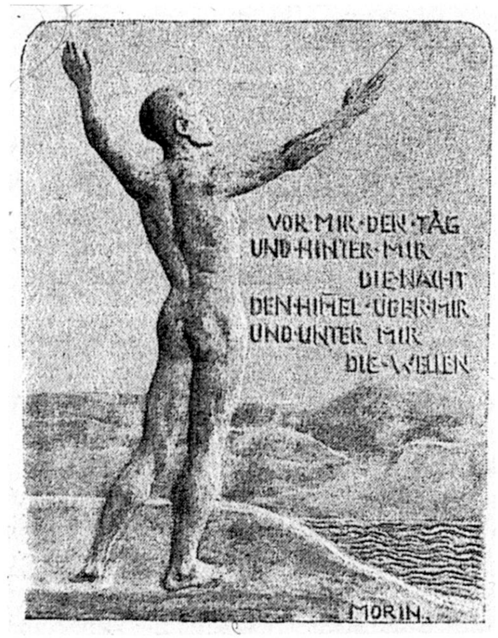

Figure 2.6 Georges Morin, Souvenir Medal from the XIV Congress on Hygiene and Demography, 1907. (Courtesy: Countway Medical Library, Harvard University.)

preparedness and improved public health in forging wholesome bodies while producing literate citizens.[28]

Such an initiative lay behind the Norwegian military physician-cum-comparative ethnologist, Carl Arbo, who surveyed military recruits, gathering anthropometric data in an attempt to explain variations in regional military readiness.[29] Arbo had sought training with the followers of anatomist and racial theorist, Anders Retzius in Stockholm, and with anthropologist, Paul Broca in Paris. By the 1890s, he theorized distinctive racial typologies within different regions of Norway based on hair and eye color, body dimensions, and especially cranial configuration. Within the nation-building movements of the nineteenth century, a search for origins had led to significant debates about the *Ur* population, focusing on two imagined racial types, the so-called "Nordics," identified with longhead shapes, white skin, and pale hair and a second "race," characterized as darker skinned, short, and round-headed.[30] In the 1890s, ethnographer Andreas Hansen proposed that the "long skulls" had primacy as the "ruling caste in Europe" and had imposed their Aryan superior culture upon the inferior and short-skulled non-Aryan peoples.[31] Hansen joined a team of scholars from the university to create a popular handbook for the Paris Universal Exposition in 1900 that consolidated these demographic statistics. Yet Hansen stated that no other country in Europe (save Ireland) had lost more population due to emigration.[32] Enlisting Arbo's biometric data, Hanson also claimed that Norwegians were the tallest, blondest, most blue-eyed, and large-chested population on the planet. This anthropological work, particularly the concerns with degeneration, hygiene, and bodily development, prompted the establishment of Norway's "Committee for Racial Hygiene" in 1908, preceded by a similar organization in Denmark and followed, in 1909, by the establishment of the Swedish Society for Racial Hygiene. As historian Jon Røyne Kyllingstad has demonstrated, the mythic notion of a distinctive Aryan race and a dual-race model in Norway depended on a taxonomy developed before Darwin and elaborated in the wake of Darwinian sciences.[33]

Racial hygiene should be understood on a continuum with other forms of national hygienism, such as bathing and other forms of physical enhancement.[34] In the summer of 1911, shortly after the opening of the Dresden hygiene exhibition, the Fifth International Congress on Therapeutic Sea Bathing opened in Kolberg, Germany, organized by committees from throughout Europe, including Norwegian physicians specializing in water cures.[35] In the later nineteenth century, a bathing culture had emerged along Norway's coasts, including the town of Åsgårdstrand, where Munch painted numerous compositions including *Bathing Boys* in 1904 (Figure 2.7) and the town of Kragerø, whose granite shoals are represented in Munch's Aula paintings. In Norway, a popular culture of swimming emerged in the 1890s, in part through the proselytizing efforts of E. Leonard Hasvold. An advocate for swimming lessons and the construction of public bathhouses, Hasvold argued that cold bathing developed muscles, morals, and hygiene – a somewhat different rational to the one operative in Parisian bathhouses, identified by Anthea Callen in Chapter 1.[36] The London Olympic Games of 1908, which offered the largest display of competitive swimming to date, became the stimulus for the founding of the International Swimming Federation.[37] By 1912, when Stockholm hosted the Olympic Games, swimming and diving had developed as central competitions, as Callen outlines in Chapter 1.[38] The new popularity of swimming as a sport and as a therapy is evidenced by the ubiquitous visual culture of bathing at the turn of the century. This was accompanied by an iconography of "practical" Vitalism throughout the Nordic countries shared by Munch's paintings.[39]

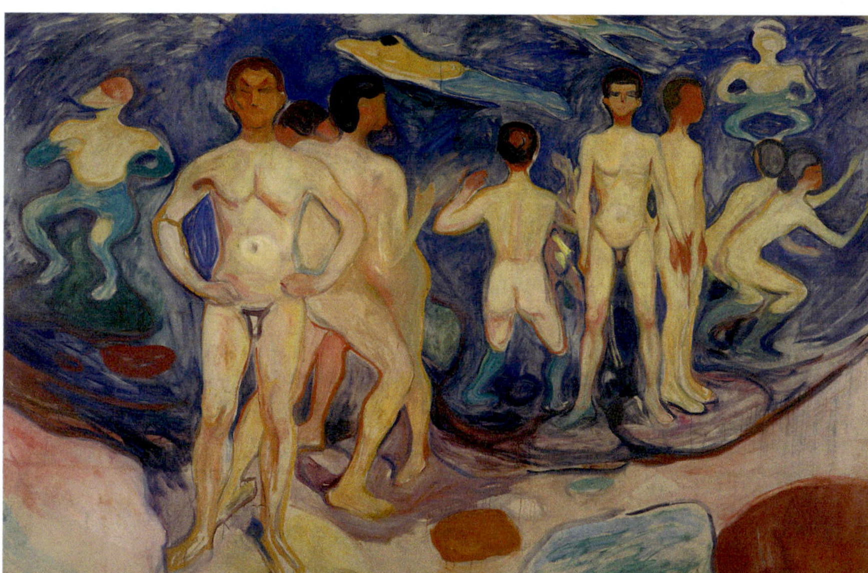

Figure 2.7 Edvard Munch, *Bathing Men*, 1904, oil on canvas, University of Oslo/Munch Museum. (Courtesy: University of Oslo, Munch Museum.)

Haeckel's anti-clerical sweeping description of a vast and mysterious web of interconnected energies and morphologies had particular resonance in Norway where Neo-Lutheran clergy had advanced the "infallible certainty" of the cycle of nature as the basis of mystery. Fueling the University's determination to reshape the curriculum to include Darwin's works and those in his wake, Haeckel's monism and his theories of evolution through use and adaptation provided for both the laboratory and social sciences, a secularizing history of human and animal descent within an animated universe.[40] Haeckel's "Darwinismus" thus offered both "traditionalists" and empiricists a means of reconciliation, legitimizing the model of the experiment as a fluid enterprise. With the notion of human origins and development, and the universe as a dynamically connected force, the university focused its attention and funding on research that would both seek the origins of life and instrumentalize its knowledge for the emerging nation.

Munch had been familiar with Haeckel's work since the time that Wilhelm Bölsch, Germany's foremost interpreter and hagiographer of Haeckel, had become an occasional member of the artist's circle of writers and scientists in 1890s Berlin.[41] Munch's library included a copy of Bölsch's biography of Haeckel, published in 1898.[42] In 1906–1907, Munch had spent considerable time in Jena and Weimar, in part to work on a posthumous portrait of Friedrich Nietzsche commissioned by his Swedish patron, Ernest Thiel.[43] The founding of the Monist League, and Haeckel's plans for his museum of zoology, occurred while Munch worked on that project. Familiar with Haeckel's comparative embryology, especially after it was contested in the scientific and popular press, from the 1890s Munch may have modeled some of his images of swimmers de-evolving in prints and paintings on Monist principles.[44]

References to Transformism (in Norwegian, known as "Stoffveksling" or metabolism), light, waves, and germination, amplifying Haeckel's Monism, appear repeatedly in Munch's writings, as they did in his visual production: "Death," Munch wrote, "is the beginning of life ... The earth and stones longed to mix with the air, and human beings and animals and trees were created and air turned into water and earth became air."[45] His speculation about the experimental sciences of Chemistry and Astrophysics merged with his interest in esoterism, as it did throughout popular and academic culture.[46]

In Munch's mural entitled "Chemistry" (Figure 2.8), a naked couple appear to engage in abiogenesis, generating life from a laboratory flask. As the male figure pours a fluid into the receptacle held by the female, this fluid rises into a gelatinous cloud to produce new human life. Below them, other substances resembling smoke and protoplasm move outward from additional flasks. In Munch's companion painting entitled "New Rays" (Figure 2.9), dynamic streaks emitted by the image of the sun at the center of the hall, and vectored lines suggesting auratic energy, converge around and interpenetrate two naked bodies. An allegory of the university's most prestigious area of research, the study of electromagnetism and sky science, this painting, along with the adjacent solar images, entangles the new objectives of the institution with the ongoing speculative work of scientists themselves. In his private journals, Munch evoked his allegories of science in his word collages:

> "*New Rays* is overflowing with light – It travels into the bodies – and in and out of the crystal. The painting is called '*Crystallization*' – there is light that travels like X-rays. The other side painting – '*Chemistry*' – represents the hidden energies –

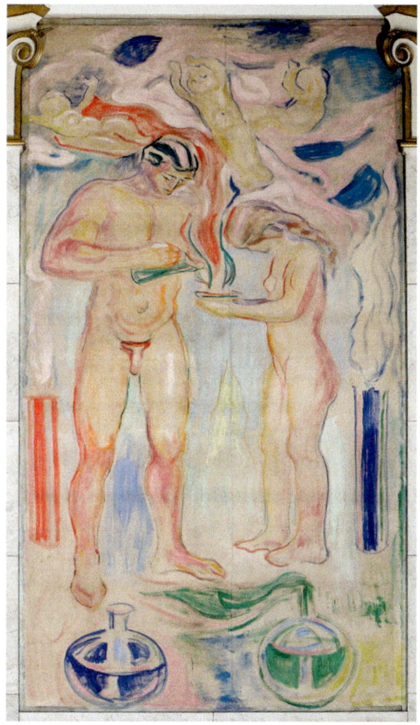

Figure 2.8 Edvard Munch, *Chemistry*, 1914–1916, oil on canvas, 450 × 225 cm, University of Oslo/Munch Museum. (Courtesy of University of Oslo, Munch Museum.)

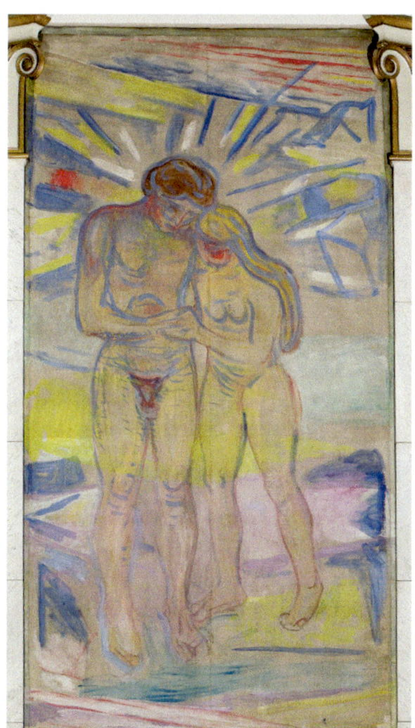

Figure 2.9 Edvard Munch, *New Rays*, 1914–1916, oil on canvas, 455 × 225 cm, University of Oslo/Munch Museum. (Courtesy of University of Oslo, Munch Museum.)

the workplace of fire and warmth. On the other side *'The Sun'* sends its rays to even greater distances ... – On the one side the energies turning inward – the blending of liquids *(Chemistry), Crystallization* and X-rays *(The two Lovers), Primitive light, the ventricle of the heart.*"[47]

In sketches for both motifs, Munch had positioned the couples over stratigraphic layers containing skeletal remains (Figure 2.10), a reference to transformism with death generating new life.

In the five-painting complex in the front of the hall (Figures 2.2a to 2.2e), from which the painted solar rays emanate, men and women are pictured embracing the light. These are all figures of regeneration, giving their bodies over to the sun in the manner of the increasingly popular arena of heliotheraphy, pioneered by the Danish physician, Niels Finsen, who had been awarded a Nobel Prize in 1903 for his work in harnessing light for therapeutic purposes.[48] Not only was his clinic in Copenhagen among the first sites in Europe to develop heliotherapeutic methods and mechanisms but his work also converged with already popular swimming and sunbathing regimes in the Nordic countries generating a Neo-Lamarkian iconography of the land giving rise to physical betterment.[49] The two paintings directly adjacent to the central mural, *The Sun*, picture light as bringing bodies into being (Figures 2.2b and 2.2d). To its right, in the painting entitled, *The Genii of Light* (Figure 2.2d) radiation from the sun becomes flesh in the form of small putti that appear to be spawned by its rays. In the paintings of men's bodies, two male figures on the right (Figure 2.2e) appear to become aroused by the light while in the painting on the left, *Awakening Men in a Flood of Light*

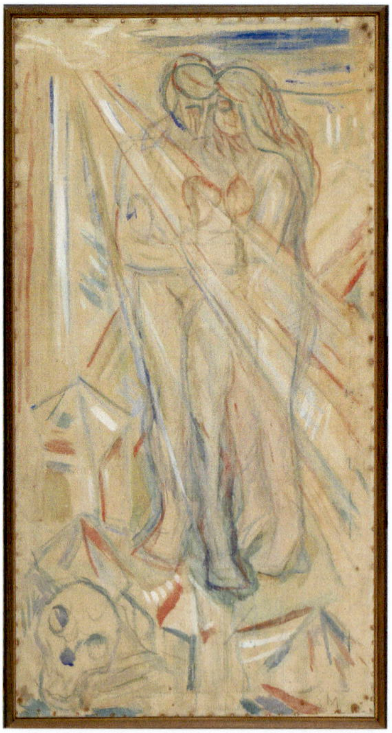

Figure 2.10 Edvard Munch, *Physics*, 1914–1916, oil in canvas, MM.M.00550. (Courtesy of Munch Museum.)

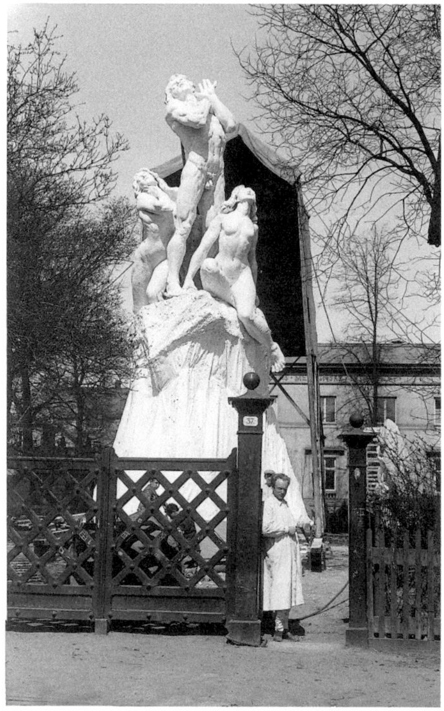

Figure 2.11 Rudolph Tegner, *Towards the Light*, 1906, bronze, intersection of Blegdamsvej and Tagensvej Copenhagen. Photograph: City of Copenhagen. (Courtesy: Tegner Museum, Sweden and the City of Copenhagen.)

(Figure 2.2b), light appears to animate two male figures who, in cinematic form, rise from a slumbering position to reach toward the sun and then toward the heavens. This painting is allied in its representation of the figural sequencing from ground to sky with the monument to Finsen installed in Copenhagen by Rudolf Tegner (Figure 2.11) in 1909, the year that Munch began to draft his mural proposals. Munch's generative sun symbolized educational enlightenment, a standard iconography for institutional art, but in this regard it also suggested the ways in which the program of national hygienism became entangled with eugenics, as articulated by Fae Brauer, for the breeding of an improved inheritance.[50] The image of a cleaning sun, the rugged coastline setting (Figure 2.2c), and the receptive bodies of Munch's "new species" (Figures 2.2 (a, b, and e), and Figures 2.7 to 2.10) crystalize this concept.

The University of Oslo had become an international center of research into the aurora borealis and electromagnetism, establishing Norway's competitive position within scientific communities.[51] The university professor, Fridtjof Nansen was not only a polar explorer and geologist but also a popularizer of Darwin's theory of natural selection.[52] Achieving renown as a humanitarian, Nansen also gained status as a national hero and an international sensation, following his three-year voyage to reach the North Pole using the theory of transpolar drift.[53] His purpose was to study the Aurora Borealis. Upon his return in 1896, he published the immensely popular book, *Farthest North,* in which he included pastel drawings he made of the Aurora in abstract waves and radiant vectors (Figure 2.12), as well as selections from his diaries.

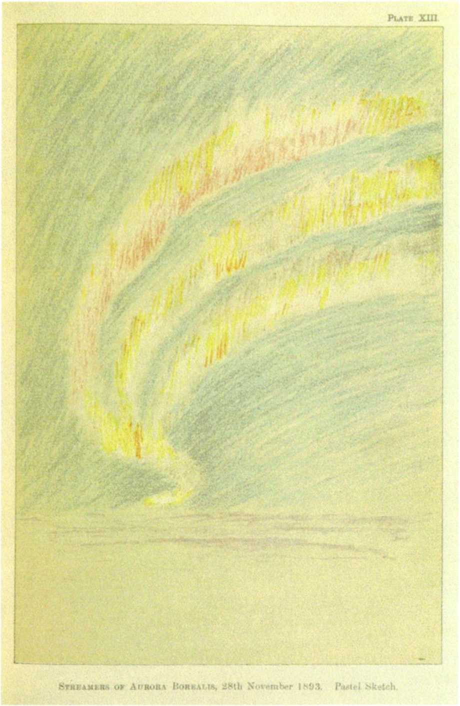

Figure 2.12 Fridtjof Nansen, Plate XIII, Farthest North, being the record of a voyage of exploration of the ship "Fram" 1893–1896, and of a fifteen months' sleigh journey by Dr. Nansen and Lieut. Johansen (New York: Harper and Row, 1897). (Courtesy: Wellesley College Library.)

Entries from 1893 move from objective descriptions of the Aurora[54] to the ecstatic voice of revealed mystery:

> Nothing more wonderfully beautiful can exist than the Arctic night. It is dreamland, painted in the imagination's most delicate tints; it is color etherealized. One shade melts into the other, so that you cannot tell where one ends and the other begins, and yet they are all there. No forms—it is all faint, dreamy color music, a far-away, long-drawn-out melody on muted strings … I see the thread of life becoming entangled in the complex web which stretches unbroken from the gentle dawn of life to the everlasting silence of the ice.[55]

Kristian Birkeland, another of the University's sky scientists, theorized that electrically charged beams moving toward the earth have an impact on its magnetic field and are guided to the polar latitudes. Visible aurora lights result from interactions between the charged particles and gases of the upper atmosphere. Birkeland originally understood these rays to be "corpuscles," and he theorized that the vacuum of space was actually a plexus, filled with these corpuscles.[56] The metaphors of a web, of light as material, form part of what Robert Brain has termed the period's "protoplasmania," the metaphor of an organic, mutable substance representing "the unified forces of the physical world fused in a … semi-fluid, elastic substance" that could be found in the cells of all organisms "with the "capacity to make the primordial rhythms of biological life perceptible to the beholder."[57] Birkeland's study of geomagnetism and the Aurora Borealis in *The Norwegian Polaris Expedition, 1902-1903*, newly published in 1908 and widely reported in the popular press, contributed to scientific as well as popular understandings of electromagnetism in the northern latitudes. Birkeland's work also coalesced with a wide network of Western esotericism, which sought to reveal the connectedness of nature and spirit, visible and invisible, as illustrated by Serena Keshavjee in Chapter 4 and Brauer in Chapter 5 of this book.[58] By the first decade of the twentieth century, Nansen and Birkeland's polar research, the aurora and magnetized particles entered into the arena of national signification decorating, for example, the Nobel Peace Prize certificates.[59] Both scientists had captured the public imagination with their "rational" mastery of the sky – of the Aurora, magnetism, and atmospheric electricity. At the same time, a romantic veneration of nature was alloyed with their scientism. Both scientists had been steeped in pedagogy inspired by Haeckel's "monism" before they developed their theories and experimental methods. In *The Riddle of the Universe*, Haeckel had characterized the instruments and discoveries of astrophysics as not antithetical to Monism, giving credence to the notion of a great chain of evolution, as he concluded:

> The monistic conviction, which we thus arrived at, of "the physical and chemical unity of the entire cosmos", is certainly one of the most valuable general truths which we owe to astrophysics ... In one part of space we perceive, with the aid of our best telescopes, vast nebulæ of glowing, infinitely attenuated gas; we see in them the embryos of heavenly bodies, billions of miles away, in the first stage of their development.[60]

Munch's sun, with its material discharges, then offers a visual poetics of the University of Oslo's most prestigious scientific work. It was through a Vitalist visual rhetoric that Munch interpreted these "new rays" that had the capacity themselves to produce life. "All living organisms, without exception," wrote Haeckel in *The Riddle of the Universe*, "are sensitive; they are influenced by the condition of their environment and react thereon by certain modifications in their own structure. Light and heat, gravity and electricity, mechanical processes, and chemical action in the environment act as *stimuli* on the sensitive psychoplasm ... an evolution of the soul."[61] Munch's many writings also echo Haeckel's monist philosophy:

> I stood on a high mountain ... I saw the little planet Earth ... I saw where the change in matter began – how the air corroded the hard matter – and the transitional forms between stone and atmosphere were created: the living – human beings, animals and plants ... I saw the single individual – from the first moment the light entered it – how its desire slowly but surely developed ... It screamed after nourishment – it saw the sun and stretched its arms towards it.[62]

Posing light as nourishment and as a form of fluid genesis, Munch approached the sky sciences in such terms.

By 1916, when these paintings had become the University's most prominent works of art, such a public image of objective sky science, as a form of solar-gestation, was anachronistic. Yet sometimes, art cuts through public rhetoric to reveal private belief and occult motivation, as illustrated by Brauer's exploration in Chapter 5 of this book of Picasso's relationship to Bergson alongside magnetism and what she calls "Occultist Cubism." The artist's "new species" identified in this chapter crystalized the national imperatives of the university's new experimentalism while both referencing and vitiating the national anxieties about decline through the promise of cultural and scientific renewal. In this Neo-Lamarckian allegory, the sun is granted the capacity for cosmic renewal in a new nation-seeking cultural distinction. Munch imagined and imaged origins: his slippery, spectral bodies make manifest these complex entangled national atavisms with Nordic Vitalisms and scientific modernisms.

Notes

1 Munch's bathers in relation to Vitalism have been the subject of numerous publications, including Gunnar Sørensen, "Vitalismens år?," *Cras* (1981) 26–42; Patricia G. Berman, "Body and Body Politic in Edvard Munch's Bathing Men," *The Body Imaged. The Human Form and Visual Culture since the Renaissance*, ed. Kathleen Adler and Marcia Pointon (New York and Cambridge: Cambridge University Press, 1993); Patricia G. Berman, "Mens sana in corpore sano: Munchs vitale kropper," *Livskraft: Vitalismen som kunstnerisk impuls 1900-1930*, ed. Ingebjørg Ydstie and Karen Lerheim (Oslo: Munch Museum, 2006) 45–64; Patricia G. Berman, "Dionysus with Tan Lines: Edvard Munch's Discursive Skin," *A Fine Regard: Essays in Honor of Kirk Varnedoe*, ed. Patricia G. Berman and Gertje Utley (Aldershot, UK and Burlington, USA: Ashgate Publishing Limited, 2008) 68–85; *The Spirit of Vitalism. Health, Beauty and Strength in Danish Art, 1890-1940*, eds. Gertrud Hvidberg-Hansen and Gertrud Oelsner (Copenhagen: Museum Tusculanum Press, 2011); Lill-Ann Körber, *Badende Männer: Der nackte männliche Körper in der skandinavischen Malerei und Fotografie des frühen 20. Jahrhunderts* (Bielefeld: transcript Verlag, 2014).

2 Edvard Munch, in *Konkurransen om den Kunstneriske Utsmykning av Universitetets Nye Festsal* (*Competition for the Artistic Decoration of the University's New Festival Hall*); (Kristiania: W. C. Fabritius & Sønner A/S, 1911) 4.

3 This model is indebted to Barbara Larsen's observation that "linked developments in the natural sciences among materialist and occult investigations"; refer Barbara Larson, *The Dark Side of Nature: Science, Society, and the Fantastic in the Work of Odilon Redon* (University Park: The Pennsylvania University Press, 2005) 180.

4 See Sven Halse, "Vitalisme – fænomen og begreb," *Kritik* (171, 2004): 1–7; Halse, "Vitalismen: flere tilløb til en begrebsdiskussion," *Kritik* (38, 2006): 119–125; and Hvidberg-Hansen and Gertrud Oelsner, "Den vidtfavnende vitalisme," 46–57.

5 The significant literature examining the paintings and the convoluted history of the commission includes Roy A. Boe, "Edvard Munch's Murals for the University of Oslo," *Art Quarterly*, vol. XXIII, no. 3 (Autumn 1960): 233–246; Otto Lous Mohr, *Edvard Munchs Auladekorasjoner i lys av ukjente utkast og sakens akter* (Oslo: Gyldendal Norsk Forlag, 1960); Patricia G. Berman, "Monumentality and Historicism in Edvard Munch's University of Oslo Festival Hall Paintings," Ph.D. dissertation (New York University, 1989); Tina Yarborough, "Exhibition Strategies and Wartime Politics in the Art and Career of Edvard Munch, 1914-1921" (Ph.D. dissertation, University of Chicago, 1995); Petra Pettersen, "Munch's Aula-Dekorasjoner," *Samlede Malerier: Catalogue Raisonne*, ed. Gerd Woll, Vol. III, 1909–1920 (Oslo: Munch-Museet; Kaare Berntsen and Cappelen Damm, 2008) 829–851; Patricia G. Berman and Peder Anker, *Edvard Munchs Aulamalerier: Fra Kontroversielt Prosjekt til Nasjonalskatt* (Oslo: Messels Forlag, 2011; and Ydstie and Lerheim, 2011).

6 Pascal Rousseau, "Radiation: Metabolizing the 'New Rays," *Edvard Munch: The Modern Eye*, eds. Angela Lampe and Clément Chéroux (Paris and London: Centre Pompidou and Tate Modern, 2011) 168.

7 See Monika Wagner, *Allegorie und Geschichte. Ausstattungsprogramme öffentlicher Gebäude des 19. Jahrhunderts in Deutschland* (Tübingen: E. Wasmuth, 1989).

8 I take the term "transluminated" from Gertrud Oeslner, "Healthy Nature," (Hvidberg-Hansen and Oelsner) 159ff.

9 Ernst Haeckel, *Die Welträthsel (The Riddle of the Universe at the Close of the Nineteenth Century)* (1899; New York: Harper and Brothers Publishers, 1905) 363. The best study of Munch's engagement with Haeckel's work is Shelly Wood Cordulack, *Edvard Munch and the Physiology of Symbolism* (Madison and Vancouver: Fairleigh Dickinson University Press, 2002) 97ff.

10 See for example Jennifer Shaw, *Dream States. Puvis de Chavannes, Modernism, and the Fantasy of France* (New Haven: Yale University Press, 2002); Serena Keshavjee, '"La Vie Renaissant de le Mort': Albert Besnard's 'Non-Miraculous' History of Creation," Chapter 4, *Picturing Evolution and Extinction: Regeneration and Degeneration in Modern Visual Culture,* eds. Fae Brauer and Serena Keshavjee (Newcastle upon Tyne: Cambridge Scholars Publishing, 2015) 61–82.

11 Jon Røyne Kyllingstad and Thor Inge Rørvik, *Universitet i Oslo, 1811-2011*, vol. 2, 1870–1911; *Vitenskapenes universitetet*, ed. John Peter Collett (Oslo: Unipub, 2011) 19ff; John Collett, *Historien om Universitetet i Oslo* (Oslo: Universitets Forlaget, 1999) 107. The debates were focused on the contest between religion and "scientism," Classical philology and modern languages, and ancient and modern history and literature as the bases of the new curriculum.

12 Kyllingstad and Rørvik, 1999, 107–114.

13 The popularization of Darwinism is cited in Hans Fredrik Dahl, *A History of the Norwegian Press, 1660-2015* (London: Palgrave MacMillan, 2016) Chapter 4; Kirsten Shepherd-Barr, *Theater and Evolution from Ibsen to Beckett* (New York: Columbia University Press, 2015) 37.

14 Kyllingstad and Rørvik, 1999, 32–38. On the racializing work that such satires perform, see Julia Voss, "Monkey, Apes and Evolutionary Theory: From Human Descent to King Kong," *Endless Forms: Charles Darwin, Natural Science and the Visual Arts*, eds. Diana Donald and Jane Munro (Cambridge: Fitzwilliam Museum; New Haven: Yale Center for British Art, 2009) 215–234.

15 Øivind Larsen, Bård Alsvik, Magne Nylenna, and Elisabeth T. Swärd, *Helse og Nasjonbyggning* (Oslo: Gyldendal Norsk Forlag, 2005); Mark Micale, *The Mind of Modernism: Medicine, Psychology, and the Cultural Arts in Europe and America, 1880–1940* (California: Stanford University Press, 2003).

16 Einer Lie and Hege Roll-Hansen, *Faktisk Talt: Statistikkens historie i Norge* (Oslo: Universitetetsforlaget, 2001) 132.

17 Willliam H. Hubbard, "Historical Demography," *Making a Historical Culture: Historiography in Norway*, eds. William H. Hubbard, Jan Eivind Myhre, Trond Nordby, and Sølvi Sogner (Oslo, Copenhagen, Stockholm, and Boston: Scandinavian University Press, 1995) 320.

18 Einar Niemi, Jan Eivind Myhre, and Knut Kjeldstadli, *Norsk innvandringshistorie*, Vol. 2: *i Nasjonalstatens Tid 1814-1940* (Oslo: Pax Forleg A/S, 2003) 21, 331ff.

19 Jon Røyne Kyllingstad, *Measuring the Master Race. Physical Anthropology in Norway, 1890-1945* (Cambridge: Open Books Publishers, 2014).

20 See Ydstie and Lerheim, 2011; Michael Hau, *The Cult of Health and Beauty in Germany: A Social History, 1890-1930* (Chicago and London: The University of Chicago Press, 2003).

21 Wilhelm Loewenthal, *L'Enseignement actuel de l'Hygiène dans les facultés de médecine en Europe* (Paris: Librairie H. Le Soudier, 1887) 117. See also the authoritative 10-volume *Handbuch der Hygiene*, ed. Theodor Weyl (Jena: Verlag von Gustav Fischer, 1883–1904); Manuel Frey, *Der reinliche Bürger: Entstehung und Verbreitung bürgerlicher Tugenden in Deutschland, 1960-1980* (Göttingen: Vandenhoeck & Ruprecht, 1997).

22 For the implications of such concerns for symbolist artists, see Barbara Larson, "Microbes and Maladies: Bacteriology and Health at the Fin de Siècle," *Lost Paradise: Symbolist Europe* (Montreal: The Montreal Museum of Arts, 1995) 387, note 9.

23 Ferdinand Hueppe, *Handbuch der Hygiene* (Berlin: Verlag von August Hirschwald, 1899) 5.

24 *Offizieller Katalog der internationalen Hygiene Austellung, Dresden, Mai bis Oktober, 1911* (Berlin: Rudolf Mosse, 1911) 378; also refer Berman, *Livskraft*, 2006, 48–49.
25 Finn Olmstad, *Norsk Idretts Historie: Forsvar, sport, klassekamp 1861-1939* (Oslo: Aschehoug, 1987) 39.
26 Benjamin Ward Richardson, *The Health of Nations: A Review of the Works of Edwin Chadwick* (London: Longmans, Green, and Co., 1887) 256.
27 Hartvig Nissen, *ABC of the Swedish System of Educational Gymnastics: A Practical Handbook for School Teachers and the Home* (Philadelphia and London: F. A. Davis, 1891) 1–2.
28 See for example the case of organized sports in Berit Skirstad and Kristin Felde, "Sports Policy in Norway," *National Sports Policies: An International Handbook*, eds. Laurance Chalip, Arthur Johnson, and Lisa Stachura (Westport, C.T.: The Greenwood Press, 1996) 320; also cited in Berman, *Livskraft*, 2006, 51.
29 Kyllingstad, *Measuring the Master Race,* 87ff.; Patricia Berman, "From Folk to a Folk Race: Carl Arbo and National Romantic Anthropology in Norway," Chapter 1, *Constructing Race on the Borders of Europe: Ethnography, Anthropology, and Visual Culture, c. 1850-1930,* eds. Marsha Morton and Barbara Larsen (London: Bloomsbury Publishing, 2021) 25–49.
30 Kyllingstad, *Measuring the Master Race*, 87ff.
31 Kyllingstad, *Measuring the Master Race*, 55.
32 Sten Konow and Karl Fischer, *Norway: Official Publication for the Paris Exhibition, 1900* (Kristiania: Aktie-Bogtrykkerit, 1900) 106–107; 118–120.
33 See Bjørg Evjen, "Measuring Heads: Physical Anthropological Research in North Norway," *Acta Borealia* (Vol. 2, 1997): 6ff.
34 *Utopianism and the Sciences, 1880-1930, Groningen Studies in Cultural Change*, eds. Maria G. Kemperinck and Leonieke Vermeer (Leuven: Peeters, 2010); Elisabet Stavenow-Hidemark, "Hygienism kring sekelskiftet," (Fataburen: Nordiska Museets och Skansens Årsbok, 1970) 47–54.
35 *Bericht über den V. Internationalen Kongress für Thalassotherapie* (Kolberg, 5–9 June 1911), eds. Prof. Dr. Dietrich and Dr. Kamner (Berlin, 1911).
36 Olmstad, 135. See also Irene Berg Godinez, "Friluftsbad Christiania – frå reinsomhed til sport og rekreasjon," *Byminner,* 97/2 (1997): 37.
37 Bill Mallon and Ian Buchanan, *The 1908 Olympic Games: Results from All Competitors in All Events, with Commentary* (Jefferson, North Carolina: McFarland & Co., 2000) 242.
38 Bill Mallon and Ture Widlund, *The 1912 Olympic Games: Results from All Competitors in All Events, with Commentary* (Jefferson, North Carolina: McFarland and Co., 2001) 153ff, 286ff, and 329ff.
39 See Holger Koefoed, "Badeliv og vitalisme i kunsten," *Byminner,* Vol. 97, No. 2 (1997): 60–66.
40 Nicholas Hope, *German and Scandinavian Protestantism 1700-1918* (Oxford: Clarendon Press, 1995) 441; 515; Kyllingstad and Rørvik, 194–197.
41 Carla Lathe, "The Group *Zum Schwarzen Ferkel:* A Study in Early Modernism" (Ph.D. dissertation, University of East Anglia, 1972) 30; for Haeckel's work on Darwin, and Bölsch's popularization of Haeckel, see Marsha Morton, "From Monera to Man: Ernst Haeckel, *Darwinismus*, and Nineteenth-Century German Art," *The Art of Evolution: Darwin, Darwinisms, and Visual Culture*, eds. Barbara Larson and Fae Brauer (Hanover: Dartmouth College Press, University Press of New England, 2009) 60–61; note 8. On Munch and Monism, see Patricia Berman, "Edvard Munch's 'Modern Life of the Soul,'" *Edvard Munch: The Modern Life of the Soul,* ed. Kynaston McShine (New York: The Museum of Modern Art Exhibition Catalogue, 2006) 34–51.
42 Wilhelm Bölsche, *Ernst Haeckel: Ein Lebensbild* (Dresden: C. Reisner, 1900).
43 On this commission, see Andreas Brändström and Hilde Rognerud, *Munch! Nietzsche, Thiel och Nordens störste konstnär* (Stockholm: Thielska Galleriet, 2013).
44 Berman, "Dionysus with Tan Lines," 2008: 68–85.
45 Edvard Munch, Sketchbook T2702, Munch Museum, translated in Poul Erik Tøjner, *Munch in His Own Words* (Munich: Prestel Verlag, 2001) 104.

46 Refer Corinna Treitel, *A Science for the Soul* (Baltimore and London: The Johns Hopkins University Press, 2004), Chapter 6 in which Treitel explores how occult techniques were used to solve modern problems; also refer Simone Natale, "A Cosmology of Invisible Fluids: Wireless, X-Rays, and Psychical Research around 1900," *The Canadian Journal of Communication*, 36/2 (2011): 263–275.
47 Edvard Munch, *Notebook,* reg. no. N55 (Munch Museum).
48 Finsen published his findings in Danish, German, French, and English beginning with his pioneering study, "Om lysets indvirkninger paa huden," *Hospitalstidende* 1 (1893): 721–728. This and other papers were published in German in 1899, and *La Photothérapie* appeared in French the same year (Paris: Georges Carré et C. Naud). Hugo Roesler's "Niels Ryberg Finsen's Disease and His Self-Instituted Treatment," *Annals of Medical History* 8 (1936): 353–356, includes a list of his publications on this topic; on Finsen and phototherapy, see Tania Woloshyn, *Soaking Up the Rays: Light Therapy and Visual Culture in Britain, c. 1890-1940* (Manchester: Manchester University Press, 2017).
49 Berman, "Dionysus with Tan Lines," 2008, 80. On heliotherapy and its implications for eugenics, see Tania Woloshyn, "Regenerative Tanning: Pigmentation, Neo-Lamarckism Eugenics and the Visual Culture of the Cure de Soliel," Chapter 10, *Picturing Evolution and Extinction: Regeneration and Degeneration in Modern Visual Culture*, eds. Fae Brauer and Serena Keshavjeee (Newcastle upon Tyne: Cambridge Scholars Publishing, 2015) 193–216.
50 Fae Brauer, "Making Eugenic Bodies Delectable: Art, 'Biopower' and 'Scientia Sexualis,'" Introduction, *Art, Sex, and Eugenics: Corpus Delecti* (Aldershot: Ashgate Publishing, 2008) 3–7; 19–21.
51 John Peter Collett, *Historien om Universitetet i Oslo* (Oslo: Universitetsforlaget, 1999) 109–114; 117–120.
52 Geir Hestmark, "Nansen – 'Vitenskaperen,'" *Nansen ved to århundreskifter*, eds. Olav Christiensen and Audhild Skoglund (Oslo: Aschehoug, 1996) 130.
53 The cultural and national importance of his public profile cannot be exaggerated, as his trip to the Arctic helped to give Norway an international scientific profile and, at home, helped to popularize the sport of cross-country skiing, a central feature of the Nordic reform movement's emphasis on bodily regeneration in nature; see Rune Slagstad, *Sporten: En Idéhistorisk Studie* (Oslo: Pax Forlag, 2008) 239–266.
54 Fridtjof Nansen, Diary entry from 8 November 1893, *Farthest North, Being the Record of a Voyage of Exploration of the Ship "Fram" 1893–96 and of a Fifteen Months' Sleigh Journey by Dr. Nansen and Lieut. Johansen* (New York: Harper and Brothers, 1897).
55 Nansen, Diary entry from 8 November 1893, *Farthest North*, 252; 302. For an analysis of this rhetoric, see Robert Marc Friedman, "Making the Aurora Norwegian: Science and Image in the Making of a Tradition," *Interdisciplinary Science Reviews*, Vol. 35, No. 1 (March 2010): 65.
56 Alv Egeland and William J. Burke, *Kristian Birkeland: The First Space Scientist* (Dordrecht: Springer Verlag, 2005) 8–14; 27–43.
57 Robert Michael Brain, "How Edvard Munch and August Strindberg Contracted Protoplasmania: Memory, Synaesthesia, and the Vibratory Organism in Fin-de-Siècle Europe," *Interdisciplinary Science Reviews*, Vol. 35, No. 1 (March 2010): 7.
58 Lynn Gambrell, *Exploring the Invisible: Art, Science, and the Spiritual* (Princeton: Princeton University Press, 2002) 163ff.
59 Robert Marc Friedman, "Civilization and National Honour: The Rise of Norwegian Geophysical and Cosmic Science," *Making Sense of Space: The History of Norwegian Space Activities*, ed. John Peter Collett (Oslo and Stockholm: Scandinavian University Press, 1995) 9.
60 Haeckel, *The Riddle of the Universe*, 369.
61 Haeckel, *The Riddle of the Universe*, 111.
62 Edvard Munch, *Notebook,* reg. N665-1 (Munch Museum, trans. Tøjner) 103.

3 «L'art et le muscle»

Robert Delaunay's *L'Equipe de Cardiff* and Pierre de Coubertin's *Internationalist Vitalism*

Pascal Rousseau

At the 1913 Salon des Indépendants where he exhibited *L'Equipe de Cardiff*, Robert Delaunay was by no means alone in representing a sporting event (Figure 3.1). In the same Salon, Albert Gleizes exhibited *Joueurs de football* (Figure 3.2). Despite Delaunay's and Gleizes' misleading titles, football was equivalent to them to a game of rugby, at that time called "football-rugby" – football then, as today, being accommodated under the name of "football-association". With *L'Equipe de Cardiff*, Delaunay offered a "hymn to sport", under the internationalist aegis of Pierre de Coubertin, the illustrious proselyte of the Olympic spirit and facilitator, in France, of Anglo-Saxon rugby clubs. Following Coubertin, Delaunay made an ideological choice in favour of a "sport-spectacle" based upon the Anglo-Saxon model of sport as "entertainment" against "la morale des sports" of Paul Adam.[1] Created only a year or so before the outbreak of the Great War, as a counterpart to the hawkish declarations of the Italian Futurists referred to in Chapter Six, *L'Equipe de Cardiff* celebrated an energizing and pacifist dimension of "modernist vitalism", particularly Baron Pierre de Coubertin's "internationalist vitalism".[2] Instead of commemorating the strategic preparation for armed conflict, Delaunay endeavoured to convey, through *L'Equipe de Cardiff*, how cultural communication and transnational competition could be engendered by sport, particularly football, and how this could, in turn, disarm the destructive impulses of warmongering nations.[3]

«Génération 1912»: Sport as a Remedy for Neurasthenic Decline

Creating a scrum in the heart of Paris may be regarded as an avant-garde strategy. Yet this aesthetic of combat needs to be counterbalanced with *fin-de-siècle* degeneration, and its urban nosological translation as neurasthenia.[4] As Anson Rabinbach puts it:

> The nation as well as the individual found itself abandoned to the vicissitudes of the will, to the emotions and to the enemies of productive order. Fatigue, manifest as entropy, was not just undermining the conservation of energy for social usage but seemed to threaten modernity itself.[5]

This was seemingly compounded by what was derided as the pervasiveness of Modernism identified as Decadent art. At the beginning of 1913, Dr. Lucien Grellety's *Névrosés et décadents* appeared.[6] Despite being published some twenty years after Max Nordau's *Degeneration* – first published in 1892 and translated into French two years later – it lambasted all the Decadent literature that had been castigated by this Hungarian physician living in Paris, while stigmatizing Symbolism

DOI: 10.4324/9781003045595-5

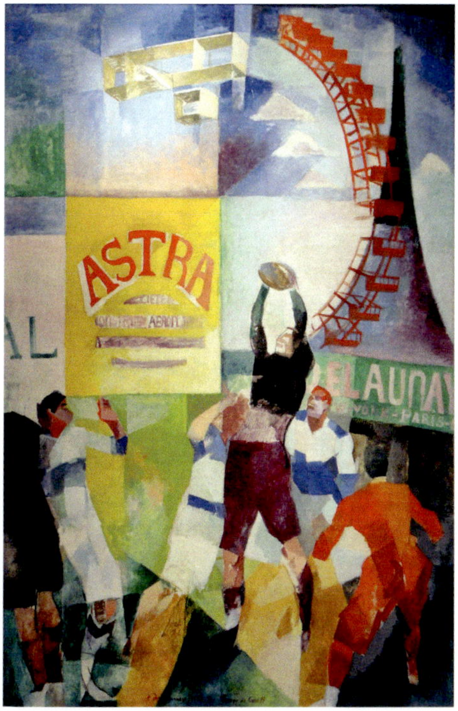

Figure 3.1 Robert Delaunay, *L'Equipe de Cardiff*, 1913, oil on canvas, 326 × 208 cm. (Musée d'Art Moderne de la Ville de Paris; Accession No. AMVP 1115. Public domain.)

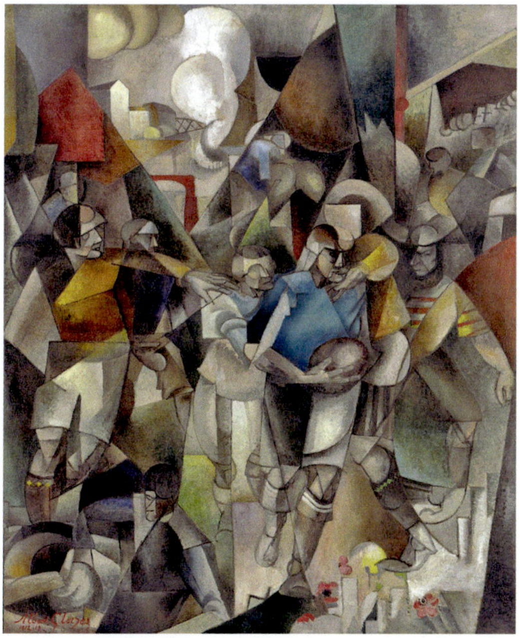

Figure 3.2 Albert Gleizes, *Les Jouers de football*, 1912–1913, oil on canvas, 225.4 × 183 cm. (National Gallery of Art, Washington D.C. Public domain.)

for its neuro-narcissistic inclinations.[7] In much the same terms as Nordau, Grellety condemned the Decadents' "refined (obsession) with intellectual pleasure" and their "morbid humanitarianism",[8] while developing new arguments for the "glorification of energy", defended by Maurice Barrès, embellished by certain medical analyses.[9] In turn, these arguments were reinforced by eugenic discourses, particularly those of the "Société française d'eugénisme".[10] Inaugurated the day after the First International Eugenics Congress closed in London in July 1912, the French Eugenic Society circulated its Galtonian and Neo-Lamarckian eugenic policies from the first issue of its journal, *Eugénique,* published from early 1913.[11]

By formation of the French Eugenic Society, the aestheticization of decadence had seemingly transpired.[12] As Dr. Grellety proudly announced: "Dilettantism is no longer fashionable: The era of aesthetes and young malingerers has passed".[13] Opposing thinking as a pragmatic concept, the new currents of the avant-garde turned towards action, boosted by the well-established philosophical tracts of Friedrich Nietzsche, Hippolyte Taine, Jean-Marie Guyau and Henri Bergson. Concern with action and energy abounded in literature on art as testified by what Charles Lalo hailed as the "feeling of vitality", and the "indefinite irradiation of a source of energy".[14] In *Essais sur la sensibilité contemporaine,* Raphaël Cor regarded the model of a "philosophy of sensitivity" as "the fundamental structure of the living self", which he regarded as integral to a "model of young, crude, heroic strength".[15] At the same time, Jean Bourdeau explored "affective philosophy" as the "instinctive will of action",[16] while his colleague, Le Tainturier-Fradin hailed sport for providing an "ordering principle" that could be applied to all aspects of social life entailing "courage, energy and mastery of the self".[17]

While each of these texts seemed to flow with the Bergsonian current,[18] as outlined in Chapter Five, the "internationalist vitalism" of Pierre de Coubertin, inherent to each of them, is the notion dear to nineteenth century psycho-physiological theorists of an *ideomotor* of movement, will and representation. This was articulated by Étienne Rey as "the natural balance between desire and act" and the end of "this fatal opposition between thought and action, which was the defect of previous generations".[19] This ideo-dynamism, defended by Dr. Toulouse in the columns of the journal *Demain,*[20] was based on a reflex chain reaction activated by empathic projection with an image illustrating an action. This image would purportedly be able to induce in the spectator an interior movement, and through its mimetic reproduction, intensify their energy. To observe an image representing a galvanizing movement was also regarded as a means of garnering strength and nourishing the momentum of decision-making. Such an "affective image", surmized Albert Bazaillas, would be at the heart of a new "art loving intensity".[21]

The visual economy of chromatic contrasts is clearly identified as the distinctive sign of this "athletic" aesthetic.[22] Georges Rozet, one of the most famous defenders of the campaign for national recovery through sport, spared himself any subtlety: "In the eyes of our athletes, the only things that count are very strong, astonishing visual sensations. The ring lights are 'dazzling': No one cares any longer whether the light is pink or lilac. And so we go back to the visual simplicity of the primitives".[23] If Delaunay chose to locate his rugby scene in a visually sensational forest of billboards, it was not only to illuminate a landscape on the side of the stadium that he would rub shoulders with in everyday life and in magazine images but also to consolidate, in colour, the two markers of the visual economy of modernity – sport and advertising. In a fictitious dialogue located at the heart of the 1913 Salon des Indépendants published in *Gil Blas* on 19 January 1913, the art critic, Gabriel Mourey, commented on this syndrome while Delaunay was still developing his painting, *L'Equipe de Cardiff:*

> We athletes have a mission of fulfilment in the time to which we belong... We represent health, strength, life. You are dull, we come to tone you up; we are toners. You languish, we come to stimulate you; we are stimulators. You are weak, we come to bring you new energies; we are regenerators. You are civilized, refined, Byzantine; we are barbarians. Conscious and organized barbarians! Barbarians! Ah! The beautiful word! Barbarians! that is to say, virginal and free imaginations, healthy senses, normal force ... How do you finally want artists today to conceive and practice art: As if they were contemporaries of Pericles, Leo X, Louis XIV or of ... Louis Philippe? ...
>
> We need to rest our eyes on images of joy, strength, health, life. We demand that only frank and vigorous colours vibrate around us, as [highly] pitched as possible; these sonorous reds which are like bugle blows; these acid greens that make you cringe, the sick people that you are, but which give us such exquisite sensations; these pure oranges which resemble the rays of the setting sun; these vigorous and bold blues which evoke the vision of the Mediterranean on windy days. And we also adore intense blacks and immaculate whites: and we cherish the brutal oppositions, the daring contrasts of which our decorators have made a specialty in the use of these colours, which remind us of the violence of our favorite sports. It is the suddenness, the unforeseen, the safety, the frankness of the gestures which make the gymnasts, the champions of football and boxing, all the sportsmen, finally worthy of their name.[24]

Implicit within Mourey's diatribe was the dichotomy that had animated aesthetic debates since the eighteenth century: The notorious drawing/colour dispute in which line and form (masculine) were correlated to the use of reason and intelligence while colour (feminine) was identified with feeling and affect.[25] Yet given Mourey's identification of consumerism with muscular effort and economic competition with sporting competition, these values were reversed: Colour became the prerogative of the athlete and the publicist, a sign of vigour and virility, persuasion and authority.[26] This is epitomized by Delaunay's L'Équipe de Cardiff. In the two major versions of his painting, Delaunay chose to dress the players in colourful jerseys. In the black and white source photograph used by the artist, two teams face off, one wearing striped jerseys that can be seen in the painting. The striped jerseys in the original photo were that of the Stade Toulousain players whose colours were black-and-white, not blue-and-white. In choosing these historic colours of Cardiff FC, Delaunay confirmed his interest in the Welsh team, as well as his engagement with the battle of emblematic colours.[27]

This battle was linked to a culture of action that Pierre de Coubertin, as a disciple of Taine, commented upon in his famous 1909 article on the "psychology of sportswear", reprinted at the beginning of 1913 in *Essais de psychologie sportive.* According to Coubertin, a "display of sports clothing causes a certain stir in the body", what he called "sport emotion". This "contact with sportswear" was conceived as a source of energy that could awaken an "idea of pleasure, intoxication, desire soon satisfied" and which could grow, like pure contagion, into "thinking with intensity, with a yearning for movement".[27] Sportswear, like sport, was also regarded as a means of vitalizing workers in a metropolis. Installing sport in the heart of cities entailed fighting, through prophylaxis, against the effects of overwork amongst the young urban tertiary class suffering the notorious "nervous fatigue" of modern life. In his *Essais de psychologie sportive*, Coubertin also questioned whether sport could "curb universal neurosis". His response, both Transformist and anti-Decadent, found an immediate echo in the choice of motifs adopted by Delaunay. Delaunay's choice reflects Coubertin's positioning of sport not just as a panacea to universal neuroses but as a vitalist form of adaptation to the incessant accelerationism and everyday vicissitudes of mass culture in "modern life", as Coubertin explained:

> Modern life is no longer localized or special: everything has an influence on everyone ... The rapidity and multiplicity of transport have made Man an essentially mobile being for whom distances are moreover insignificant to cross and which, consequently, require frequent changes of location ... Inside and out, the brain is kept in a state of incessant boiling. Points of view, the aspects of things, combinations, possibilities, as much for individuals as for collectives, follow one another so quickly that it is necessary to use them whenever needed, to keep always alert and remain in a state of permanent mobilization.[28]

When contextualized within Coubertin's philosophy of sport, Delaunay's paintings, *L'Equipe de Cardiff* may be then perceived unequivocally as a vitalist plea in favour of this new spirit of adaptation. In its most biological and Nietzschean sense, it was to be orientated towards an "incorporéisation" or embodiment of subjectivity in terms of "will-power" and the "living body", whose faculties of assimilation were largely dependent on a physiology of arousal.[29] The capacity to welcome and absorb external stimuli was especially stressed for the "artist" in their quest for a "force plastique" closely associated with what Nietzsche calls "Lebensreiz", an "excitation which erupts into life".[30] That team sports were correlated to the aesthetic biologism of Nietzsche was demonstrated by many other authors. In his pre-war memoirs published under the title *Joueur de balle*, Joseph Jolinon explicitly associated the philosophy of rugby with the lessons of Nietzsche.[31] This author, who described the state of mind of the "young athlete of 1913-1914", delivered descriptions of scrums and touchdowns, which seem to have been inspired by the dynamogenic assembly of *L'Equipe de Cardiff*. As Jolinon wrote:

> To each red-black torso, a red torso hangs on. Exalted acrobatics. The spring of their legs, the mass of their heads inflict on the blond meteor the trajectories of arrows. They are silent, the movement intoxicates them, they detach themselves from the ground, they fall back, their ardour multiplies, surrounded by watching eyes, rays of sun, molecular dustings, waves of joy. The modulations of waiting make the crowd vibrate with pathos. The action precipitating thought floats like a kind of hypnosis of which they will only retain a memory, confused and magnificent.[32]

More specific is the writer Paul Adam with whom Delaunay rubbed shoulders in artists' circles at the Closerie des Lilas. In 1904, Adam had made an initiatory trip to the United States from which emanated his *Vues d'Amerique* in which he saluted the spirit of competition across the Atlantic in his so-called 'cult of action'.[33] For Adam, collective Anglo-Saxon team sports like "football-rugby" constituted a pragmatic adaptation to the "struggle for life" that was necessary to negotiate the vicissitudes of modern life.[34] As he explained:

> When Americanism and Nietzscheanism will dominate the next era, there will be no ideas able to prevail against these social phenomena ... Let us ask sport to arm our character in order to be able to take our place in the first rank of those who will manifest their *Will to Power* with glory.[35]

While Delaunay absorbed Adam's "Americanism and Nietzscheanism", *L'Equipe de Cardiff* demonstrates more the pragmatic, voluntarist and liberal model that Delaunay encountered in the writings of Pierre de Coubertin,[36] particularly his promotion of Anglo-Saxon sports in France.[37] Sport was conceived by Coubertin as uniting those of different origin, dissolving class prejudices and destroying social categories: "Let the bourgeois and proletarian youth drink from the same source of muscular joy", he wrote, "from which source will flow, for both of them, the good social mood".[38] Coubertin, whose father, a Baron and academic painter, and whose mother, née Mirville, both belonged to a very old nobiliary branch, very quickly sought to distance himself from this "class spirit", particularly those dubbed the "morons of Faubourg Saint Germain". Robert Delaunay was not dissimilar. Brooded over by his mother, the Countess Berthe de Rose, a *demi-mondaine* frequenting Parisian salons and circles of influence, Sonia Delaunay would tell, in her memoirs, how much he sought, very quickly, to get out of this "pseudo-aristocratic" mold.[39]

Team sports appeared an escape from this mold while also offering "international" competition. For Coubertin, it was vital to engage in "foreign competition", as it was only through openness to the world that he envisaged peace could be sustained. "Let's export rowers, runners, fencers", he declared. "This is the free trade of the future, and the day it shall be introduced into the customs of Old Europe, the cause of peace shall have received new and powerful support".[40] The war of tomorrow would no longer be an armed conflict, according to Coubertin, but an economic battle in a new world given over to permanent circulation – what Coubertin called the "fruitful instability of modern societies" to which a "resourceful philosophy" would have to respond. As Coubertin stressed:

> The resourcefulness that this time needs will come neither from a bit of a lad, nor an upstart but simply a boy skillful with his hands, quick with effort, supple with muscles, resistant to fatigue, quick with glances, firm with decisions and trained in advance to bear changes of place, profession, situations, habits and ideas made necessary by the fruitful instability of modern societies.[41]

Additionally, Coubertin proposed a "muscular diploma" for this young generation active in a "democratic and cosmopolitan" era, not only "distinguished by its instability" but also by "tolerance and liberalism".[42] Aspiring "to unite in future festivals ... muscles and thought",[43] Coubertin pointed out that during the first Olympics, "poets came to read their unpublished works while painters exhibited their recent paintings".[44] This rapprochement between physical culture and artistic awakening became the central argument of the International Olympic Committee conference at the Comédie Française under the chairmanship of Jules Claretie. To bring athletes and artists together, Coubertin proposed the launching of five competitions in architecture, music, literature, painting and sculpture. The winning artworks were to be exhibited at the London Olympic

Games in 1908 with medals awarded for architecture, literature, music, painting and sculpture, a prospect wholeheartedly supported by the Royal Academy of Art and its President, Sir Edward Poynter, but which was eventually deferred until the 1912 Olympic Games in Stockholm. At the same time, Coubertin insisted: "Artists must attend sporting events".[45] Installing works of art in stadiums would be of mutual benefit, as Coubertin explained: "The gymnasium would not only provide painters and sculptors with original models but also suitable locations for their works and these works, in turn, would contribute to the eurythmic education and development of young athletes".[46] It was then essential to facilitate, in Coubertin's words, "a union that would bring these former divorcees together again: Muscle and spirit".[47]

The "Revenge of Stockholm": Youth Surveys and the Model of *l'artiste complet*

Before examining how this happened, we need to return to the cultural contexts which clearly motivated Delaunay's and Gleizes' choice of rugby scenes animated by "acts of bravery", addressed to the public due to attend the 1913 Salon des Indépendants. Their synchronous choices may be interpreted as their reaction to the public debate on the social role to be played by sport in training new generations, mobilized by the Parisian press in the second half of 1912. This debate was not new: Its roots lay in the defeat of 1870 and the need to ensure that the nation had the means to succeed. This was the debate on "regeneration". Initially galvanized by the publications of Max Nordau and Paul Bourget, the "regeneration" debate experienced a clear resurgence during the summer of 1912 following the crushing defeat of French athletes at the Stockholm Olympics. Hence, while the year 1912 had begun with a series of surveys on the state of French youth, it ended with a surge of responses addressing "l'athlète complet" to which Delaunay responded with his painting.

This flurry of surveys and frenzy of responses were part and parcel of what Philippe Bénéton and Robert Wohl have called the "1912-1914 generation".[48] These surveys had been initiated in March 1912 by *La Revue hebdomadaire,* asking a panel of "young people" to define their ideals.[49] Two months later, *Le Gaulois* invited a series of personalities to give their opinion on the horizons offered to "the youth of today".[50] In *L'Opinion*, Alfred de Tarde and Henri Massis, united under the pseudonym, Agathon, led an investigation into "young people of today".[51] Not only did *Le Temps* rally,[52] but so did *La Revue de la jeunesse, Les Annales politiques et littéraires, La Revue des Français* and *La Guerre Sociale.*[53] In this discursivity, the "youth of today" seemed repeatedly opposed to the dilettante, pessimistic and passive generation of late nineteenth century. Vigorous, energetic, enamoured of action, they were also identified as "anti-intellectualist",[54] a key-term in these discourses revealing how impregnated the new generations had become with Bergsonian "intuitionism".[55]

The French defeat at the 1912 Olympic Games in Stockholm provided the catalyst for this impetus. Even though France achieved seven gold medals, only one was awarded in athletics to Jean Bouin. Not only did France's failure ignite a controversy in the press but also a national campaign in favour of "physical culture", led by the influential sports journalist, Georges Rozet, a staunch defender of the "integral athlete".[56] Immediately Rozet launched an Enquête in *L'Opinion*, tellingly entitled "Let's prepare the Stockholm revenge by Creating a College of Athletes", as he explained:

> Since we were beaten in Stockholm and very badly beaten in the arduous field of athletic sports, and since ... the Americans, conquered as much by their meticulous scientific training as by the intrinsic qualities of their race, by their "class" ... for all these reasons, perhaps we should consider creating in France a true elite of muscle, educated and perfected in the American way, in an athletic college ... In short, we should think of founding – with a view to international competition and as an example to the rest of the country – a school of "phenomenons", as the United States champions have been called.[57]

Over September and October 1912, this Enquête became a "national and nationalist event",[58] reaching a climax at the beginning of 1913.[59] With the 1916 Olympic Games scheduled in Berlin, this event focused the urgency of this crisis. Those who rushed to respond included such runners as Géo André and Marcel Bal; such sociologists as André Lichtenberger; such writers as Marcel Boulanger; such physiologists as Raoul Baudet; such journalists as Pierre Laffitte and Frantz Reichel, and such politicians as Joseph Paul-Boncour. The Enquête ended with the publication of a manifesto in favour of a "College of Athletes, School of Physical Renaissance". Signed by Rodin,[60] amongst other artists, it was regarded as a means of developing not only «l'homme complet» but also "creative force" through a combination of muscle and culture, strength and invention.[61] The curriculum of English and American colleges was cited as a model.

In his *Exercices physiques et le développement intellectual*, published in 1904, Angelo Mosso, whose "energy theory of the mind" had inspired Futurism,[62] proposed the term "agonistic" to qualify this corrective use of physical education as the counterpart to intellectual development epitomized by the "classic" spirit at the Sorbonne. In order to compete in major international sporting events, team sports needed to be prioritized, particularly their spirit of "antagonism".[63] American universities, which Mosso hailed as the factory of "champions", represented his model, as he explained:

> In America, all universities have a gymnasium ... Americans possess, to such a high degree, the art of making intellectual and physical education more intense; they breathe it, with so much solicitude ... that they seem to remain young longer than Europeans ... *More life and fuller*".[64]

Drawing upon the social theories of Frédéric Le Play, Coubertin's plea for the educational model of English colleges pursued the same hypotheses. This appeared to be absorbed by Delaunay.

It was not entirely fortuitous that the press clipping Delaunay tore from the pages of *La Vie au grand air* depicted a meeting between the Stade Toulousain and the SCUF, a university team.[65] Through his choice of the title, *L'Equipe de Cardiff FC*, Delaunay signalled the "natural" domination of the Anglo-Saxons in this sport; through his motif, he signalled the debate on its relationship to French youth. Yet of the two sides in this polemic, one vigorously supporting gymnastic methodologies, philo-militaristic and revanchardist; the other ardently promoting collective sports, liberal and pacifist, Delaunay's choice of this rugby match appears consistent with the latter position and Coubertin's *Essais de psychologie sportif.* Published in the spring of 1913, Coubertin's treatise coincided with exposure of *L'Equipe de Cardiff* to the Parisian public and the tense polemic on "today's youth" debated between the gymnophiles and the collective sporting adherents. Within this debate, Coubertin's treatise incarnates collective sports, as does Delaunay's painting.

Cosmopolitan Sport: *"La touche de rugby"* versus *"le pas gymnastique"*

By the time that Delaunay began his painting, the proponents of these two political philosophies of sport vigorously opposed one another.[66] While "military gymnastics" represented the established order and its social mechanisms, "sport" was aligned with individual freedom and instinct,[67] competition and novelty.[68] This dichotomy was extended to a distinction drawn between natural order and artificial performance. The purportedly "natural" method of Lieutenant Hébert's physical education designed to virilize youth, prepare them for military action and channel their energies into moral and national ideals was pitted against what Hébert called "the fairground exhibitionism" of sport.[69] Yet it was precisely the "spectacle of sport" that the illustrated reports of *La Vie au grand air* commemorated and on which Delaunay drew for his sporting motif. In this conflict between *L'Education physique* and *La Vie au grand air*, Delaunay made his choice: The defense of sport in the form of rugby. Its popularity was demonstrated by 20,000 spectators attending the international France-Scotland meeting on 1 January 1913, and its staging at the Parc des Princes, which had become the modern arena for demonstrations of skill and exaltation of competition.

Unlike "Le Douanier" Rousseau who positioned his *Joueurs de football* in a bucolic undergrowth, Delaunay set up his football game in the Parc des Princes under the *Eiffel Tower* – the icon of the 1889 and 1900 Universal Expositions – and the *Grande Roue* – the totem of the funfair. While these two metal structures were emblematic of festive architecture integral to theatricalization in the *City of Light*, the stadium and the amusement park carried comparable significations.[70] As Pascal Ory notes, the "festival of machines" had replaced a "machine of festivals" with the cultural advent of mass entertainment, largely derived from the American model.[71] Viewed from this perspective, Delaunay does not seek to destabilize painting as a medium through his incorporation of popular culture but, on the contrary, optimizes it so that the spectator may become carried away by the intoxication of entertainment. He does so by immersing the spectator within "the cultural industry" in his painting, following Theodor Adorno and Walter Benjamin, rather than separating them from it, and enabling them to identify with such models of urban subjectivity and competition as the footballer and the aviator. Delaunay does this also through "cosmopolitanization".

The model for Coubertin's conception of the Olympic Games was that of World Fairs.[72] Coubertin had sought to align the development of collective sports under the aegis of a modernity in which art would become intertwined with technique, invention with its spectacular diffusion and globalization with universalism epitomized by the "cosmopolitization of Sport".[73] Delaunay regarded his *Equipe* as a visual condensation of this "cosmopolitization". The biplane that flies over the players in his painting signifies that aeronautics will become the new weapon of global "reconciliation", despite it being more commonly represented as the "fourth weapon" in France's preparation for revenge against Germany, as ironically signified by inclusion of the military leaflet in Picasso's 1912 collage, *Notre avenir est dans l'air.*[74] Given the internationalism identified with the Eiffel Tower, it is not surprising to find it overlooking the rugby players, especially after the major conference convened there in October 1912 to abolish borders by establishing an "international unification of the hour".[75] Since all of these signifiers converge in this painting, they make *L'Equipe de Cardiff* a hymn to universal harmony, where the competitive arena of sport becomes a genuine derivative of warlike instincts in what Alain Ehrenberg has called "the peaceful drift of general mobilization", as he explained:

> Competition embodies the utopia of a dream war in a peaceful society down to the smallest aggressive manifestations. Its staging aims to establish a relationship with the masses of spectators that is none other than that established between war and people in a democracy: A general mobilization ... The athletic project at work in the competition is nothing other than the peaceful drift towards general mobilization.[76]

Yet what Delaunay excluded from his painting of this football match is as strategic as what he chose to include. Delaunay does not portray the hysterical and gregarious crowd. Instead he pays tribute to major international sporting meetings as events able to act as a reconciliation of countries. In so doing, he salutes the virtues of collective sport in the fragile construction of a competitive but peaceful globalized world, in which the authoritarian paradigm of the war of nations and peoples becomes displaced by the liberal model of a "peaceful struggle" between commercial and economic interests.[77] Delaunay's concept then dovetails with what Pierre de Coubertin called in 1910 the "democratic future" of sport, as he explained:

> Sport should not be considered only in relation to physical education, nor even muscle culture, because then the aphorism would no longer be correct; it is sport alone, that is to say this way of feeling and practicing exercise, which constitutes an act of free will, a constant quest to do better in regular competition with one's comrades ... Sport, moreso than any other manifestation of human activity, needs to strive for internationalism.[78]

In February 1913, this optimistic concept of sport internationalism in which the curses of repeated warfare would be finally eradicated, was satirized by *Fantasio* propelling its readers into this future world:

> Finally, the battles are over: Sport is God and Frantz Reichel is its prophet. Football-rugby has replaced religion, the army, the navy. According to an old prophecy, once misinterpreted, the inter-na-tio-na-a-ales teams constitute the elite of the *gen'rhumain!* All being well, the budgets of the great European nations need only welcome it. International football matches have replaced the battles of yesteryear.[79]

Nevertheless a few months later, the declaration of war would surprise the French in the middle of summer. Once the thirst for "revanche" became a quest for demographic slaughter, no longer would many continue to raise the "flower to the gun". Delaunay remained an exception. Already he had left for Spain. On news of the outbreak of the First World War, he and his wife, Sonia, decided not to return to France and to settle in Madrid. Branded a "deserter", only when the Delaunays moved to Portugal a year later was Robert declared unfit for military duty by the French Consulate in Vigo. Afterwards, the Delaunays left for Barcelona where Robert was able to join his friend, Gleizes. A committed pacifist and anti-militarist, Gleizes had also escaped from the Front and, like Delaunay, had sought refuge from the carnage of the First World War that he hoped would never arise, in his art.

Translation from French into English: Fae Brauer and Justin Fleming

Notes

1 Paul Adam, *La Morale des Sports* (Paris: La Librairie mondiale, 1907); "Athletic Education", *The International Journal of the History of Sport,* 23/3–4 (20 August 2006): 446–482.

2 Baron Pierre de Coubertin, *Les Batailles de l'éducation Physique : Un Campagne de vingt-et-uns ans (1887–1908)* (Paris: Librairie de l'Éducation Physique, 1909).

3 Pierre de Coubertin, «L'avenir des sports au point de vue psychologique, international et démocratique», repb. «Psychologie, internationalisme, démocratie», *Revue Olympique*, (July 1910): 103. Pascal Rousseau, *Robert Delaunay : l'invention du pop* (Paris: Éditions Hazan, 2019).

4 Fae Brauer and Serena Keshavjee (eds.), *Picturing Evolution and Extinction. Regeneration and Degeneration in Modern Visual Culture* (Lady Stephenson Library, Newcastle upon Tyne: Cambridge Scholars Publishing, 2015).

5 Anson Rabinbach, *The Human Motor: Energy, Fatigue and the Origins of Modernity* (Berkeley: University of California Press, 1992).

6 Dr Lucien Grellety, *Névrosés et décadents* (Mâcon: Protat Frères, 1913).

7 Max Nordau, *Dégénérescence,* Tome Premier : *Fin de siècle : Le Mysticisme* (Paris: Félix Alcan, Éditeur, 1894); Robert A. Nye, "Degeneration, Neurasthenia and the Culture of Sport in Belle Époque France", *Journal of Contemporary History,* 17/1 (January 1982): 51–68.

8 Grellety, *Névrosés et décadents,* 11.

9 Etienne Rabaud, «Origine et transformation de la notion de dégénéré», *Revue de l'école d'anthropologie* (vol. 17, 1907): 36–46.

10 Fae Brauer, "L'Art Eugénique: Biopower and the Biocultures of Regeneration", *Genetics and French Culture,* eds. Louise Lyle, Douglas Morrey; *L'Esprit Créateur,* 52/2 (2012): 42–58; Fae Brauer, "Eroticizing Lamarckian Eugenics: The Body Stripped Bare during French Sexual Neoregulation", *Art, Sex and Eugenics. Corpus Delecti*, eds. Fae Brauer, Anthea Callen (Hampshire, UK and Burlington, VT: Ashgate Publishing, 2008) 129.

11 William Schneider, "Toward the Improvement of the Human Race: The History of Eugenics in France", *The Journal of Modern History,* vol. 54 (June 1982): 268–291; also refer Brauer, "L'Art Eugénique", 42–58.

12 Michel Winock, *Décadence fin de siècle* (Paris: Gallimard, 2017).

13 Grellety, *Névrosés et décadents,* 111.

14 Charles Lalo, *Les Sentiments esthétiques* (Paris: Alcan, 1910) 152–153.

15 Raphaël Cor, *Essais sur la sensibilité contemporaine* (Paris: H. Falque, 1912) 8.

16 Jean Bourdeau, *La philosophie affective* (Paris: Alcan, 1912).

17 Le Tainturier-Fradin, *L'activité sportive, germe d'une philosophie pratique*, 1913, quoted in *Recherches*, 43, (April 1980) 131.

18 Mark Antliff, *Inventing Bergson. Cultural Politics and the Parisian Avant-Garde* (Princeton, NJ: Princeton University Press, 1993).

19 Etienne Rey, *La renaissance de l'orgueil français* (Paris: B. Grassat, 1912) 111.

20 Dr. Toulouse, "L'action est le but supérieur de la vie", *Demain*, 1 (April 1912): 1–7.

21 Albert Bazaillas, *Musique et inconscience. Introduction à la psychologie de l'inconscient* (Paris: Alcan, 1908) 239, 241.

22 Henry de Montherlant, "Football et rugby Olympique", *Demain* (July 1924); repb. Françoise Lionel et Serge Laget, *Rugby en toutes lettres. Anthologie* (Biarritz: Atlantica, 1999) 48.
23 Georges Rozet, *La défense et Illustration de la Race Française* (Paris: Felix Alcan, 1911) 35.
24 Gabriel Mourey, «L'art et le muscle», *Gil Blas* (19 janvier 1913): 1, in response to the survey, "Art and Muscle".
25 Jacqueline Lichtenstein, *La couleur éloquente. Rhétorique et peinture à l'âge classique* (Paris: Flammarion, 2003).
26 David Cottington, *Cubism in the Shadow of War. The Avant-garde and Politics in Paris, 1905-1914* (New Haven and London: Yale University Press, 1998); Robert Lubar, "Cubism, Classicism and Ideology: The 1912 Exposicio d'Art Cubista in Barcelona and French Cubist Criticism", eds. Elisabeth Cowling, Jennifer Mundy, *On Classic Ground* (London: Tate Gallery, 1990) 309–324.
27 Pierre de Coubertin, «La psychologie du costume sportif» (February 1909); *Essais de psychologie sportive* (1913); (Grenoble: Jerôme Million, 1992) 63.
28 Pierre de Coubertin, «Le sport peut-il enrayer la névrose universelle?», *Essais de psychologie sportive,* 120.
29 William H. Schneider, *Quality and Quantity. The Quest for Biological Regeneration in Twentieth-Century France* (Cambridge: Cambridge University Press, 1990).
30 Barbara Stiegler, *Nietzsche et la biologie* (Paris: Presses Universitaires de France, 2001): 32–43.
31 Joseph Jolinon, *Le joueur de balle* (1924) (Paris: Rieder, 1929) 75.
32 Jolinon, *Le joueur de balle*, 131.
33 Paul Adam, *Vues d'Amérique* (Paris: Ollendorf, 1906).
34 Préface de Louis Dedet à l'ouvrage de Gondouin et Jordan, *Le Football* (Paris: Pierre Lafitte, 1910) 18: *«La vie universelle n'est autre chose qu'une lutte (..). Apprenons donc la lutte et le triomphe. Prenons-en l'habitude et le goût sur les terrains gazonnés où se jouent les grands jeux.»*
35 Adam, *La Morale des Sports*, 17.
36 E. Seillière, *Un artisan d'énergie française, Pierre de Coubertin* (Paris: Henri Didier, Librairie Éditeur, 1917) 22–23.
37 Pierre de Coubertin, *Une campagne de vingt et un ans La bataille de l'éduction physique.* (Paris: Librairie de l'Éducation Physique, 1909) 220.
38 Pierre de Coubertin, *Pédagogie sportive* (Paris: Les Éditions G. Crès et Cie, 1922) 145.
39 Sonia Delaunay, *Nous irons jusqu'au soleil*, 26.
40 Coubertin, *Une campagne de vingt et un ans*, 90.
41 Coubertin, *Une campagne de vingt et un ans*, 183.
42 Pierre de Coubertin, «La philosophie du débrouillard», discours prononcé dans le grand amphithéâtre de la Sorbonne, le 30 juin 1907, lors de la fête organisée par la Société des Sports Populaires pour la «distribution du Diplôme des Débrouillards», repb. *Une campagne de vingt et un ans*, 217; 219.
43 Coubertin, *Une campagne de vingt et un ans*, 192.
44 Coubertin, *Une campagne de vingt et un ans*, 192.
45 Coubertin, *Une campagne de vingt et un ans*, 198.
46 Coubertin, *Une campagne de vingt et un ans*, 198.
47 Coubertin, *Une campagne de vingt et un ans*, 200.
48 Philippe Bénéton, «La génération de 1912-1914. Image, mythe et réalité?», *Revue Française de Science Politique,* XXI/5 (Octobre 1971) 981–1009; Robert Wohl, *The Generation of 1914* (Cambridge: Harvard University Press, 1979).
49 Fernand Laudet, «Enquête sur la jeunesse», *La Revue hebdomadaire* (23 March–20 July 1912).
50 Jules Bertaut, «La jeunesse d'aujourd'hui», *La Gaulois* (1, 15 juin 1912).
51 Agathon (Alfred de Tarde, Henri Massis), «Les jeunes gens d'aujourd'hui», *L'Opinion* (13 April–29 June 1912); repb. Agathon, *Les jeunes gens d'aujourd'hui* (Paris: Plon, 1913).
52 Émile Henriot, «À quoi rêvent les jeunes gens?», *Le Temps,* 23, 24 April 1912 (7, 13); 27 May; 4 June 1912.

53 Pierre Baudin, «Nos enfants. A quoi rêvent-ils, Que rêvons-nous pour eux?» (national survey directed by Henri Mazel), *Revue des Français*, 6 (25 June 1912): 113.
54 Agathon, «Consciemment ou d'instinct, elle est anti-intellectualiste», Agathon, *Les jeunes gens d'aujourd'hui* (Paris: Plon, 1913) 19. In his survey, *Revue hebdomadaire* (n°29, 20 juillet 1912, 302), Émile Faguet highlighted the influence of Bergson, Bourget and Barrès.
55 In his response to the survey of Jules Bertaut, *La Gaulois* (15 juin 1912), Bergson commended the regeneration of energy among youths. On the relationship between these surveys and Bergson's engagement with youth, refer André Colomer, «Aux sources de l'héroïsme individualiste. M. Bergson et les Jeunes gens d'aujourd'hui», *L'Action d'art*, 2 (1 March 1913): 1.
56 Georges Rozet, *Défense et illustration de la race française* (Paris: Alcan, 1911).
57 Georges Rozet, «Préparons la revanche de Stockholm. Il faut créer un Collège d'athlètes», *L'Opinion* (17 August 1912): 1.
58 Jean Paul Delaplace, *L'Opinion* (17 August 1912) 78.
59 Anon., «Le collège d'athlètes», *L'Intransigeant* (1 January 1913): 3.
60 Auguste Rodin, *L'Opinion*, 37 (14 September 1912): 335–336: le «projet du collège d'athlètes (l') intéresse»
61 J. R. Guasco, *Le Collège d'athlètes* (Paris: Audin, 1913) 17.
62 Vincent Pidoux, «Sur les traces d'une théorie énergétique de l'esprit. Angelo Mosso et le travail des émotions», *PSN (Psychiatrie, Sciences Humaines, Neurosciences)*, 13/2 (2015): 43–69.
63 Angelo Mosso, *Exercices physiques et le développement intellectuel* (Paris: Alcan, 1904) 76.
64 Mosso, *Exercices physiques,* 156–157.
65 Thierry Gervais, «L'invention du magazine. La photographie mise en page dans *La Vie au grand air* (1888-1914)», *Études photographiques*, 20 (June 2007): 50–67.
66 Eugen Weber, "Pierre de Coubertin and the Introduction of Organized Sport", *My France. Politics, Culture, Myth* (Cambridge: Harvard University Press, 1991) 207–225.
67 G. Hébert, *L'éducation physique ou l'entraînement complet par la méthode naturelle* (Paris: Vuibert, 1912) 51.
68 Pierre de Coubertin, «Une nouvelle formule d'éducation physique», *Revue mensuelle du Touring Club de France* (20 March 1902) 146–151; *L'Education des adolescents au XXème siècle. I. L'Education physique : la gymnastique utilitaire. Sauvetage-Défense-Locomotion* (Paris: Alcan, 1905).
69 Hébert, *Le Sport contre l'éducation physique,* 98.
70 John F. Kasson, *Amusing the Million: Coney Island at the Turn of Century* (New York: Hill & Wang, 1978).
71 Pascal Ory, *L'exposition universelle* (Bruxelles: Complexe, 1989) 111.
72 In 1889, Pierre de Coubertin introduced to the Exposition Universelle a «Congrès des exercices physiques», and curated the exhibition «L'exposition athlétique»; refer *Revue athlétique*, 25 May 1890.
73 Pierre de Coubertin, «La Haute Ecole» (March 1911); repb. *Essais de psychologie sportive* (1913), (Grenoble: Jerome Million, 1992) 144.
74 D'Estournelles de Constant, «Pour l'aviation», *La Revue bleue*, 9 (27 février 1909) 271–273: Despite Estournelles de Constant's articles on "unexpected aviation", *Bulletins of the International Conciliation,* and his pleas for pacifist aviation in *Revue Bleue,* the aeroplane was not represented as a tool of cross-border fraternity.
75 Alphonse Berget, «Heure», *Larousse mensuel,* 71 (January 1913) 609–611.
76 Alain Ehrenberg, «Aimez-vous les stades. Architecture de masse et mobilisation», *Recherches*, 43 (April 1980) 33–34.
77 In August 1912, Coubertin had triumphed in the Stockholm Literature Olympics with his pacifiste bilingual French-German poem, «Ode au sport».
78 Pierre de Coubertin, «L'avenir des sports au point de vue psychologique, international et démocratique»; repb. «Psychologie, internationalisme, démocratie», *Revue olympique* (July 1910) 103.
79 Legrogneux, «Complications footballkaniques. Extrait d'un correspondant de guerre en l'an 1950», *Fantasio*, 157 (1 February 1913): 475.

Part II OCCULTIST VITALISM

Magnetism, Parapsychology, Spiritism and Theosophy

4 Visualizations of the Vital-Psychic Force

Serena Keshavjee

Henri Bergson's term *élan vital* in *L'Évolution créatrice* to signify an impulse that directs the evolutionary development of both individuals and species, became the best-known term for the animating force thought responsible for the creation of all forms – a never-ending creative momentum at the core of a unitary universe. Strategically, Bergson posed Vitalism as a philosophical challenge to the positivist world view upheld during the French Third Republic.[1] Alongside Bergson were the scientists who applied experimental and observational methodologies to prove that such an animating force actually existed. Thus, by the beginning of the twentieth century and before publication of *Creative Evolution*, it was generally accepted that the psychic force was a variation of the vital force, which became the popular explanation for the physical manifestations witnessed in séance rooms.[2]

In his 1922 *Traité de métapsychique*, Nobel Prize winning physiologist and psychical scientist, Dr. Charles Richet, pointed out that such new discoveries and inventions as electricity, the telephone, x-rays, bacteria and the airplane were challenging the scientific orthodoxy about the physical world and opening up areas of study into "metapsychics", otherwise known as parapsychology or, as it was commonly called at the turn of the century, psychic studies.[3] Starting with this premise, this chapter explores how a group of leading psychical scientists visualized an invisible force, typically described as directed and dynamic that purportedly transcended the material world. In trying to represent this vital force, scientists in Germany, Britain and France, active between 1850 and 1940, produced a series of striking scientific illustrations that reveal a confluence of biology and psychic studies while utilizing experimental methods and photographic technologies.[4] Research-oriented aesthetics into the vital-psychic force contributed to changes in the formal conventions for images of apparitions, from the radiations of light of Victorian era pictorialism to the modernist ectoplasmic materializations of the early twentieth century.

Conflating the Vital and Psychic Forces

In French scientific circles the concept of an animating force was widely debated, both before, during and after the appearance of Bergson's *L'Évolution créatrice* in 1907.[5] That same year, Camille Flammarion, scientific popularizer, astronomer and

DOI: 10.4324/9781003045595-7

a forerunner in psychical research, declared that the standard suppositions about physical matter were open to question:

> In reality, there is only one principle in the universe, and it is at once intelligence, force, and matter, embracing all that is and all that could possibly be. *That which we call matter is only a form of motion.* At the basis of all is force, dynamic and universal mind or spirit.[6]

As contemporary scientists learned more about what constituted matter, Flammarion proposed that the mechanistic understanding of life was indefensible.[7] While defining himself as a Vitalist, Flammarion used terminology different from Bergson's, emphasizing "dynamism, the vital energy" and more readily embracing a teleological Vitalism that proposed an intelligent and designed universe.[8] For the astronomer, the life force transcended the known material world, as Flammarion explained: "There exists in our cosmos a dynamic element, imponderable and invisible, diffused through all parts of the universe, independent of matter, visible and ponderable, and acting upon it; and in that dynamic element there is an intelligence superior to our own".[9]

Although Director of the Institut Métapsychique International of Paris, Dr. Gustave Geley, referred to Bergson throughout his 1919 book, *De l'inconscient au conscient*, he adopted more fully Flammarion's notion of a directing dynamism. Reciting Flammarion, Geley wrote that "the purely mechanical concept of nature is insufficient", arguing that the concept of a directed "dynamo-psychicism" was challenging materialism.[10] Vitalist doctrines helped Geley develop a naturalistic explanation for the supernatural phenomena occurring through mediumship. Despite their regular attendance in séances over the years, neither Flammarion nor Geley believed in ghosts, with Geley empathically stating that ghosts do not exist: "There is no question of phantoms of the dead or of the living; of spirits or of genii", he declared. "Materialisation is today no longer the strange half-miraculous manifestation that was discussed in early spiritistic literature".[11] Many psychical researchers rejected the notion of discarnate spirits and looked to energies and natural laws for an explanation of the strange phenomena, especially the telekinetic activity and materializations witnessed in the séance room – what Geley labelled supernormal physiology.

Interest in the vital force was fuelled somewhat by renewed interest in Lamarckian evolutionary theory.[12] Historian, Sebastien Normandin, explains that by 1900, Vitalist theories boosted the emerging sciences of French Neo-Lamarckian evolutionary theory, Haeckelian embryology, dynamic psychology of the unconscious mind and, I would add, psychic studies.[13] French Neo-Lamarckism conferred a sense of agency to the evolutionary process.[14] Seven years before Bergson laid out his updated version of Vitalism in *Creative Evolution*, Flammarion had widely disseminated his teleological Vitalist theory of evolution whereby "the law of progress ... regulates all life".[15] Similarly Geley describes a vital, Transformist impulse, upending Bergson's resistance to teleology:

> This essential factor is a kind of interior impulse, an original and undefined "vital surge," (*élan vital*). This vital impulse pertains to an immanent principle which is life, intelligence and matter. It transcends them all, in the past, present and the future.

> It presupposes them, contains them and precreates them. This immanent principle, however, has no final completeness itself; it comes into existence progressively as it creates the evolving universe. It constitutes what M. Bergson calls 'Duration'. This 'Duration' is not very easily understood.[16]

Within the fields of psychology and dynamic psychiatry, where many psychical researchers were trained, dissatisfaction with the limitations of materialism was especially keen. As Geley hints, Flammarion's Vitalism was also easier to understand than Bergson's. At the turn of the twentieth century, a good deal of testing of the vital force was carried out by scientists who were investigating the "open" unconscious mind through altered states of consciousness exhibited by "sensitive" patients, including somnambulists, mediums and hysterics.[17] Noting the extraordinary feats of intelligence that mediums sometimes exhibited at séances, playwright Maurice Maeterlinck described the unconscious mind as the font of all knowledge. Maeterlinck poetically labelled the unconscious the "hidden guest". As he explained: "This brings us back once more to the omniscience and perhaps omnipotence of our hidden guest, to the brink of the mysterious reservoir of every manner of knowledge".[18] Further, the unconscious is "a spot in which everything is known, in which everything is possible, to which everything goes, from which everything comes, which belongs to all, to which all have access, but of which long-forgotten roads must be learnt again by our stumbling feet".[19]

As science historian Peter Bowler notes, around 1900 the animating force became conflated with the mind or willpower as a nonmaterial "life force" that transcends the known material world.[20] Some of the most respected psychical scientists working in Europe, including Flammarion and Geley, attributed the insight that mediums revealed in a trance-state, to the repository of the open unconscious mind. To explain the physical happenings inside the séance room, especially telekinesis and excreted materializations, these psychical scientists also theorized that the vital force could be projected outside of the body, almost like a lever. Both Flammarion and Geley applied Vitalist and Lamarckian characteristics to their understanding of psychic phenomena, accepting that all life forms progress from simple to complex, directed by some sort of "intelligence"; intrinsically creative, the life force contains an inherited knowledge that is essential both for the individual organism to develop and for a species to thrive.[21] Psychic researchers adapted the mechanism of the vital force to explain the physical happenings and the ectoplasmic spirit extras that they were seeing in the photographs taken in the darkened séance room.

In 1922, Maeterlinck, Flammarion and Geley's colleague, Nobel-prize winner Charles Richet, produced one of the most comprehensive compendiums on the state of research on materializations and séance happenings. Richet's weighty scientific reputation contributed to an understanding of the unconscious mind as able to access foreknowledge. In *Thirty Years of Psychical Research: Treatise on Metapsychics*, Richet noted that entranced mediums sometimes demonstrated heightened levels of intelligence. To support this claim he pointed to investigations into therapeutic applications of the trance-state, including the early work of the mesmerist, Marquis de Puységur whose research, Richet felt, afforded examples of second sight, which he identified with lucidity.[22] In a matter-of-fact statement explaining the physical feats seen in séances, such as long-distance movement, Richet proposed that the vital force could manifest itself as a mechanical force and exteriorize as "rigid rods", as he explained: "The substance that produces these telekinetic movements is a kind of lever, cantilever or rod which emerges from the body of the medium and is reabsorbed into it".[23]

Even more spectacularly, the rods could develop into the simulated body parts, typically hands and miniature faces, that appeared during the séance and, later, on the photographic plate, becoming the central focus of psychical researchers at the début-du-siècle. According to historian Carlos Alvarado, it was not unusual for the vital force to be conceptualized as the mechanism of the unconscious mind. By the end of the nineteenth century, a range of "new" energies, variously called "animal magnetism, neomagnetism, nervous force, neuric force, Od, psychic fluid, psychic force, vital energy and vital force" were equated to the discoveries of electricity, light and heat.[24] Richet referred to these up-to-date theories about energy, including that of Einstein, to support his ideas:

> Materialization is a mechanical projection: we already know the projection of light, of heat and of electricity; it is not a very long step to think that a projection of mechanical energy may be possible. The remarkable demonstrations of Einstein show how close mechanical and luminous energy are to one another.[25]

Interest in the vital force and the open unconscious mind remained on the margins of orthodox science. Yet, even so, between 1850 and 1940, credible scientists endeavoured to create and constantly update a set of pictorial codes to visualize these forces. Over this period, visualizations of the vital force transformed from glowing luminous energy to projected mechanical forces, following scientific trends.

"Spirit Lights": Rendering the Vital Force as Light

Richet and Maeterlinck begin the story of the modern effort to discover the vital principle, or life force, with Mesmerism, a treatment that synthesized ancient theories and contemporary science into a compelling performance of healing.[26] Franz Antoine Mesmer had claimed to be able to direct a force that ran through animated bodies, comparing it to a magnetic effect. Utilizing the principles of the etheric universe, Mesmer theorized that an individual's animal magnetism was connected to a universal fluid that was everywhere and surrounded all beings. In the eighteenth century, the first generation of mesmerists produced hand-drawn instructional diagrams that initiated the scientific effort to depict the vital force, laying down formal conventions to depict an invisible force. Typically, a mesmerist passed their hands over the patient to unblock their vital fluids for curative purposes. Instructional diagrams show the doctor's or healer's hands guiding their animal magnetic energy through the patient with directional lines of force, which is explored in the next chapter of this book. The hand continued to be a consistent motif in the visualization of the vital force in the twentieth century.[27]

An important example of the conflation of the vital force with the emerging idea of a dynamic directed universal force was put forth by the German chemist, Karl van Reichenbach, in *Researches on Magnetism, Electricity, Heat, Light, Crystallization and Chemical Attraction in Their Relations to the Vital Force*.[28] Through a series of controlled experiments during the 1840s, Reichenbach claimed that when a strong magnet was passed over "sensitive" patients who were then placed in a dark room, they were able to see light glowing from a variety of materials, including metal, crystal and vegetal items, especially from their own hands.[29] Reichenbach decided that this was a new force, the odic force, reflecting the "power permeating the whole of nature", and correlated it with the ancient conception of an animating force, the "vital force" of his treatise's title.[30]

Because he was not "sensitive" himself, Reichenbach never saw the luminous force with his own eyes but relied on descriptions by his patients. Luckily, his more artistically inclined patients provided him with illustrations for his publication. Barbara Reichel, "suffering" from nerves after a severe fall as a child, sketched out the simple diagrams used as plates in Reichenbach's treatise.[31] The chemist discovered that Reichel had a propensity to see nuanced coloured lights shooting out as flames: She drew "fiery bundles of light" that flowed and flickered from the finger-tips, some ten inches long, as if poles of a magnet".[32]

Even more visually compelling were the faintly coloured drawings of a glowing hand, bouquet, face and crystals created by the Viennese artist, Gustav Anschuetz, a "healthy" man who could nonetheless see glowing lights.[33] Reichenbach explains how Anschuetz unveiled his artwork illustrating the odylic lights (Figure 4.1):

> One morning when I visited him, he surprised me with a black picture: at the first moment I saw nothing in it in the light in which he exhibited it to me; but when

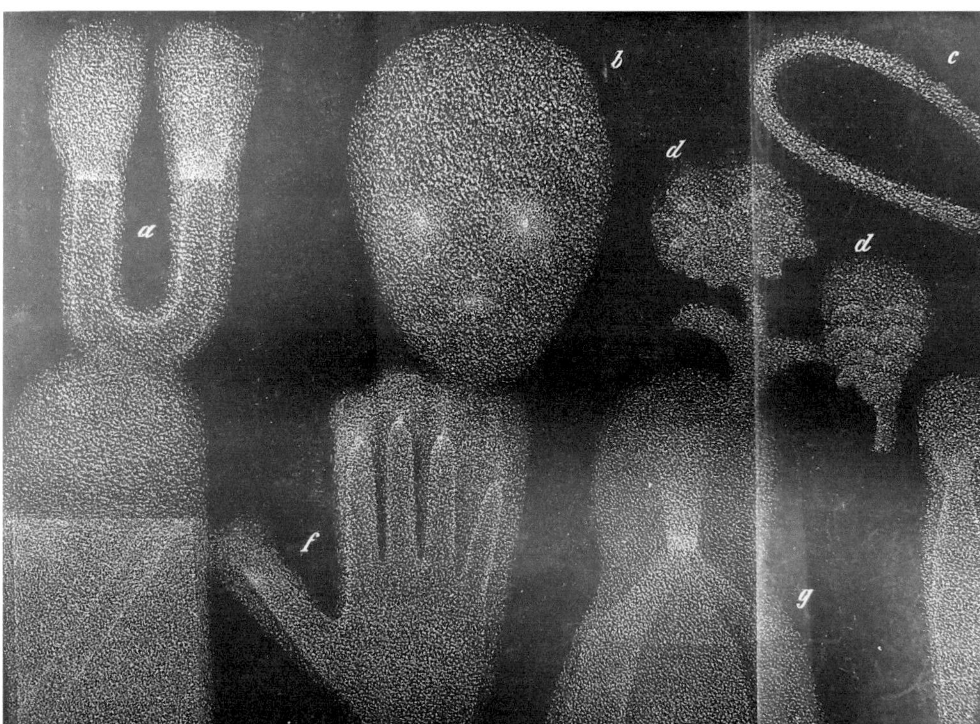

Figure 4.1 Gustav Anschuetz, Plate 1, reproduced in Karl von Reichenbach, *Researches on Magnetism, Electricity, Heat, Light, Crystallization and Chemical Attraction in Their Relations to the Vital Force*, trans. William Gregory (London: Taylor, Walton and Maberly, 1850 edition). (Photographed by the author in the History and Special Collections for the Sciences at UCLA Library. The author thanks Maxwell Zupke for permission to reproduce.)

> he turned it, a cloudy figure rose delicately and *phantom-like from the darkness*: it was the face of his beautiful wife, as it had appeared to him in its odic light in deep darkness. It was surrounded by crystals, magnets, flowers and hands, and I saw before me a picture of natural phenomena such as human eyes had never met before.[34]

In his 1851 translation of Reichenbach's book, John Ashburner chose the metaphor of a phantom materializing. In so doing, he anticipated the challenge for the next wave of researchers of the vital impulse – psychical scientists, who were quite literally trying to portray a barely visible, variable ectoplasmic substance as it exteriorized out of the medium's orifices in a darkened séance room. Reichenbach had set up a controlled space with blacked-out windows to test his patients' abilities to see radiant light and these conditions became the standard for observing the vital force for psychical researchers.[35] In his book published in 1896, *L'Extériorisation de la sensibilité*, Albert de Rochas d'Aiglun contributed to the revival of mesmerism by republishing Gustav Anschuetz's painting in a small black-and-white detail alongside vibrant, colourful versions of Reichel's handmade diagrams (Figure 4.2).[36] Even as modern Spiritualism grew in popularity through the last thirty years of the

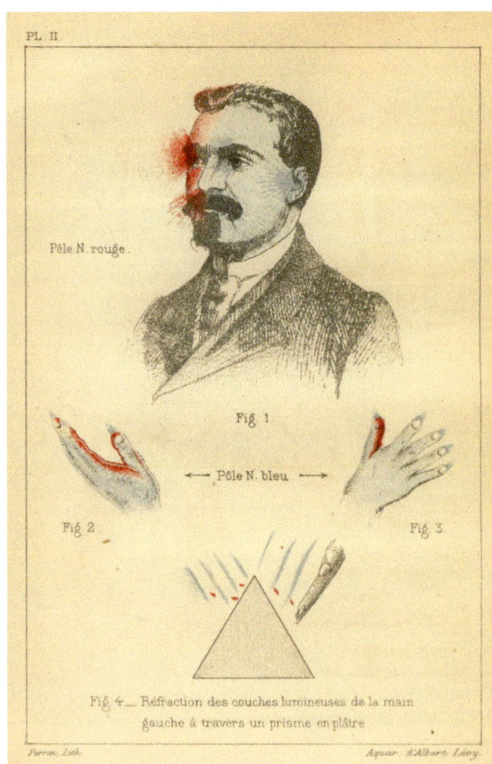

Figure 4.2 Barbara Reichel, Plate II: Figs. 1, 2, 3 and 4, reproduced in Albert de Rochas d'Aiglun, *L'Extériorisation de la sensibilité: Étude expérimentale & historique* (Paris: Bibliothèque Chacornac, 1896). (Courtesy of Medical Historical Library, Harvey Cushing/John Hay Whitney Medical Library, Yale University.)

nineteenth century while the term "odic force" became replaced by the term "psychic force", Reichenbach's experiments continued to have a lasting effect on laboratory conditions and attitudes towards "sensitives".

Interest in Spiritualism developed throughout the nineteenth-century, especially in serving many purposes as a reformist religion, a comfort for the bereaved and a source of entertainment.[37] The curiosity that arose about ghosts is reflected by the revival of The Ghost Club in the same year as inception of the Society for Psychical Research in 1882 followed by one of the earliest exhibitions dedicated to ghost imagery: *The Phenomena of Materialisation,* held at St. James Hall in London in 1886, in which were displayed 28 images of ghostly materializations, the term materialization purposely used to create the sense of scientific visualizations.[38] These included Gabriel von Max's print, *Spirit Greeting;* James Tissot's celebrated mezzotint, *Apparition Mediunimique* [sic], *Dark seances d'Eglinton du 20 May 1885, Londres,* showing the materialized ghost of Tissot's lover who died in 1882, Kathleen Newton with a Spirit Guide (Figure 4.3), as well as the set of watercolours that John Gerrard Keulemans made for *Twixt Two Worlds: A Narrative of the Life and Work of William Eglinton* (1886), written by John Farmer.[39]

J. G. Keulemans, an illustrator of birds and a practicing Spiritualist, carefully indicated the variable nature of the substance through which the ghost materialized using pictorial codes that were standard in the nineteenth-century for depicting apparitions powered by luminous energy. His colour plates show a light source,

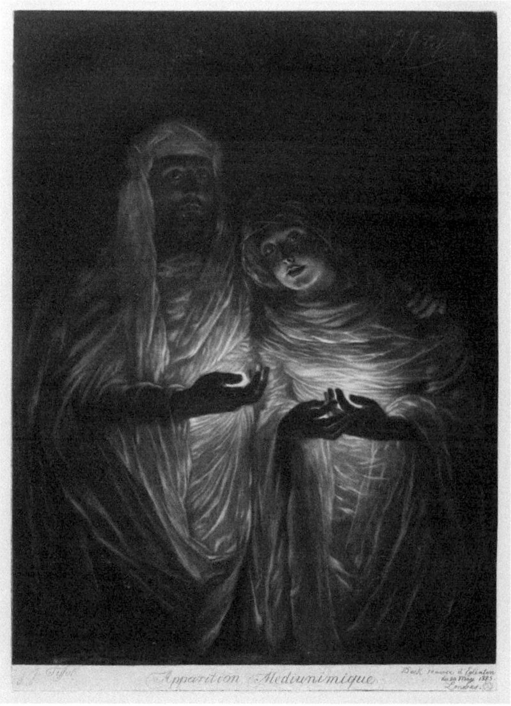

Figure 4.3 James Tissot, *The Mediumistic Apparition,* 1885, mezzotint on chine applique on Woven paper, 71.7 × 54.5 cm. (Courtesy of Art Gallery of Ontario, Toronto: Gift of Allan and Sondra Gotlieb; Image@Art gallery of Ontario.)

moving into a gaseous form, with wispy, linear cords emanating from the medium, William Eglinton's navel, morphing into a fully formed phantom. (Figure 4.4) During the 1880s, it is notable that Tissot, Keulemans and artist Albert Besnard depicted the essence of the discarnate spirit as a hand-held orb of bright light that energized the formation of a full apparition, as illustrated by Figures 4.5 and 4.6.[40] According to an eyewitness account of Eglinton's séances, the intensity of the orb was reminiscent of the brilliant light of a "phosphorescent miner's lamp" but incongruously, it was limited in range: "It throws no light into the room, though it is in itself of a more or less brilliant phosphorescent quality".[41] Keulemans' bright ovoid forms in Figure 4.5 might be a reference to one of the most popular texts about a ghostly apparition – the eminent chemist William Crookes's encounter with the ghost of Katie King. Crookes described the scene in detail:

> Holding up his handmade, bright phosphorescent lamp, a "small corked glass bottle" with a little phosphorated oil in it, to see the ghost better, and "still kneeling, I passed the lamp up and down so as to illuminate Katie's whole figure and satisfy myself thoroughly that I was really looking at the veritable Katie whom I had clasped in my arms a few minutes before and not the phantasm of a disordered brain.[42]

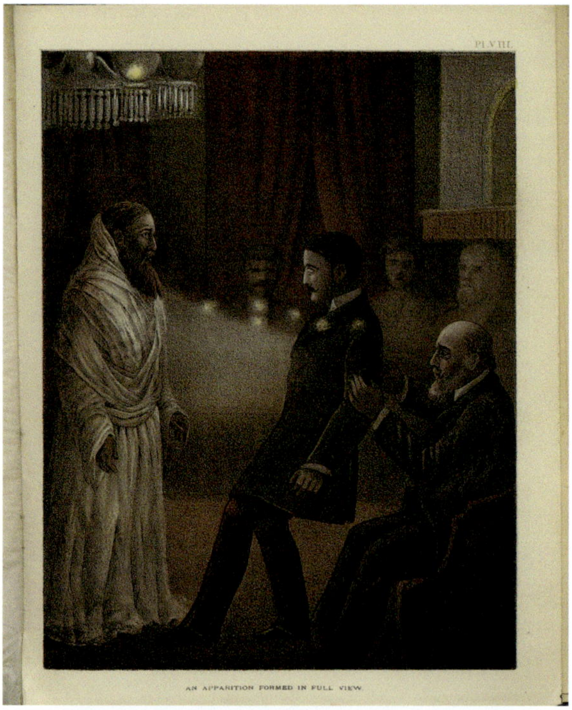

Figure 4.4 John Gerrard Keulemans, *"An Apparition Formed in Full View"*, c.1885–1886, chromo-lithograph, Plate VIII, reproduced in John S. Farmer, *Twixt Two Worlds: A Narrative of the Life and Work of William Eglinton* (London: The Psychological Press, 1886) 192. (Courtesy of University of Manitoba Archives and Special Collections.)

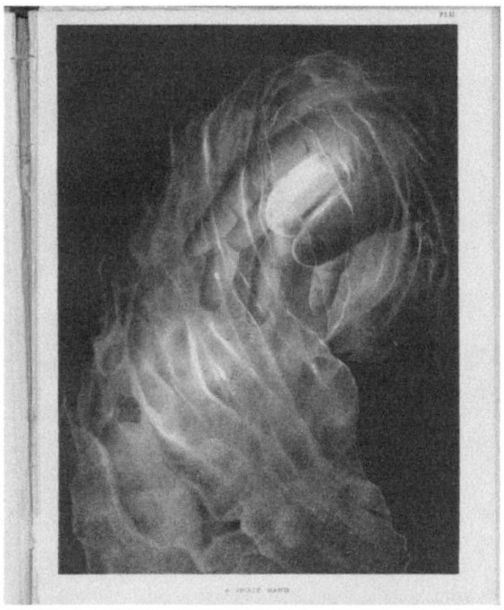

Figure 4.5 John Gerrard Keulemans, "*A Spirit Hand*", c.1885–1886, chromo-lithograph, Plate II, reproduced in John S. Farmer, *Twixt Two Worlds: A Narrative of the Life and Work of William Eglinton* (London: The Psychological Press, 1886) 48. (Courtesy of University of Manitoba Archives and Special Collections.)

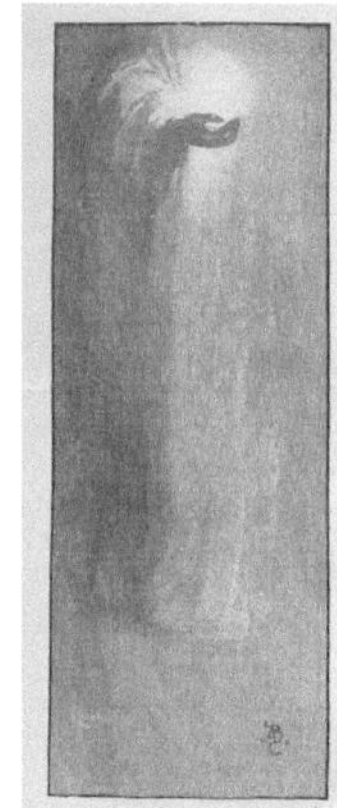

Figure 4.6 Albert Besnard, "*Nous vîmes la silhouette sombre d'une main*", n.d. Drawing reproduced as a wood engraving by Frédéric Florian in Yveling Rambaud, *Force Psychique* (Paris: Ludovic Baschet, 1889) 43. (Collection of the author.)

In this case, Crookes' well documented experiments maybe one of the inspirations for these prints.

One of the most interesting illustrations of the discarnate spirit as a radiant light source is Plate I in *Twixt Two Worlds*, entitled "Spirit Lights" (Figure 4.7). Keulemans drew stages of subtly coloured orbs in a grid framework, reflecting Reichenbach's and Anschuetz's pseudoscientific layout of odic light radiations. While Farmer uses the term "substance" to describe Eglinton's spirits lights, Keulemans may have been familiar with the emerging biologically oriented research in psychical science, because his use of the term "bioplasm" seems to derive from the title of Lionel S. Beale's 1872 book, *Bioplasm*. As Farmer observes, "Mr. Keulemans records that he has noticed also slight, but rapid, vibrations in the ramifications referred to, as though the substance (*bioplasm*) still retains its vitality".[43] A British physician, Beale carefully chose the term bioplasm to reinforce the vital characteristics of protoplasm, a primal, structureless jelly found in elementary animal and vegetal cells. Farmer's explanation of "Spirit Lights" in *Twixt Two Worlds*, based on his discussions with Keulemans, mentions "blood" and "brain" matter inside the glowing orbs: "These two lines are of a pale greyish red, exactly the colour of living brain matter. The red spot is the colour of human blood".[44] While this corporeal description of the light-filled globes does not match the visuals in the book, it alludes to changes in the research aesthetics of the vital force towards a biological substance. By the turn of the century in Britain, Germany and France, the substance that formed this apparition was being described less as a light source and more as a mechanical projection of primal matter called variously "the substance", "ectoplasm", "teleplasm" or "bioplasm", reflecting the rise of the science of biology.[45] Coincidently around this period, scientists were rejecting hand-drawn illustrations for photographic technologies, also contributing to stylistic changes for the scientific rendering of materializations.

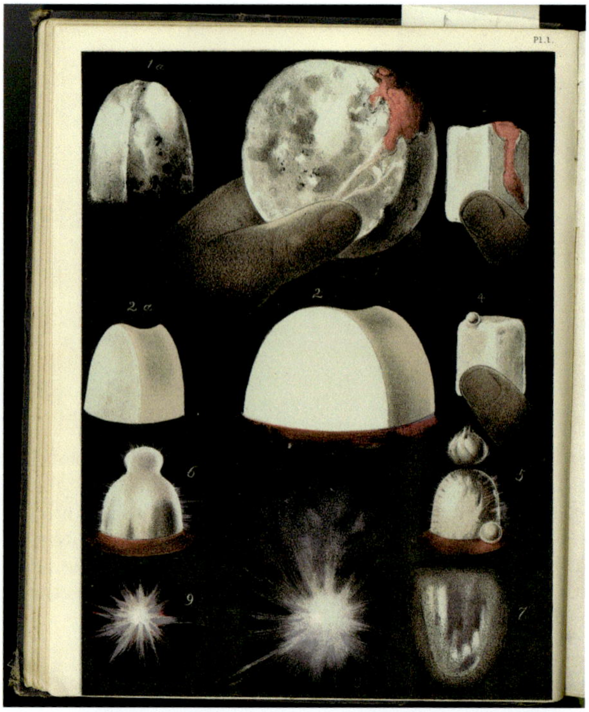

Figure 4.7 John Gerrard Keulemans, *"Spirit Lights"*, colour drawing reproduced as a chromolithograph, Plate I, in John S. Farmer, *Twixt Two Worlds: A Narrative of the Life and Work of William Eglinton* (London: The Psychological Press, 1886) 24–25. (Collection of the author.)

Nineteenth-century spirit photographers relied on conventions of Victorian portrait painting, as Sarah Willburn explains, especially in their use of props, poses and the stylistic trends of salon Impressionism signifying the dominant pictorialist style of spirit photographs in the nineteenth-century.[46] In the prints of ghostly apparitions by such respected artists as Tissot, Gabriel von Max and Besnard, emerging spirit photographers found excellent models for their photographs. As scientists' interests moved way from ether theory to biology, visualizations of the psychic force also changed. Within some parapsychology circles, the next stage in visualizing the vital force upended Victorian pictorialism and Victorian science with a pared-down image that may be described as both minimalist and modernist. [47]

"Rudimentary and Crude Productions": The Vital Force in the Age of Biology [48]

There was ample evidence of how easily a spirit extra could be faked on a photographic plate, as demonstrated by the fraud trials of Édouard Isidore Buguet and William Mumler, as well as by magician Harry Houdini's pictorialist efforts to debunk Spiritualism with amusing forgeries (Figure 4.8). However, nothing dissuaded psychical scientists from turning to photographic evidence.[49] By 1910, the formal elements and standard compositions of evidentiary photographs of phantoms

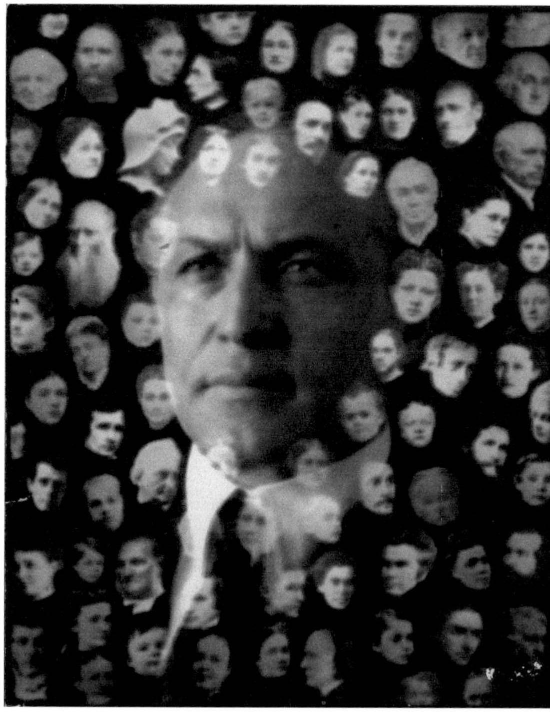

Figure 4.8 Harry Houdini, "*Harry Houdini with Spirits*", Photograph, c.1924, M2014.128.703.24. (Courtesy © McCord Museum.)

were transforming in style, as illuminated in Chapter Five of this book, as a consequence of improvements in photographic technologies and developments in biology. Adopting experimental methodologies, some psychic scientists set up laboratories to test the telekinetic and teleplasmic abilities of mediums, such as table turning, making imprints in clay or wax and exteriorizing materializations. Specialized lenses, high speed emulsion glass plates and faster flash powders emphasized sharply focused details of protoplasmic prolongations, shown frozen at a moment in time. While spirit extras continued to be conveyed as communicating during the twentieth century, they were less likely to be represented as naturalistically rendered figures floating in the air around a sitter, as in the Houdini example. Instead "miniature face forms" began to pop out of excreted ectoplasmic clumps, exactly like an insect breaking free of its chrysalis, reflecting the grounding of materializations in biology.[50]

The best example of the formal changes for rendering these materializations are the series of photographs of the performative medium Eva Carrière, née Marthe Béraud (1886–1969). Eva C., as she became known, dominated scientific testing in the early twentieth century. Nevertheless she was only ever able to excrete incomplete, "undecided" formations as "rudimentary" projections.[51] Yet no one was disappointed with her production. The rendering of fully formed phantoms, often swathed in veils, such as those photographed by William Crookes of Katie King, increasingly fell out of favour in some research circles, for close up views of the medium exuding formless paste, or as Houdini described them, views of "horrible, revolting viscous substance".[52]

Nevertheless Geley thought Carrière's ability to hold onto the fragmented solid state of the excretion to be "astonishing", and he was excited that her ectoplasmic simulacrum of amputated limbs and rods seemed to have the "appearance of life" in the biological sense.[53]

One of the most photographed mediums at the beginning of the twentieth century, Carrière worked with all of the important continental scientists. This included Richet whose medical research into hysteria, hypnotism and the unconscious mind dates back to the late 1880s, well before he met Carrière in 1905 when he witnessed her infamous materialization of "Bien Boa".[54] During the summer of 1894, Richet had invited respected scientists from Germany, England, Italy and France to his Mediterranean retreat to join in séances with the Italian medium, Eusapia Palladino. This group included the sexologist and medical doctor Schrenck-Notzing, who had trained in therapeutic hypnotism in France; Dr. Frederic W. H. Myers; engineer and Spiritualist, Gabriel Delanne; professor of psychology and photographer, Julien Ochorowicz and the respected British physicist, Oliver J. Lodge.[55] From this series of controlled séances emerged the Vitalistic biological rationale for materializations that reconceptualized the luminous energy into an organic "substance" and contributed to standardizing a new set of pictorial codes better suited for photography.

Lodge's careful account of the sittings of 1894 reveal that this group grappled with their experiences in séance experiments, working out parallels with the physical world. Best known for his promotion of Ether theory to explain how ghosts communicate from the beyond, Lodge tentatively reckoned that table turning might be due to "vitality at a distance" and that the "prolongations" he saw were reflections of the "power of protoplasmic activity".[56] Similarly Richet turned to the properties of protoplasm, taking credit in the 1923 edition of *Thirty Years of Psychical Research* for coining the widely used term "ectoplasm", as he explained:

> The word "ectoplasm," which I invented for the experiments with Eusapia, seems entirely justified. This ectoplasm is a kind of gelatinous protoplasm, formless at first, that exudes from the body of the medium and takes form later. This embryo-genesis of materializations shows clearly on nearly all the photographs. In the early stages there are always white veils and milk patches and the faces, fingers and drawings are formed little by little in the midst of this kind of gelatinous paste that resembles moist and sticky muslin.[57]

Declaring the vital-psychic force of the medium to be the controlling mechanism for the materializations, Richet moved away from describing luminous, fully formed materializations in favour of organic extrusions. He worked hard to explain how the prolongations changed shape and state, with a range of descriptors such as "undecided form", stipulating that the "mysterious" phenomena, "which we call *materialisation,* is accompanied by a sort of disaggregation of the pre-existing matter, so that the new matter formed is formed at the expense of the old, and that the medium *empties herself*, so to speak, in order to constitute the new being which emanates from her …".[58] Richet, Lodge and later Geley's research introduced a decidedly solid quality to ectoplasm. Finding the hard photographic evidence of the gelatinous substance is credited to researcher, Juliette Bisson, when she took a flash photograph of ectoplasm in 1909.[59]

Science historian, Robert Brain has thoroughly contextualized how and why Richet's ectoplasmic theory came to underwrite psychical research for scientists, philosophers and artists, by the early twentieth century. As Brain outlines, Lodge and Richet promoted biological "amoeboid theories of neuronal mobility" to describe how the medium, Eva C's cellular "substance", exteriorized from her body while in a trance-state, entailed shape shifting pseudopodal extensions. Lodge and Richet considered that all telekinetic and teleplastic activity was directed by a vital force, or as some described it, a "directing idea".[60] Bergson's *Creative Evolution* contains a number of passages characterizing the essential property of basic animal life symbolized as asymmetrical movement, directed by a "tremendous internal push", which causes cells to send out pseudopodia in order to grow.[61] These theories illuminate how the discovery of rudimentary amoebae captured the imagination of psychical scientists.

The Limits of Photography: Protoplasm, Ectoplasm and Biocentrism

The evolutionary theorist and biologist, Ernst Haeckel, was a key figure in generating excitement about the elemental properties of intra-cell protoplasm that would be reflected in the writings of Bergson, Geley, Richet and twentieth century Biocentrism philosophers. Throughout the 1860s, Haeckel explored how higher life forms could evolve from a structureless plasma or primordial slime called *Urschleim*.[62] Thomas Huxley contributed to the scientific debate about this ubiquitous, gelatinous medium with his apparent discovery of "Bathybius Haeckelii".[63] Named after Haeckel, Huxley used art potters' clay to explain the facility of protoplasm to create forms in his publication "The Physical Basis of All Life":

> Protoplasm, simple or nucleated, is the formal basis of all life. It is the clay of the potter which, bake it and paint it as he will, remains clay, separated by artifice, and not by nature, from the commonest brick or sun-dried clod. Thus it becomes clear that all living powers are cognate, and that all living forms are fundamentally of one character.[64]

Although Huxley retracted his discovery ten years later, Haeckel developed the concept into monera, the keystone for his nature-centric world view. Monism, which in 1906 Haeckel cofounded with Wilhelm Ostwald, the promoter of "energetism", presented a "unifying conception of nature as a whole", a pantheistic Vitalism rooted in German nature Romanticism.[65] Art historian, Oliver Botar, positions Monism as a subset of Biocentrism (from *Biozentrik*), a fin-de-siècle, nature-centric philosophy updated by biology.[66] The term "Biocentric" was employed by both the ecologist, Raoul Francé and critical Vitalist philosopher, Ludwig Klages, who alongside his colleague, biologist, Hans Driesch, another Biocentric and vitalist philosopher, sat in séances with Dr. Schrenck-Notzing between 1921 and 1922.[67] Thus these Biocentric philosophers became familiar with materialized protoplasm, called ectoplasm, in the well-equipped laboratory of Schrenck-Notzing.[68] Brain explains that by the twentieth century protoplasmic matter, central to the most basic cells, was characterized with autoplasticity and hereditary memory, and came to signify generative forces for philosophers, artists and scientists alike.[69]

Active in debates over protoplasm between evolutionists and creationists during the 1860s and 1870s, both Haeckel and Huxley proposed that a chemical process initiated biological life.[70] While Dr. Lionel Beale held onto a divine origin, he chose the term *bioplasm* to indicate a living matter that "grows and moves" by an external force.

Yet, as already indicated, bioplasm is the very term that Farmer and Keulemans used in 1886. Beale further described protoplasm as "life, force and matter", anticipating Flammarion's phrase of 1900, "intelligence, force and matter".[71] This demonstrates how complex were the relationships between Christian concepts, evolutionary theory, psychical science and Spiritualism, as Bowler acknowledges in *Reconciling Science and Religion*.[72] For this chapter, it is enough to point out that many twentieth-century psychical scientists accepted the Vitalistic conception that protoplasm could be expelled out of the body into an endless production of creative forms, as articulated in the writings of Bergson, Flammarion and Biocentric philosophers, Driesch and Raoul Francé.

In 1909, this was demonstrated by a small team of psychical researchers including the French sculptor, Juliette Bisson, Eva Carrière and Schrenck-Notzing, who took up Vitalistic protoplasmic concepts and translated them into evidentiary photographs. It is this group that dramatically changed the look of apparitions from electromagnetic luminous radiations to viscous protrusions. Historian of photography, Martyn Jolly observes how visualizations of ectoplasm were reformulated by these three players, who "developed a new form of materialization, not the theatrical unveiling of a full-bodied personality from behind the cabinet curtains but the slow painful extrusion of wet organic matter from the visible body of the medium, which gradually formed into an entity".[73] Between 1909 and 1912, Bisson organized regular scientific séances in her Parisian sculpture studio, turning a niche into a curtained-off laboratory space and taking advantage of the characteristics of black and white photography, designed a knitted garment of bloomers and stockings for Eva C. Bisson and invited the key French psychical researchers into her domain, including Richet; the academic painter, Léon Chevreul; Delanne and professional photographer Guillaume de Fontenay, all of whom had been involved for years in photographing materializations.[74] However, it was Schrenck-Notzing who was Bisson's main collaborator. Bisson set up a flash apparatus in her studio in 1909. Although tricky and dangerous to use, she conquered the technique so well that Geley asserted that it was Bisson who proved through flash photography that the protoplasmic "substance" existed.[75] In 1914, Bisson and Schrenck-Notzing both published books, respectively entitled *Les Phénomènes dits de matérialisation*, and *Der kampf um die materialisations-phänomene,* generously recognizing each other's contributions.[76] These staged black-and-white images of Carrière published in these books created a new and influential style for scientific photographs of materializations.

Schrenck-Notzing travelled between Munich and Bisson's studio to participate in her scientific séances. He also hired the Munich-based, academic artist, Karl Gampenrieder to draw materializations by hand when flash photographs were spoiled or imperfect. A séance held in Bisson's studio on 13 March 1911 included some of the most important psychical researchers in Europe, including Schrenck-Notzing, Richet and de Fontenay. There they witnessed Carrière excrete about ten inches of a whitish, jelly-like material endowed with mobility. Remarkably, a set of 30 different hand-like forms of various sizes were exteriorized in that session, some of which had the ability to flex fingers and one of which had a fingernail.[77]

These spectacular hands were illustrated in Schrenck-Notzing's *Phenomena of Materialisation,* in two different media, reflecting two different moments

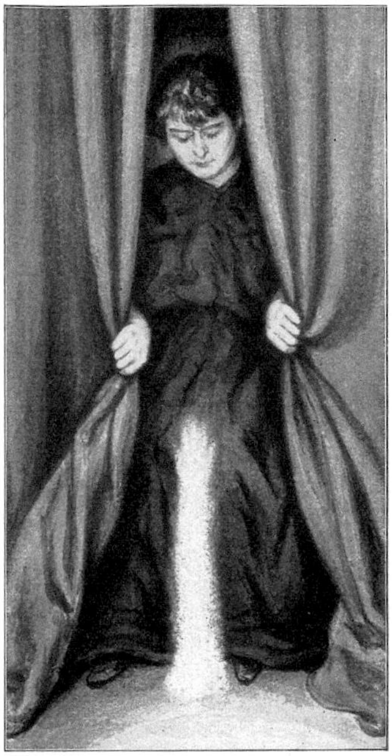

FIG. 26. DRAWING AFTER RECORD OF SITTING OF 13 MARCH 1911.

Figure 4.9 Karl Gampenrieder, Drawing after record of sitting of 13 March 1911; reproduced in Albert von Schrenck Notzing, *Phenomena of Materialisation: A Contribution to the Investigation of Mediumistic Teleplastics,* Fig. 26; trans. E. E. Fournier d'Albe (London: Kegan Paul, Trench, Trubner and Co., New York, P. Dutton, 1920). (Collection of the author.)

in the sequence of externalizing protoplasm. While Figure 4.9 is a drawing by Gampenrieder, Figure 4.10 is a photograph directed by Schrenck-Notzing and Bisson. Geley, Richet and Schrenck-Notzing's on-site research indicated that ectoplasm was exteriorized in different states, emerging in gaseous or vaporous form, "instinctually" moving into a liquid or a solid state, with the tendency to morph from a formless shape to a formed one.[78] In Gampenrieder's drawing, a "white column of smoke curled upward" appears to form a basic forearm and hand shape.[79] The artist utilized Victorian pictorialist leitmotifs of radiating light to illustrate the gaseous state of the substance, depicting the variable nature of ectoplasm. While drawings made by hand may easily show sequential time, a flash photograph, even with multiple cameras, freezes moments in time and cannot reflect changes of state.[80] Still photographs were then the empirical foundation for Schrenck-Notzing's research. Through photographs, he emphasizes the solid nature of the hand-like excretions describing them as if an "unfinished" sculptural work, but substantive, "like fur or wadding, composed of a bundle of stripes and cords, resembling lint or wool, intermixed with transparent veiling".[81] When Geley praises the solid characteristics of Carrière's excretions, he appears to be coming to terms with the limits of photography for capturing movement, as well as accepting the new style of static,

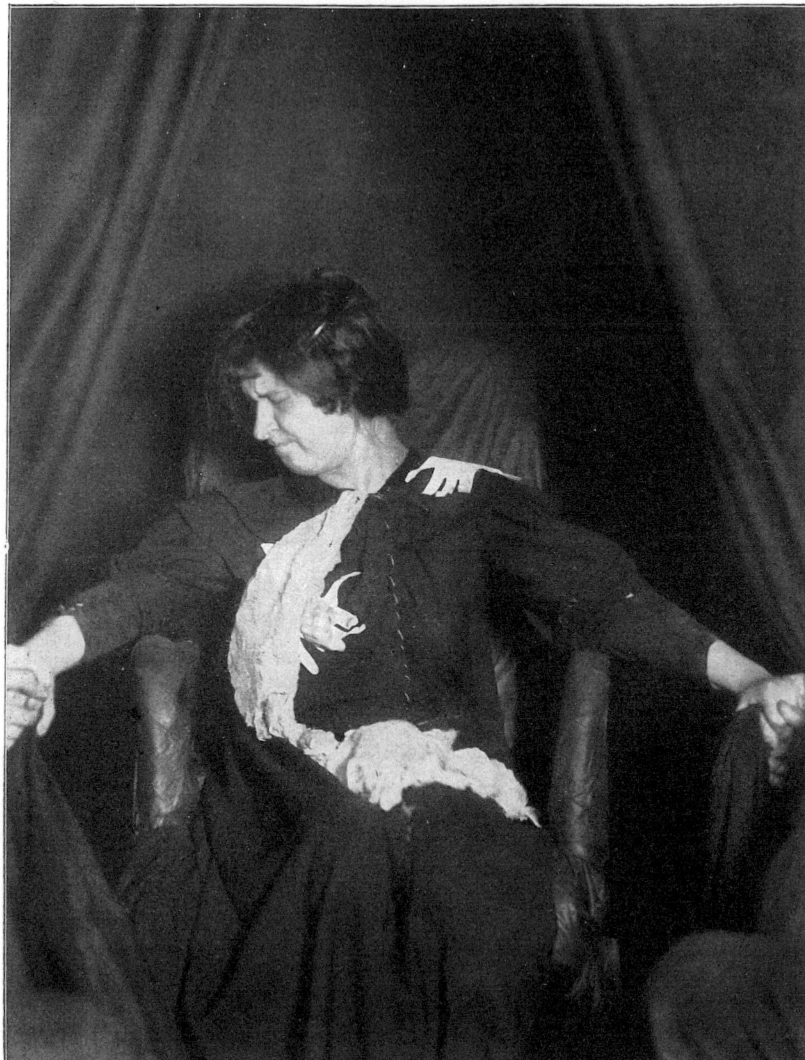

Fig. 27. Flashlight photograph by the author, 13 March, 1911.

Figure 4.10 Karl Gampenrieder, Flashlight photograph by the author, Albert von Schrenck Notzing, 13 March 1911; reproduced in Albert von Schrenck Notzing, *Phenomena of Materialisation: A Contribution to the Investigation of Mediumistic Teleplastics,* trans. E. E. Fournier d'Albe (London: Kegan Paul, Trench, Trubner and Co., New York: P. Dutton, 1920) Fig. 27. (Collection of the author.)

organically shaped ectoplasmic photographic images that Bisson and Schrenck-Notzing initiated. As Geley explains:

> The substance of materialization is more amenable to examination under its liquid or solid forms... It has been observed under this form, from several mediums,

especially from the famous medium Eglinton. But it is from the medium Eva that this *solid substance is generated with astonishing completeness.*[82]

Harry Houdini and even art historians have parodied these photographs of materializations, suggesting they represent a decline in the aesthetic standards set by the Victorian spirit photographers.[83] Nevertheless, when Bisson in 1909 conquered the magnesium flash and fixed a still photograph of the shape-shifting ectoplasmic excretion, she encouraged the scientists around her to adapt their research aesthetics entailing the vital force. With their photographs, this group based in Munich and Paris moved away from delicate renderings of light or veils typical of Victorian pictorialism towards the modernist style, emerging for example in Lázló Moholy-Nagy's "New Vision" photography.[84] Yet while Biocentric scientists, artists and photographers were interested in the ectoplasmic materializations, psychical researchers endeavoured to measure the primal forms and the primal matter of life itself.[85] Geley summed up what was at stake for these scientists: "Ectoplasm would thus be able to give us the key to human and animal biology such as the origin of the species. It would really offer in itself the explanation of the mysteries of life".[86] With their ectoplasmic photographs, Bisson, Schrenck-Notzing and Carrière charted one route of vitalist modernism reflecting the fascination with the biomorphic origins of life.

Thank you to Fae Brauer, Oliver Botar, Tim Pearson, Paul Dutton, Geneviève Riou, Murray Leeder and Emma Dux for their comments on this chapter. This chapter was researched and written with the support from the Social Science and Humanities Research Council.

Notes

1 Henri Bergson, *L'Évolution créatrice* (Paris: Alcan, 1907); for this chapter, the English edition is referenced: Henri Bergson, *Creative Evolution* (London: Macmillan, 1922). On Vitalism, psychical research and Bergson, see Justin Sausman, "'It's Organisms That Die, Not Life': Henri Bergson, Psychic Research, and the Contemporary Uses of Vitalism," *The Machine and the Ghost: Technology and Spiritualism in Nineteenth to Twenty-First-Century Art and Culture*, eds. Sas Mays and Neil Matheson (Manchester: Manchester University Press, 2013) 16–36, especially, 20. Refer also Sebastien Normandin and Charles T. Wolfe, *Vitalism and the Scientific Image in Post-Enlightenment Life Science, 1800–2010* (New York: Springer, 2013) and Sebastien Normandin, "Visions of Vitalism: Medicine, Philosophy and the Soul in Nineteenth Century France" (PhD diss., McGill University, 2003).

2 In *Great Secret* (New York: The Century Co., 1922) 226, Maurice Maeterlinck speculated that the life force was akin to heat, electricity, light, the od, the etheric fluid, and the Akasha; Charles Richet, *Traité de métapsychique* (Paris: Alcan, 1922); republished, *Thirty Years of Psychical Research: Being a Treatise on Metapsychics,* trans. Stanley de Brath (New York: The MacMillan Company, 1923) 103.

3 Richet, *Thirty Years of Psychical Research,* 9–10.

4 On ectoplasmic photographs, see Andreas Fischer, "The Reciprocal Adaptation of Optics and Phenomena: The Photographic Recoding of Materializations," *The Perfect Medium: Photography and the Occult*, eds., Chéroux Clément and Andreas Fischer (New Haven: Yale University Press, 2005) 171–216; see also Neil Matheson, "Ectoplasm and Photography: Mediumistic Performances for Camera," *The Machine and the Ghost*, 2013, 78–102. For this chapter, I have concentred on photographs by psychical researchers and naturalistic drawings with an observational affect, such as those by Tissot and Besnard. While automatic drawings and mediumistic art by, for example, Georgina Houghton and Hilma af Klint, is an equally interesting topic that I have researched,

it is outside the scope of this particular chapter; see Julia Voss et al., *World Receivers: Georgiana Houghton, Hilma af Klint, Emma Kunz, and John Whitney, James Whitney, Harry Smith* (München: Lenbachhaus, 2018); Tracey R. Bashkoff et al., *Hilma af Klint: Paintings for the Future* (New York, NY: Guggenheim Museum Publications, 2018) and Kurt Almqvist, *Hilma af Klint: the Art of Seeing the Invisible.* (Stockholm, Sweden: Axel and Margaret Ax:son Johnson Foundation, 2015).

5 See review in Normandin and Wolfe, *Vitalism and the Scientific Image in Post-Enlightenment Life Science,* chapter 1, and Sausman, "'It's Organisms That Die, Not Life,'" 18.

6 Camille Flammarion, *Mysterious Psychic Forces: An Account of the Author's Investigations in Psychical Research, Together with Those of Other European Savants* (Boston: Small Maynard & Co., 1907) 429 (italics mine).

7 Flammarion, *The Unknown* (New York: Harper and Brothers, 1900) x. In *Mysterious Psychic Forces*, 449, Flammarion states that the "purely mechanistic concept of nature is insufficient and that there is in the universe something else than so-called matter. It is not matter that rules the world: it is a dynamic and psychic element".

8 Flammarion, *Mysterious Psychic Forces,* 1907, 430. Within psychic scientific literature, Flammarion is more often cited than Bergson. While Bergson represents "critical" Vitalism, which resisted teleological assumptions, Flammarion is perceived by Normandin as a "naive" Vitalist, whose version of its philosophy appears more optimistic and compelling; for "critical" versus "naive" Vitalism," see Normandin, "Visions of Vitalism," 4, 30, 217. Normandin points out, 218–219, that Bergson and Driesch, the primary critical Vitalists, both served as directors of the SPR in 1913 and 1926, respectively.

9 Flammarion, *The Unknown*, 1900, xi; similarly, Flammarion, *Mysterious Psychic Forces*, 1907, 429. Flammarion states that mysterious forces followed the laws of mechanical forces but were invisible and immaterial, dynamic and intelligent.

10 Gustave Geley, *From the Unconscious to the Conscious*, trans. Stanley De Brath (London: William Collins, Sons & Co. Ltd., 1920) 66–67, originally published as *De l'inconscient au conscient* (Paris: Félix Alcan, 1919).

11 Geley, "Ectoplasm," *ASPR* (April 1924): 272–279; 273, trans. from "L' Ectoplasmie," *Revue Métapsychique,* no. 7 (1921): 355-361. For Geley's supernormal physiology as a parallel to normal physiology, refer *From the Unconscious to the Conscious,* 1920, Chapter 2, Part 2.

12 Peter J. Bowler, *Reconciling Science and Religion: The Debate in Early-Twentieth-Century Britain* (Chicago: University of Chicago Press, 2001) 378.

13 Normandin, "Visions of Vitalism," 22, discusses the differences between Neo-Lamarckian and Darwinian versions of evolutionary theory; he also identifies how Vitalism thrived within Transformist France, 208–209.

14 Peter J. Bowler, *The Non-Darwinian Revolution: Reinterpreting a Historical Myth* (Baltimore: John Hopkins University Press, 1988).

15 Flammarion, *The Unknown*, 1900, x.

16 Geley, *From the Unconscious to the Conscious,* 1920, 163–64. In the translator's note, vii, Stanley de Brath states that Geley was updating Bergson's *élan vital* with a "concrete energy," defining "energy as an influence forming all the varieties of cellular tissue out of one primordial substance and moulding those tissues into organic form under the impulsion of a Directing Idea".

17 Henri Ellenberger, *The Discovery of the Unconscious: The History and Evolution of Dynamic Psychiatry* (New York: Basic Books, 1970); Brady Brower *Unruly Spirits: The Science of Psychic Phenomena in Modern France* (University of Illinois Press, 2010) xxii.

18 Maurice Maeterlinck, *The Unknown Guest,* trans. A. Teixeira de Mattos (London: Methuen and Co., 1914) 81.

19 Maeterlinck, *The Unknown Guest*, 1914, 81–82.

20 Bowler, *Reconciling Science and Religion*, 2001, 135–136.

21 Flammarion, *Mysterious Psychic Forces*, 1907, 429 and 431. See also Geley, *From the Unconscious to the Conscious*, 49. For an excellent contextualization of ectoplasm, see Robert Brain, "Materialising the Medium: Ectoplasm and the Quest for Supra-Normal Biology in *Fin-de-Siècle* Science and Art," *Vibratory Modernism,* eds. A. Enns and S. Trower (London: Palgrave Macmillan, 2013) 112–141, specifically 128–129.

22 Richet, *Thirty Years of Psychical Research*; on lucidity, see chapter 1; also, Charles Richet, *Our Sixth Sense* (London: Rider and Co., 1926) 22 and 31.

23 Richet, *Thirty Years of Psychical Research*, 1926, 442. Richet calls them rigid rays, based on Ochorowicz's observations, 440.

24 See Carlos S. Alvarado, "Human Radiations: Concepts of Force in Mesmerism, Spiritualism and Psychical Research," *JSPR*, 70.3: 884 (2006): 138–162, quotation 138.

25 Richet, *Thirty Years of Psychical Research*, 1926, 468; these forces "appear to have intellectuality, will and *intention*, which may not be human, but which resemble human will and intention. Intellectuality—the power of choice, intention, and decision conformably to a personal will — characterizes all metapsychic phenomena" (5).

26 The term "Spirit Lights" in the heading comes from John S. Farmer, *Twixt Two Worlds: A Narrative of the Life and Work of William Eglinton* (London: The Psychological Press, 1886). For a history of mesmerism see Betsey van Schlun, *Science and the Imagination: Mesmerism, Media, and the Mind in Nineteenth-Century English and American Literature* (Berlin: Galda & Wilch Verlag, 2007) 36–37; Richard Noakes, *Physics and Psychics: The Occult and the Sciences in Modern Britain* (Cambridge: Cambridge University Press, 2019) chapter 1, 35, and Richet, *Thirty Years of Psychical Research*, 1926, 21–24. Sofie Lachapelle discusses the rise in neo-magnetism in France in the 1890s. *Investigating the Supernatural: From Spiritism and Occultism to Psychical Research and Metapsychics in France, 1853–1931* (Baltimore: Johns Hopkins University Press, 2011) chapter 2, 52–58, as does Brower, *Unruly Spirits*, 2010, 56, and Fae Brauer, "Magnetic Modernism: František Kupka's Mesmeric Abstraction and Anarcho-Cosmic Utopia", *Utopia: The Avant-Garde, Modernism and (Im)possible Life*, eds. David Ayers et al (Berlin, Boston: De Gruyter, 2015) 123–152.

27 For mesmerist images, see William Davey, *The Illustrated Practical Mesmerist: Curative and Scientific* (London: J. Burns, 1889) especially Plate 8, with the healer's hand (Wellcome Collection; also *From Energy to Information: Representation in Science and Technology, Art, and Literature*, eds. Bruce Clarke and Linda Dalrymple Henderson (Stanford: Stanford University Press, 2002), especially Henderson, "Vibratory Modernism: Boccioni, Kupka and the Ether of Space," 126–149.

28 There are two different English translations of Karl van Reichenbach's book: Reichenbach, *Researches on Magnetism, Electricity, Heat, Light, Crystallization and Chemical Attraction in Relation to the Vital Force*, trans. William Gregory (London: Taylor Walton and Maberly, 1850; (New Hyde Park, N.Y.: University Books, 1974) and Reichenbach, *Physico-Physiological Researches on the Dynamics of Magnetism, Electricity, Heat, Light, Crystallization, and Chemism, in Their Relations to Vital Force*, trans. H. John Ashburner (London: H. Baillière, 1851). The translation used in this chapter is indicated by dates.

29 See Reichenbach, *Researches on Magnetism* [1850] 1974 edition, 1–3, for the explanation of the process and methodology. Sensitive patients were described as diseased, having weak nerves, being easily agitated, terrified, sensitive to smell, hysterics, and spontaneous somnambulists. Noakes, *Physics and Psychics*, 39, points out that there were critiques of Reichenbach's dependence on young women describing them as "instruments of research," akin to automatic recording devices.

30 See Reichenbach, *Researches on Magnetism* [1850] 1974 edition, Foreword (np).

31 It seems that Barbara Reichel's drawings are the basis for Plate 1, figs. 1, 2, 9, and 10; see *Researches on Magnetism* [1850] 1974, quotation 10, explanation of drawings, 9–11; 89.

32 Reichenbach, *Researches on Magnetism* [1850; 1851], 1974, 89.

33 For Reichenbach on Gustav Anschuetz, see *Researches on Magnetism*, 1974, 255–258.

34 Reichenbach, *Researches on Magnetism*, 1851, 328. Reichenbach laments that the published engraving does not do justice to the coloured drawing; 255). The 1850 translation does not use the phantom analogy; only the 1851 edition by Ashburner uses that word. In the preface to the 1850 edition, ghostly sightings of the past are explained by these lights.

35 Reichenbach, *Researches on Magnetism* [1850] 1974, 6; xl–xli, describes the methods and conditions for his experiments. Reichenbach's inability to see the odic light parallels psychical scientists' experience because ectoplasm only appeared in a dark room.

36 Albert de Rochas, *L'Extériorisation de la sensibilité: Étude expérimentale et historique* (Paris: Chamuel, 1896). For the continued interest in magnetism and development of avant-garde art, see Nicolas Pethes, "Psychicones: Visual Traces of the Soul in Late Nineteenth-Century Fluidic Photography," *Medical History*, 60.3 (2016): 325–341; also refer Brauer, "Scientistic Magnetism and Hauntological Metarealism: The Phantasmatic Doubles of Duchamp and Durville", *Realisms of the Avant-Garde,* eds. David Ayers et al (Sixth volume, European Avant-Garde and Modernism Studies, Walter de Gruyter, 2020) 42–75.

37 On the range of ways that séances were received, see Lynn L. Sharp, *Secular Spirituality: Reincarnation and Spiritism in Nineteenth-Century France* (Lanham: Lexington Books, 2006); Simone Natale, "The Medium on the Stage: Trance and Performance in Nineteenth Century Spiritualism", *Early Popular Visual Culture,* 9: 239–25, and Esyllt Jones, "Spectral Influenza: T. G. and Lillian Hamilton, Interwar Spiritualism, and Pandemic Disease," *Epidemic Encounters: New Interpretations of Pandemic Influenza in Canada, 1918–1920,* eds. Esyllt Jones and Magda Fahrni (Vancouver: University of British Columbia Press, 2012) 193–221.

38 *The Phenomena of Materialisation* catalogue of drawings and prints is described in *Light,* 6 (16 January 1886): 34. I thank Krystyna Matyjaskiewicz for informing me that this would have been a one-day event.

39 John Gerrard Keulemans's images are published in Farmer, *Twixt Two Worlds*, 1886.

40 Melissa E. Buron, Krystyna Matyjaszkiewicz, et al., *James Tissot* (San Francisco: Fine Art Museums of San Francisco, 2020) especially "The Visions of Tissot," 66–71.

41 M.A., "Notes by the way", *Light* (27 February, 1886) 97–98; also refer William Crookes, "The Last of 'Katie King.' "The Photographs of 'Katie King,' by the Aid of the Electric Light", *The Spiritualist* (5 June 1874): 270–271.

42 W. Crookes and C. G. Helleberg, "Remarkable Spirit Manifestations" [1874] (Cincinnati: C.G. Helleberg, 1891) 8–9; quotation, 6. The discovery of the painted version of Tissot's *Apparition* with its bluish spirit lights enabled Dr Melissa Buron to compare Tissot's rendering of the light orbs with Keulemans' versions of spirit lights. Tissot described the spirit lights as having the quality of phosphorus light, perhaps referencing the well-known phosphorous lamp that Crookes used to light the studio; refer Melissa E. Buron, "*'Twixt Two Worlds': The Visions of James Tissot* (PhD. diss., Birkbeck College, University of London, 2021) 67–72. Serge Drigin depicted Crookes and his hand-held phosphorus lamp in Arthur Conan Doyle, *The History of Spiritualism* (New York: Arno Press, 1975) 270.

43 Farmer, *Twixt Two Worlds*, 1886, 164 (question mark in the original; italics mine). Farmer uses the term "substance" throughout the text and thus seems to be quoting from Keulemans when he explains the plates, 164–166, using the term "bioplasm". For more on bioplasm, refer Lionel S. Beale, *Bioplasm: An Introduction to the Study of Physiology & Medicine* (London: J. & A. Churchill, 1 January 1872).

44 Farmer, *Twixt Two Worlds,* 1886, 164.

45 On ectoplasm, see Brain, "Materialising the Medium", 2013; Karen Redrobe Beckman, *Vanishing Women: Magic Film and Feminism* (Durham, North Carolina: Duke University Press, 2003) 78–92; Brower, *Unruly Spirits*, 2010, 117–122.

46 Sarah Willburn, "Viewing History and Fantasy through Victorian Spirit Photography", *The Ashgate Research Companion to Nineteenth Century Spiritualism and the Occult*, eds., Tania Kontou and Sarah Willburn (Farnham: Ashgate, 2012) 359–381. On Victorian photography, see Jennifer Tucker, *The Social Photographic Eye: Brought to Light: Photography and the Invisible, 1840-1900*, ed. Corey Keller (San Francisco: San Francisco Museum of Modern Art, 2009) 37–50. Melissa Buron, in conversation with the author, February 2020, noted that Tissot's choice of mezzotint for his ghost print was highly unusual in the 1880s.

47 John Harvey, *Photography and Spirit* (London: Reaktion, 2007) 82. Some psychic researchers, including Conan Doyle with William Hope, mediumistic photographer Ada E. Deane and Hector Durville continued to research within the parameters of electromagnetism and light radiations, photographing fully formed phantoms

rendered in a pictorialist photographic style. See Brauer, "Scientistic Magnetism and Hauntological Metarealism", *Realisms of the Avant-Garde,* 2020, 42–75. After Schrenck-Notzing and Bisson published their photographs in 1914, they were frequently imitated in Europe and North America setting a trend towards a modernist scientific image of ectoplasm.

48 These descriptors are taken from n.a. [Eric J. Dingwall], "Report on a Series of Sittings with Eva C.," *Proceeding of the Society for Psychical Research,* 32 (1922): 209–343, quotation 209–210; also Eric J. Dingwall, "The Plasma Theory," *JASPR,* 15, 5 (1921): 207–219.

49 For a survey of commercial-studio ghost photographers including William Mumler, John Beattie and Jean Buguet, see Fred Gettings, *Ghosts in Photographs: The Extraordinary Story of Spirit Photography* (Montreal: Optimum, 1978); also *The Perfect Medium,* 2005.

50 Geley, *From the Unconscious to the Conscious,* 65, used the chrysalis as an example of how materializations work in his conception of a supernormal physiology.

51 Eva C.'s excretions were described as "flat" by Albert von Schrenck-Notzing, *Phenomena of Materialisation: A Contribution to the Investigation of Mediumistic Teleplastics* [1914] trans. E. E. Fournier d'Albe (London: Kegan Paul, Trench Trubner, 1920) 90. Originally published as Albert Schrenck-Notzing, *Der Kampf um die materialisations-phänomene* (Munich: Ernst Reinhardt, 1914), Richet, *Thirty Years of Psychical Research,* 1923, 279, describes them as "undecided" and in a state of "flou" (cloudy form); see also "Concerning the Phenomenon Called Materialisation," *Annals of Psychical Science* 2 (Oct. and Nov. 1905), part 1, 207–210; part 2, 269–289. For "rudimentary" productions, see n.a.[Eric Dingwall], "Report on a Series of Sittings with Eva C.," 1922, 209–210; Geley used the term "incomplete" about Eva's excretions, *From the Unconscious to the Conscious,* 61. For historical information on ectoplasm, see Carlos S. Alvarado, "Musings on Materializations: Eric J. Dingwall on 'The Plasma Theory'", *Journal of Scientific Exploration* 33, no. 1 (2019): 73–113.

52 See Harry Houdini, *A Magician Among the Spirits* (New York: Harper and Brothers 1924) 179, particularly his experiences with Carrière and Bisson, chapter 10.

53 Geley, *From the Unconscious to the Conscious,* 1920, 52; 60–61. Also see the section on materializations, 51–73.

54 For Richet's details about the 1905 séance with Carrière, see "Concerning the Phenomenon Called Materialisation," parts 1 and 2; Schrenck-Notzing, *Phenomena of Materialisation,* 6–7. Carrière's career falls off in the 1920s, and in 1954 Rudolf Lambert writes a scathing report linking her with fraud.

55 Noakes, *Physics and Psychics,* 234, states that Palladino's lost her credibility during a series of séances in 1895. Andreas Sommer, "Policing Epistemic Deviance: Albert Von Schrenck-Notzing and Albert Moll," *Medical History,* 56, no. 2 (2012): 255–76, reveals that Richet and Schrenck-Notzing were friends from 1888 and that they developed ideas about materialization collaboratively; see Heather Wolffram, "In the Laboratory of the Ghost-Baron: Parapsychology in Germany in the Early 20th Century," *Endeavour,* 33, no. 4 (2009): 152–157.

56 Oliver J. Lodge, "Experience of Unusual Physical Phenomena Occurring in the Presence of an Entranced Person (Eusapia Paladino)", *Journal of the Society for Psychical Research,* 6 (1894): 306–336; quotations 334. In this group, Lodge became a believer in discarnate spirits; see Noakes, *Physics and Psychics,* chapters 3-4. Geley, "Ectoplasm," 276, quotes Schrenck-Notzing, who states that Lodge "was the first to advance, in 1895 [pseudopods] as an explanation of telekinesis".

57 Richet, *Thirty Years of Psychical Research,* 1923, 515. The term ectoplasm comes to dominate, although Schrenck-Notzing prefers "teleplasm" and Bisson uses the word "substance"; for more on this term, see Michael Granger, "D'où vient le mot "ectoplasme" dans son acception Spirite et Métapsychique," *Revue Spirit,* 157: 2 (2014): 15–17.

58 Richet, "Concerning the Phenomenon Called Materialisation," 1905, 288; italics and question mark in original quotation.

59 Geley credits Bisson with discovering the substance in "Ectoplasm," *ASPR*, 274–275. Schrenck Notzing, *Phenomena of Materialisation*, Fig. 1, 1909, describes how Bisson took the early flash photographs of the "substance," 47-49. See Bisson, *Les Phénonmènes dits de Materialization*, figs 2 and 3, for her first ectoplasm photographs.

60 Brain, "Materialising the Medium," 2013, 122; Alvarado, "Musings on Materializations," and Eric J. Dingwall on "The Plasma Theory,'" 107. Geley, *From the Unconscious to the Conscious*, quotation: vii.

61 Bergson, *Creative Evolution*, 118. He goes on to state "the amoeba deforms itself in varying directions"; 280; 266. Scientific images of asymmetrical protoplasm represented primal forms and caught the attention of psychic researchers and some Modernist artists. On Henry Moore's biomorphic sculpture and interest in Vitalism, see Edward Juler, "Life Forms: Henry Moore, Morphology and Biologism in the Interwar Years", *Henry Moore: Sculptural Process and Public Identity* (London: Tate Research Publication, 2015).

62 For Ernst Haeckel's conception of protoplasm, see Brain, "Materialising the Medium," 2013, 124–128. On Haeckel and art, see Marsha Morton, "From Monera to Man: Ernst Haeckel, Darwinismus, and Nineteenth Century German Art," *The Art of Evolution: Darwin, Darwinisms, and Visual Culture*, eds. Barbara Larson and Fae Brauer (Hanover and London: The University Press of New England, 2009) 59–91.

63 "Bathybius Haeckelii" was cast as a ubiquitous gelatinous medium found universally in all life forms. Although by 1878–79, Huxley retracted his discovery that Bathybius Haeckelii was urschliem, the substance continued to be championed by Haeckel; see Philip F. Rehbock, "Huxley, Haeckel, and the Oceanographers: The Case of Bathybius Haeckelii," *Isis*, 66:4 (1975): 504–533, especially 531–533; also Gerald L. Geison, "The Protoplasmic Theory of Life and the Vitalist-Mechanist Debate". *Isis*, 60: 3 (Autumn, 1969): 272–292; Lionel Smith Beale, Protoplasm: or, Life, Force, and Matter (London: J. Churchill, 1870); Lionel S. Beale, *Bioplasm: An Introduction to the Study of Physiology & Medicine* (London: J. & A. Churchill, 1872).

64 Thomas Henry Huxley, *On the Physical Basis of Life* (New Haven: College Courant, 1869), quotation 16–17; also "Inventor of Protoplasm". "A great Med'cine Man, among the Inquiring Redskins". *Vanity Fair* (January 28, 1871).

65 Haeckel, quoted in Oliver A. I. Botar, "Prolegomena to the study of Biomorphic Modernism: Biocentrism, Lászlo Moholy-Nagy's "New Vision" and Ernõ Kállai's Bioromantik" (PhD diss., University of Toronto, 1998) 191. Ostwald popularized a dynamic conception of the universe that supported Flammarion's conception; on Ostwald's "energetism," the "idea that energy is the substrate of all phenomena," and its relationship to critical Vitalism, see Botar, 194.

66 Botar, "Prolegomena to the Study of Biomorphic Modernism," 1998; see *Introduction*, 190–192, for the contextualization of Monism.

67 On Biocentric philosophers see Botar, "Prolegomena to the study of Biomorphic Modernism," 1998, 6–7; Klages, 230–232, and Raoul Francé, 233–246.

68 Robert Brain aligns Schrenck-Notzing's photographs with Baraduc's light images; he lays out how Symbolist artists skirted around Vitalism, especially through synesthesia; see Brain, "Protoplasmia: Huxley, Haeckel and the Vibratory Organism in Late Nineteenth Century Science and Art", *The Art of Evolution: Darwin, Darwinisms and Visual Culture*, 2009, 92–123. On synesthesia and art see Maurice Tuchman *The Spiritual in Art: Abstract Painting 1890-1985* (Los Angeles: LACM, 1987)

69 On the scientific properties of protoplasm, see Brain, "Materialising the Medium," 2013, 125; Geison, "The Protoplasmic Theory of Life and the Vitalist-Mechanist Debate," 1969, 288.

70 Florence Raulin-Cerceau, "The Concept of Chemical Evolution before Oparin", *Genesis–In the Beginning: Precursors of Life, Chemical Models and Early Biological Evolution*, ed. Joseph Seckbach (Dordrecht: Springer, 22 (2012) 891–906, especially 894–895; it was fundamental for evolutionary theory that life could begin without the input of an external agent.

71 Flammarion, *The Unknown*, 1900, 307; *Mysterious Psychic Forces*, 1907, 430; Beale, *Protoplasm*, and Beale, *Bioplasm*, 206, where he explains that bioplasm was controlled by an independent agency.

72 Bowler, *Reconciling Science and Religion*, 2001: Introduction.
73 Martyn Jolly, *Faces of the Living Dead: The Belief in Spirit Photography* (London: British Library Publishing Division, 2006) 64.
74 In 1909, Bisson and her husband, the vaudeville playwright, Alexandre Bisson, met the 23-year-old Eva Carrière, who had replaced Paladino as the most important subject for testing by the twentieth century. In 1910, Schrenck-Notzing brought a second flash apparatus and a stereoscopic camera to Bisson's laboratory. Bisson's setup eventually included two flash systems and between seven to nine cameras. On the early flash system, see Kate Flint, "'More Rapid than the Lightning's Flash': Photography, Suddenness, and the Afterlife of Romantic Illumination," *European Romantic Review*, 24: 3 (2013): 369–383. Little is known about Bisson's artistic career: She exhibited at the Salon des Artistes Français between 1898 and 1902, receiving an honourable mention in 1901 for her plaster, *Misère*.
75 Geley, "Ectoplasm," 274–275. Bisson made the case for using the word, "substance," in her lecture to the Metaphysical Congress in Copenhagen in 1910. The notion that Bisson discovered proof of ectoplasm is repeated by Arthur Conan Doyle, *The History of Spiritualism*, 107. Women who were active as psychical researchers were downplayed in the literature, and often treated disrespectfully by their male counterparts; see Beth Robertson, *Science of the Séance: Transnational Networks and Gendered Bodies in the Study of Psychic Phenomena, 1918–40* (Vancouver: UBC Press, 2016). In *Phenomena of Materialisation*, 149, Schrenck-Notzing outlines the controls that were placed on Carrière, including vaginal scrutiny and forced regurgitation. Even Bisson, as the lead researcher, was subjected to body searches on occasions.
76 Juliette Alexandre Bisson, *Les Phénomenès dits de matérialisation: étude expérimentale* [1914] (Paris: Alcan 1921). Schrenck-Notzing, *Der kampf um die materialisationsphänomene* (Munich: Ernst Reinhardt, 1914). Bisson's book is more substantial and better illustrated than Schrenck-Notzing slimmer volume of 1914. His 1920 English translation is much expanded with photos and new experiments. On Schrenck-Notzing's training in France, and his close relationship with Richet, see Sommer, "Policing Epistemic Deviance," 255–276. Flash photographs were directed by Bisson and Schrenck-Notzing, with help from G. Delanne and G. de Fontenay, among others, who were familiar with using the flash to photograph phantoms in dark room experiments with Flammarion; see Matheson, "Ectoplasm and Photography," 89–90.
77 Descriptions from Schrenck-Notzing, *Phenomena of Materialisation*, 87–90.
78 Geley, *From the Unconscious to the Conscious*, 1920, 52.
79 Karl Gampenrieder was not working from life, but probably from the spoilt photographs until 1911; refer Schrenck-Notzing, *Phenomena of Materialisation*, 274; on this image, 88.
80 On this point, see Beckman, *Vanishing Women*, 81–98, especially 87.
81 Schrenck-Notzing, *Phenomena of Materialisation*, 89.
82 Geley, *From the Unconscious to the Conscious*, 1920, 52 (italics mine).
83 See Houdini, *A Magician Among the Spirits*, 1924, 179; for his experience meeting Carrière and Bisson, see chapter 10; for negative assessments of these twentieth century photographs, refer Harvey, *Photography and Spirit*, 82.
84 Oliver A. I. Botar, "New Vision and the Aestheticization of Scientific Photography in Weimar Germany", *Science in Context* 17 (4) (2004): 525–556. Lázló Moholy-Nagy included 12 of Schrenck Notzing's materialization photographs in his *Film and Foto* exhibition in Stuttgart (1929). See Karl Steinorth et al., *International Ausstellung Des Deutschen Werkbunds: Film und Foto*, (Stuttgart: Deutsche Verlags-Anstalt, 1979), figs. 948–959. On the twentieth century reception of ectoplasm, see Leigh Wilson, *Modernism and Magic: Experiments with Spiritualism, Theosophy and the Occult* (Edinburgh: Edinburgh University Press, 2013) Chapter 5, especially 142–144; also Tom Gunning, "Phantom Images and Modern Manifestations: Spirit Photography, Magic Theater, Trick Films and Photography's Uncanny," *Cinematic Ghosts: Haunting and Spectrality from Silent Cinema to the Digital Era*, ed., Murray Leeder (London: Bloomsbury 2015) 17–38; Karl Schoonover, "Ectoplasms, Evanescence, and Photography", *Art Journal*,

62: 3 (Autumn, 2003): 30–43; also Karen Redrobe Beckman, *Vanishing Women*, 81–98, and L. Anne Delgado, “Bawdy Technologies and the Birth of Ectoplasm,” *Genders*, no. 54 (September, 2011) where ectoplasm is compared to childbirth.

85 I am developing the argument laid out by Brain in “Materialising the Medium”, 2013, 113: “Ectoplasm became a special instance of protoplasmic investigation, and therefore of Life in its most fundamental operations, yet within supernormal settings”. He continues on the mediums, 126: “materialisation presented an opportunity to study the workings of protoplasm under special conditions. The mediums’ materialisation of limbs, heads and amorphous forms showed distinct similarities to the projection and retraction of pseudopodia from the cells of protozoa and the regeneration of limbs from certain organisms”. On Baraduc and visual culture, see Anthony Enns, “Vibratory Photography,” *Vibratory Modernism,* eds. Anthony Enns, Shelley Trower (London: Palgrave Macmillan, 2013) 117–197; Pethes, “Psychicones: Visual Traces of the Soul in Late Nineteenth-Century Fluidic Photography,” 325–341.

86 Geley, “Ectoplasm,” 277.

5 Vitalist Picasso

Bergson's "Psychic States", Phantasmatic Luminescence and Occultist Cubism

Fae Brauer

> There is no work of art that does not indicate an opening for life, a path beneath the cracks. Everything I have written is vitalist, at least I hope it is.
>
> – Gilles Deleuze[1]

Spurred on by the vital force to experience states of "becoming simian" and "becoming tribal", Pablo Picasso may be regarded as a perpetual vitalist from the time he started living in Paris.[2] Yet the vitalisms he engaged constantly shifted, as is revealed by those he pursued during the different phases of his Cubisms. During his Analytical Cubism, Picasso had begun to explore the dissolution of matter and invisible realities conveyed by the new sciences of electromagnetic rays, x-rays, radioactivity and non-Euclidean geometry. Following Wilhelm Röntgen's discovery of invisible rays, Picasso seems to have recognized, following Linda Dalrymple Henderson's observation, "the inadequacy of human sense perception and raised fundamental questions about the nature of matter itself".[3] By his Hermetic phase of Cubism, Picasso appears to have abandoned any confinement of the human body to a contour with a wholesome surface and a clear distinction between its interior and exterior, let alone a distinct separation from the world it inhabits. Instead, Picasso seems to have reconceived its unconstrained phenomenology as surmized by Gilles Deleuze: "As there is no surface, the inside and outside, the contained and the uncontained no longer have a precise limit".[4] The phenomenological ramifications of this reconception are captured by Akira Mizuta Lippit:

> No longer inside nor out, within nor without, body and world form a heterogeneous one (a one that is not one but together, side-by-side, a series of contiguous planes and surfaces, plateaus). You are in the world, the world is in you.[5]

Amidst laicization of the Radical Republic, the imposition of standardized time and the introduction of Taylorism, Picasso's Hermetic Cubism became associated with *Bergsonisme*,[6] inscribed as *Anarchiste*,[7] aligned with Néo-Mallarméan evocation and referred to by the Neo-Symbolist journal, *La Phalange*, as "*l'Art négatif*",[8] and "*Le Mysticism Contemporain*".[9] In the literary newspaper, *Gil Blas*, it was identified as the product of "*une secte mystérieuse*".[10] It became construed as, according to André Salmon,"*les marques de l'occulte, du symbole ou de la mystique*".[11] These correlations arose during the vibrant discursive network ignited by Picasso's close relationships with Max Jacob, Guillaume Apollinaire and Salmon.

From the moment Picasso moved in with Jacob in 1901, he was exposed to Camille Flammarion's paranormal phenomena plus Dr Gustave Geley's *L'Être subconscient*

DOI: 10.4324/9781003045595-8

and "*dynamo-psychism*", mentioned by Serena Keshavjee in Chapter 4. He was also exposed to Bergsonism, Mallarméism, magnetism, mediumism, mysticism, spiritism and occultism, as well as astrology, fortune-telling, magic, palmistry, the Tarot, opium nights and neo-symbolist poetics, as Fernande Olivier vividly recalls:

> Ce fut l'époque ou Max Jacob se vit menace d'acquerir une celebrité spéciale. Cartomancien, astrologue, chiromancien, voyant. Nouvelle fantaisie! Se prenait-il au serieux? Était-il sincere? Pour ma part je n'ai jamais pu evaluer la part de sincerité de Max. Enfin cela fut un nouvel amusement. La superstition s'en mela. On le consultait sur tout.[12]

By no means did this abate with the rapid growth from 1905 of what Peter Read calls Picasso and Apollinaire's "fraternal complicity" and development of their "interdisciplinary laboratory".[13] Once joined by Salmon later that year, they formed "*la bande à Picasso*".[14] So often did these writers gather at Picasso's that Salmon vividly recalled the sign Picasso drew over his studio door: *Le rendez-vous des Poètes*.[15] Their fervent interdiscursivity intensified with the Bergsonism they explored alongside the Neo-Symbolist journals they read and to which they contributed: *La Phalange, Les Marges, Le Mercure de France, L'Occident, La Plume* and *Vers et Prose*.

Following their engagement with Bergson's *Time and Free Will, Matter and Memory* and *Creative Evolution,* by late 1909 their interdiscursivity entailed an exploration of Bergson's "*l'élan vital*", particularly intuition, sensory memory, empathy, "felt time" and the "turn of experience" in relation to Bergson's concepts of "psychic states", "sensory memory" and "becoming".[16] Simultaneously their explorations encompassed "*la force vitale*", particularly some of the revelations of magnetism, occultism and spiritism. This entailed investigation of mediumistic visions, spectres and the ghosts of their friend, Henri 'le douanier' Rousseau. "Everyone who knew Rousseau remembers his liking for ghosts", recalled Guillaume Apollinaire, "teasing him, thumbing their nose at him and breaking wind in a way so foul that it made the functionary feel quite ill".[17] Their investigations also appeared to entail the ways "a completely invisible emanation could manifest its presence on a photographic plate",[18] as exposed by the phantasmatic photography of Hector Durville and Albert de Rochas, as well as Hippolyte Baraduc's *psychicones*.

Through this network of mediation, "*la bande à Picasso*" was also exposed to the occultist research unleashed by Gerard Encause: "Papus", his Kabbalistic Order of the Rose-Croix and its Librairie du Merveilleux with its monthly review, *L'Initiation*. This included Papus' books on occultism, magic and illustrated journals, as well as those he edited with stage magnetist, Donato: *Almanach de la Chance* and *La Vie Mysterieuse*. Launched in January 1909, the first issue of *La Vie Mysterieuse* entitled "Les Tables parlants" was followed by vividly illustrated articles on *Magnétisme, Cartomancie, Spiritisme,* as well as poltergeists, psychism, séances, spiritism and telepathy. Its November 1911 issue, "Le Mediumnisme et l'Art", was illustrated by Rousseau's paintings. *La bande à Picasso* was also aware of the Société Magnétique de France founded by Hector and Henri Durville, their *Journal du magnétisme et psychisme experimental* and their publishing house, *Librairie du magnétisme*, with treatises illuminating "la force psychique", phantasmatic doubles and magnetism as an "agent lumineux": A luminous force.[19] With Baraduc and Madame Blavatsky founding members, the Société Magnétique was aligned with Parapsychology and Theosophy, while Durville drew upon the theosophical treatises of Annie Besant in translation from 1896.

To unravel the relationship of Picasso's Vitalism to *la force vitale* and Bergson's *l'élan vital,* his Hermetic Cubism shall be explored from four multidimensional perspectives. To navigate the rupture with positivist empiricism and scientistic realism represented by Cubism and its relationship to *l'élan vital*, Picasso's durational portrait of Daniel-Henry Kahnweiler (Figure 5.1) shall be initially compared with Léon Bonnat's *Portrait of the President of France, Armand Fallières* (Figure 5.2) and Jacques-Émile Blanche's study for a portrait of Henri Bergson (Figure 5.3). After contextualizing these portraits within world standard time, Taylorism and the ergonomic reconception of the body as *"le moteur humain"*, their differences shall be examined within a new psychology of time, Bergson's concept of psychic states, sensory memory and becoming. In the second part, Kahnweiler's portrait shall be rescrutinized through the lens of spiritist photography of *la force vitale* and *le corps vitale fluidique* captured by Baraduc, as well as the magnetist photography of phantoms and uncanny luminescence captured by Durville and Rochas, particularly in relation to its auratic luminosity and dissolution of the body in space. In the final part, the fusion of *l'élan vital* and *la force vitale* shall be explored through the conjunctions of Bergsonism with hermeticism, mysticism and occultism in Picasso's illustrations for Jacob's semi-autobiographical prose-poems, *Saint Matorel* and *Le Siège de Jérusalum*, to illuminate how they manifest Occultist Cubism.

World Standard Time, A New Psychology of Time and Psychic States: Picasso, Bonnat and Bergson's *L'Élan Vitale*

In 1884 when representatives of 25 countries at the Prime Meridian Conference in Washington proposed that the earth be divided into 24 time zones one hour apart and Greenwich be established as zero meridian, uniform time was established worldwide.[20] Five years later, Bergson's *Essai sur les données immédiates de la conscience* was published.[21] There he questioned whether the experience and memory of sensations, feelings and passion within the continual flow of "psychic states" that create inner life could be quantified within standardized time and measured in terms of units that impersonally regulate capitalist societies. Undeterred, President Raymond Poincaré hosted the *International Conference on Time* in Paris in 1912 to establish a uniform method of maintaining accurate time signals and transmitting them world-wide. At 10 a.m. on 1 July 1913, when the Eiffel Tower sent the first time-signal to be transmitted world-wide, World Standard Time became, as Stephen Kern observes, "the most momentous development in the history of uniform public time since the invention of the mechanical clock in the fourteenth century".[22] Immediately corporations and institutions adopted it to monitor workers, rank students, gauge productivity, accelerate performance, enhance competition and increase profits. It became integral to implementation of Taylorism in France and the development of ergonometry in which the body was reconceived as "*le moteur humain*".[23]

This standardization of time was, according to Josef Conrad, "authoritarian time" in which temporal precision became an obsession.[24] The ability to monitor work facilitated by the clock was blamed for dramatic increases in nervous breakdowns and what American neurologist, George Beard, termed neurasthenia: "New American nervousness".[25] In France, it seemed manifest by increasing numbers of male and female patients admitted to Salpêtrière, Bicêtre and La Charité suffering nervous breakdowns, manic depression and hysteria from disastrous working and living conditions.[26] For Bergson's close friend, the French psychologist, Pierre Janet, a new psychology of time was required, which was endorsed by Bergson. Drawing upon neurology, psychology and metaphysics, Bergson pointed out that the intensity of sensations, feelings and passions could not be monitored through standardized measurements of time, let alone conceived in a mathematically measured instant as a point in time. Since standardized time disrupted the secure feeling of an affective and indivisible continuity emanating from what Bergson called "the complex symphony" of pleasurable and painful experiences played out in memories – in which past and present seemed synchronized with intuitions of future time – he envisaged that it could lead to a traumatizing fragmentation of the psyche.[27]

In *Essai sur les données immédiates de la conscience*, Bergson argued that while the positivist, mechanistic and scientistic conception of time may facilitate regulation of the political economy, the very idea of a homogeneous and measurable time was artificial. It was antithetical to the inner 'felt' experience of time realized in memories of time past intermingled with the intensity of sensations during contemporary time and intuitive anticipation of future time, as he explained in *Time and Free Will:*

> It is usually admitted that states of consciousness, sensations, feelings, passions, efforts, are capable of growth and diminution. We are even told that a sensation can be said to be twice, thrice, four times as intense as another sensation of the same kind.... Sometimes the feeling which is suggested scarcely makes a break in the compact texture of psychic phenomena of which our history consists; sometimes it draws our attention from them, but not so that they become lost to sight; sometimes, finally, it puts itself in their place, engrosses us and completely monopolizes our soul.[28]

Since the felt experience of time was considered to arise from empathy with every state of the object or subject felt intuitively, what Bergson theorized was an empathic form of consciousness reconceptualising experience. As he explained: "We call intuition here the sympathy by which one is transported into the interior of an object in order to coincide with what there is unique and consequently inexpressible in it".[29]

Intuition and empathy were then able to lead to what Bergson called "the very inwardness of life".[30] In this case, Bergson's theory of "psychic states" entailed an empathic identity with persons and objects, especially artworks, through the psychological experience of time as duration, with fluctuating sensations that "engrosses us and completely monopolizes our soul".[31] To explain how this happens, he added: "There are thus distinct phases in the progress of an aesthetic feeling, as in the state of hypnosis.... the psychic states whose intensity we have just defined are deep-seated states".[32] Bergson then explained how these "deep-seated states" were able to change perception:

> But little by little it permeates a larger number of psychic elements, tinging them, so to speak, with its own colour; and lo! Your outlook on the whole of your surroundings seems now to have changed radically. How do you become aware of a deep passion, once it has taken hold of you, if not by perceiving that the same objects no longer impress you in the same manner? All your sensations and all your ideas seem to brighten up; it is like childhood again. We experience something of the kind in certain dreams....[33]

As Gilles Deleuze highlighted from his readings of Bergson, in the realms of matter and space where life is liberated from organic barriers, divisions between persons and things may dissolve. As thresholds disappear, human subjects may be conceived in constant states of becoming. These are relative states of deterritorialization without boundaries, entailing a multiplicity of movement and interpenetrating positions in space over a duration of time. "Matter or mind/spirit, reality has appeared to us as a perpetual becoming", Bergson explained. "This is the non-organic life which grips the world".[34] Becoming is, as he elaborated, "the substantial state of being always on the verge of transforming itself". It is, Deleuze concluded as if reiterating Bergson, "the powerful inorganic life that grips the world".[35] How this happens may be illuminated by Picasso's *Portrait of Daniel-Henry Kahnweiler* (Figure 5.1).

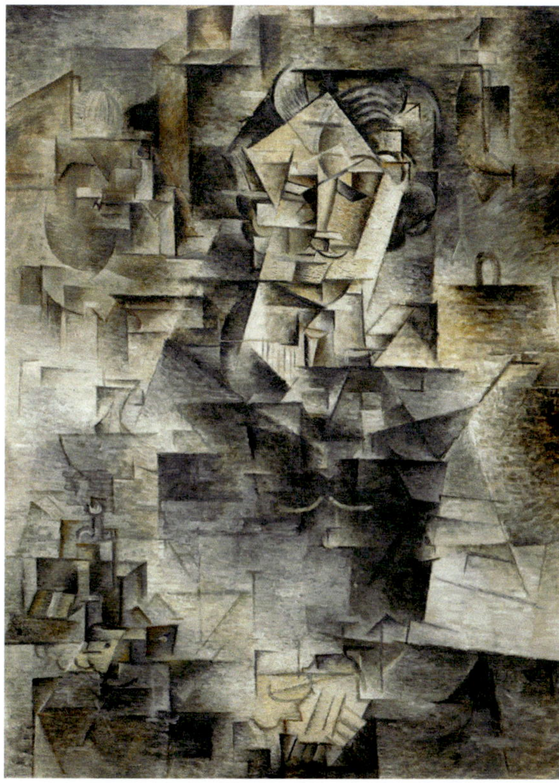

Figure 5.1 Pablo Picasso, *Portrait of Daniel-Henry Kahnweiler,* Autumn 1910, oil on canvas, 39 9/16 × 28 9/16 in. (100.4 × 72.4 cm). (Courtesy of The Art Institute of Chicago, Gift of Mrs. Gilbert W. Chapman in memory of Charles B. Goodspeed, 1948.561; © 2016 Estate of Pablo Picasso/Artists Rights Society (ARS), New York. Public Domain.)

In Chapter 2, as Pat Berman identifies, Munch's bodies are boundaryless, in the process of becoming. Their slippery contours move in and out of the surrounding space to signify their imbrication within a creative energy or vitalist force in terms of Bergson's concept of *L'élan vital.* The same may be said of Picasso's dissolution of boundaries between persons and things in his Hermetic Cubist paintings and the disintegration of their solidity into interpenetrating fragments, captured over time, from multiple perspectives. This is illuminated by Picasso's portrait of Kahnweiler (Figure 5.1) and the challenges its reconceptions of the body posed to ergonometry, Taylorism, Hippolyte Taine's scientific exactitude and the scientific rationalism of Ernest Renan's *L'Avenir de la Science.*[36] It also posed a challenge, as Mark Antliff argues, to the Cartesian rationalism exalted by Charles Maurras.[37] These rationalisms were seemingly epitomized by portraits painted by the most commissioned and acquisitioned of all artists in the Radical Republic, Léon Bonnat, caricatured

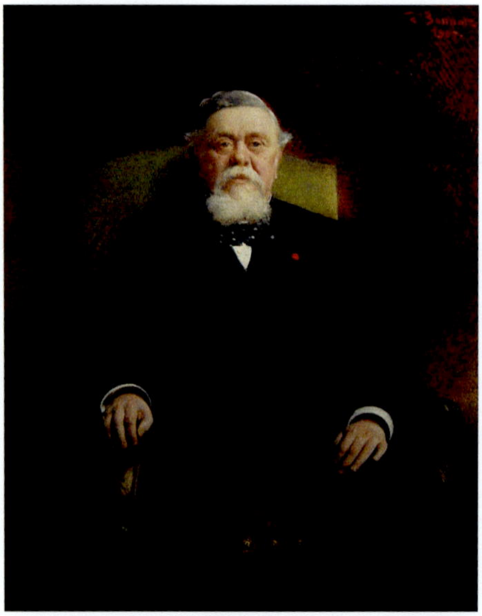

Figure 5.2 Léon Bonnat, *Portrait of Armand Fallières, Président de la République Français (1906–1913)*; oil on canvas, 1907; 143- 114.5 cm. (RMN-GrandPalais/Muséed'Orsay 20401; Hervé Lewandowski. Public Domain.)

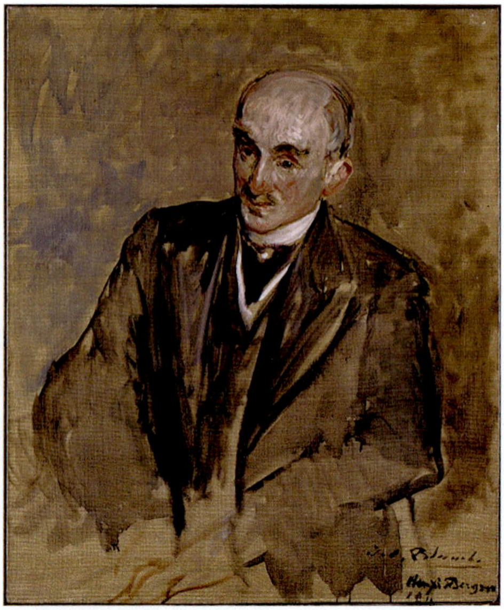

Figure 5.3 Jacques-Émile Blanche, *Étude pour le portrait d'Henri Bergson*, oil on canvas, 1911: Musée des Beaux-Arts de Rouen. (Photo © RMN Grand Palais/Agence Bulloz. Public Domain.)

by *L'Assiette au beurre* as "L'Artiste Officiel"[38] (Figure 5.2). They were even evident in the impressionistic portrait of Bergson painted by the Parisian society painter, Jacques-Émile Blanche (Figure 5.3).

From the time he exhibited his portrait of the first President of the Third Republic, Adolphe Thiers, at the 1877 Salon, Bonnat had become, according to the *Gazette des Beaux-Arts*, "the painter-laureate with the right to paint all the celebrities of the world, in politics, art, science and letters". Not without a touch of irony it surmized: "Everyone illustrious, as much in foreign countries as in France, was painted by Bonnat".[39] Drawing upon the sophisticated portraiture and spontaneous brush-marks of John Singer Sargent, Blanche was more concerned with capturing fashionable society and celebrities ranging from Comte Robert de Montesquiou and the Duchess of Rutland to Aubrey Beardsley, Charles Condor, Claude Debussy, Henry James, Vaclav Nijinsky, Anna de Noailles and Marcel Proust – the cousin of Bergson's wife. Such a celebrity had Bergson become, as indicated in the *Introduction* to the present volume, that Blanche insisted upon adding him to his retinue of celebrity portraits. During the painting of his portrait, Bergson had even asked Blanche about Cubism.[40] Yet far from this portrait arising from any interest in Bergson's philosophy or indeed Cubism, it seemed to pertain more to Blanche's status, his "sole ambition" becoming, according to Proust, "a much sought-after man of the world".[41] While Blanche was a founding exhibitor at the Salon National des Beaux-Arts, Bonnat was Honorary President of the Salon des Artistes Français and more concerned with scientific rationalism and empirical exactitude in art arising from direct visual observation : and anatomical scrutiny.[42] This is demonstrated by Bonnat's portrayal of the sixty-six-year-old President of the French Republic, Armand Fallières (Figure 5.2), popularly known as « le père Fallières », father of the nation, painted four years before Blanche's portrait of Bergson and three years before Picasso's portrait of his twenty-six-year old German-born, Paris-based art dealer (Figure 5.1).

A Senator with staunch allegiances to the *Gauche démocratique* who became President of the Senate from 1899 until 1906, Fallières remained in power as President of the Republic until 1913, supported by a union of left-wing parties. Three months after his election to the Presidency in 1906, his portrait was commissioned from Bonnat by l'État des Beaux-Arts for 10,000 francs.[43] An avant-garde outsider of 'official' art, reliant upon his dealers, Picasso created a strategic series of portraits of his art dealers and collectors, the last of which was his portrait of Kahnweiler. The first was one of Picasso's earliest collectors, Wilhelm Uhde. Uhde had opened his own gallery on rue Notre-Dame-des-Champs by 1908 with a small exhibition of Picasso's artwork followed by Picasso's Cubist portrait of him in Autumn 1909.[44] Simultaneously Picasso began producing similar portraits of his two other Paris-based art dealers, Clovis Sagot and Ambroise Vollard. The first art dealer in Paris to exhibit as many as 65 of Picasso's artworks in 1901 at his rue Lafitte gallery, Vollard had consistently bought large numbers of Picasso's paintings from 1906 and had continually advanced him generous sums.[45] As Gary Tinterow surmizes, "Vollard remained Picasso's most loyal – and most prestigious – dealer, even if prices continued to be low".[46] Hence Picasso's series of dealer portraits seems to have been planned as leverage. Even though Picasso regarded somewhat immodestly his "Cubist Portrait" of Vollard "as the best one" of all those done of this avant-garde art dealer – by "almost everybody" as Picasso put it, including Pierre Bonnard, Paul Cézanne, Auguste Renoir and Félix Vallotton – ultimately it failed to curry favour.[47] Even though Vollard hastily mounted a large solo retrospective exhibition of Picasso's artwork after it in December 1910, he never offered Picasso the security of a contract.[48]

Despite Picasso not signing an exclusive three-year contract with Kahnweiler until 18 December 1912, his portrait may have been instrumental in securing it, as well as other deals. Given Picasso's outright rejection of the Paris salon exhibition system, his portrait may have been instrumental to their agreement to exhibit Picasso's Cubist paintings continually at the Galerie Kahnweiler.[49] It may have proven instrumental to the contracts Kahnweiler was negotiating with his European network of art dealers for their exhibition and sale of Picasso's artwork.[50] It may have also proven instrumental to the contracts that Kahnweiler had begun negotiating with poets, composers and his artists for the publication of collaborative cross-disciplinary *livres d'artistes*, as elaborated in the third part of this chapter. Serendipitously, as Picasso would have been aware, Kahnweiler had visited Jacob to negotiate one of these contracts the day after the poet had experienced his apparition of Jesus on 22 September 1909. It was then that Kahnweiler had reportedly asked Jacob:

> "Je voudrais éditer les livres illustrés par mes peintres. Apollinaire sera le premier, Derain illustrera l'Enchanteur Pourissant. As-tu quelque chose?"[51]

In light of these strategic motivations, as well as the aspiration to create a Bergsonian spatio-temporal reconception of Kahnweiler's "psychic state" over "felt time", Picasso's portrait differed substantially from that of the President of France by Bonnat.

Focusing his portrait upon three-quarters of Fallières' body, Bonnat pictured Fallières seated and facing the beholder in a conventional anatomo-chronological scheme.

Despite Fallières' august position, Bonnat portrays him with no signifier of ostentation or luxury to break his sombre attire, save for the crimson insignia glowing from his button-hole signifying his distinction as *Grand maître de la Légion d'honneur.* Painting him from direct observation and the same stationary position, as signified by the photograph of Bonnat perched on a four-legged wooden stool directly in front of the portrait holding his palette (Figure 5.4), Fallières is rendered from a single perspectival vanishing point. As the singular light source travelling from the left-hand side of his body illuminates Fallières' physiology, the three-dimensional mass and solidity of the President's head and body seems to loom out of the darkness. Although rendered over some twenty sittings, due to these pictorial strategies, Bonnat's portrait appears as if Fallières were captured at a single moment in time and from the same position in space. Even though Blanche's portrayal of Bergson in oil paint appears Impressionistic (Figure 5.3), his treatment of space and time is no different. Arising from some thirty visits made by Kahnweiler to Picasso's studio from mid-September 1910, Picasso's portrait appears almost the opposite.[52] Even though Kahnweiler was photographed sitting on a settee in Picasso's studio (Figure 5.5), unlike Bonnat's location of himself in front of his portrait for the camera (Figure 5.4), Picasso is not in the photograph. He is the photographer who took this photograph.

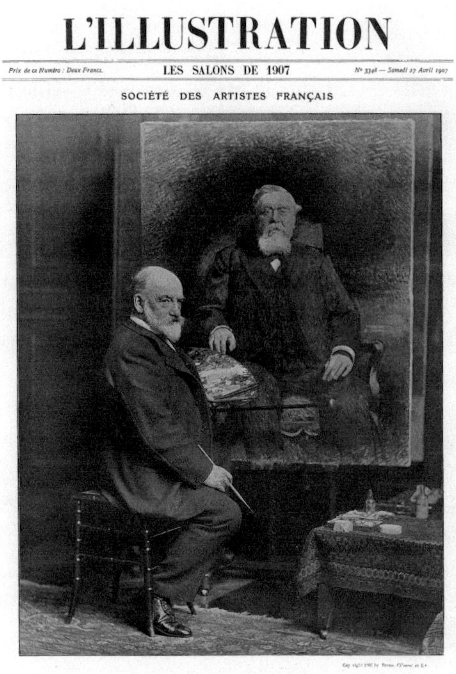
L'ILLUSTRATION

Prix de ce Numéro : Deux Francs. LES SALONS DE 1907 *N° 3348 — Samedi 27 Avril 1907*

SOCIÉTÉ DES ARTISTES FRANÇAIS

M. Léon Bonnat travaillant au portrait de M. Fallières.

Figure 5.4 Braun et Cie, "M. Léon Bonnat travaillant au portrait de Armand Fallières", *L'Illustration,* No. 3348, 27 April, 1907. (Author's collection and photograph.)

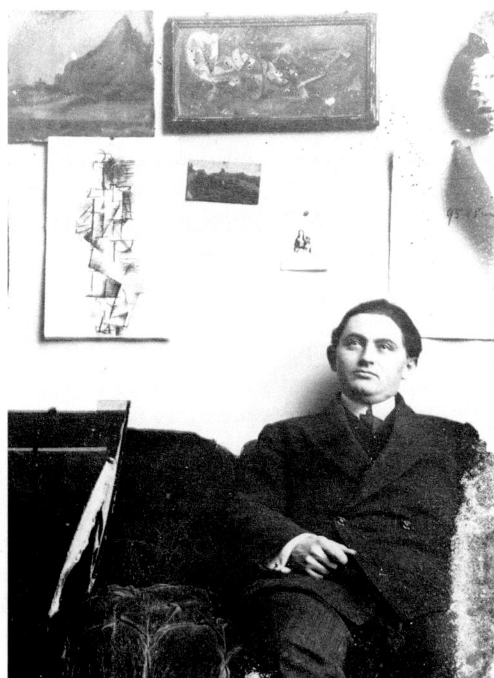

Figure 5.5 Pablo Picasso, *Daniel-Henry Kahnweiler in Pablo Picasso's Studio,* 11 Boulevard de Clichy, Paris, as photographed by Picasso, Autumn 1910; Succession Picasso – Gestion droits d'auteur. (Musée Picasso de Paris Photo©RMN-Grand-Palais. Public Domain.)

Unlike Bonnat and Blanche, Picasso did not render Kahnweiler's portrait from life, in the same stationary position with a single light source, let alone from direct observation of his art dealer sitting in the same position every time they met. Instead, his portrait appears to have been captured from Maurice Merleau-Ponty's phenomenological concept of embodied perception, drawn from Bergson – before affective consciousness becomes measured, analysed and rationalized as in Bonnat's and Blanche's portraits.[53] This affective state entailed Picasso drawing upon his intuition and sensory memory of Kahnweiler during his mobile interactions with him from different spatial perspectives over time and facilitated by his transformation of form into multidimensional interpenetrating facets. As Bergson pointed out, "there is no perception which is not full of memories".[54] To elaborate this point in relation to image memory, Bergson added: "Perception is never a mere contact of the mind with the object present; it is impregnated with memory-images which complete it as they interpret it".[55] From their first meeting late in 1907 at Kahnweiler's Art Gallery by the Madeleine, arranged through their mutual friend, Uhde, Picasso's memory images of Kahnweiler would have accumulated and, following Bergson's analogy, 'melted' into one another to form his durational being.

By Autumn 1910, Picasso and Kahnweiler had become close friends, as their correspondence reveals. In meeting and embracing almost daily, unlike Bonnat's more formalised contact with the President of France, Picasso would have accumulated in his memory many sensory images of this friend for his portrait. While the mass and solidity of Fallières' entire head and body are conveyed through finely gradated tonal modelling, rendered from the darkest dark to the lightest parts from what seems like a singular moment in the same space, Picasso has disregarded Kahnweiler's three dimensionality. Only fragments of Kahnweiler's head and body appear to be recalled from different perspectives over many moments and in the various spaces where he and Picasso would embrace, touch, chat, joke, laugh, listen to one another while tasting and smelling coffee, rolling tobacco, inhaling, exhaling and smelling cigarette smoke and absorbing their body odours when closely interacting with one another. Only fragments recalled in Picasso's sensory memory are rendered as indexical signifiers. These include the five wavy strands of Kahnweiler's dark hair as remembered through sight, smell and touch, juxtaposed with the straight shaft of his nasal bone and the rectangular plane of light stretching from his left ear to his chin.

In contrast to Bonnat's clear rendering of Fallières' crisp white shirt front and white spotted navy cravat, and Blanche's loose rendering of Bergson's white neck collar and shirt amidst his purple professorial cravat, the whereabouts of Kahnweiler's stiff white collar is imprecise. A corner of it may be hinted at by the white angles above what could be the knot of his tie, above what could be shirt pleats, and well above what could be an indexical signifier of his fob-watch chain looping across his suit coat. Further down the canvas where Kahnweiler's lap might be appears an indexical sign of Kahnweiler's clasped hands. Yet unlike Bonnat's rendering of the metacarpal bones and phalanges by the ligaments, arteries and veins throbbing within Fallières' hands, as well as the growth and patina of the President's nails, Picasso provides no comparable indication of Kahnweiler's anatomy. Following Deleuze's extrapolation from Bergson, Picasso appears to have captured his art-dealer as a body without organs.[56]

Slightly displaced from the centre of Kahnweiler's portrait, in the space far left of his photograph (Figure 5.5), Roland Penrose had discerned a pair of wooden statues from New Caledonia that Picasso had acquired two years earlier, with the head appearing to face towards Kahnweiler.[57] Yet as these fragments seem to be viewed from different perspectives in space, it seems as if Picasso has rendered his memories of Kahnweiler's "psychic states" from his interaction with Kahnweiler's "whole personality" over time. As Albert Gleizes pointed out in *Du Cubisme*, "To establish pictorial space, we must have recourse to tactile and motor sensations, indeed to all our faculties. It is our whole personality which, contracting or expanding, transforms the plane of the picture".[58] In capturing Picasso's felt experience of Kahnweiler over time, following Bergson, Picasso's painting may then be regarded as a durational portrait, not a positivist view rendered through empirical exactitude in what Marcel Duchamp dubbed "retinal painting".[59] Instead of a single moment or the countable units of chronometric-time, Picasso seems to have endeavoured to capture Kahnweiler's "psychic states" in non-sequential time and "tactile" space as conceived by Bergson's duration. Hence Bergson's concept of "psychic states" offered artists, as Patrick Jones surmizes, "an alternative vocabulary to defend aesthetic experience in the face of an ever-encroaching, life-denying scientism".[60]

While the head and body of the French President can be clearly disinterred from the surrounding space in Bonnat's painting (Figure 5.2), as can the head and upper torso of Bergson in Blanche's portrait (Figure 5.3), Picasso's portrait of Kahnweiler appears the opposite (Figure 5.1). Not only can Fallières' head and body be identified absolutely but also clearly distinguished from the objects depicted in the painting: The muted green coloured velvet back-rest of the chair, the metal-studs securing the velvet to the seat and the carved wooden armrests (Figure 5.2). By contrast, Kahnweiler's body and the objects surrounding him in Picasso's portrait can only be identified relatively in relation to one another. With no clear distinction between organic and non-organic matter in this continuum, life in Picasso's portrait appears liberated from organic barriers to convey an inorganic vitalism comparable to Bergson's *l'élan vital* (Figure 5.1). By discontinuing lines and leaving geometric facets open, Picasso's painting of Kahnweiler's head and body, together with the surrounding objects, seems to flow from one into the other and melt into the surrounding space through interconnecting passages. It is as if, following Bergson, "matter is dissolved into numberless vibrations, all linked together in uninterrupted continuity, all bound up with each other, and travelling in every direction like shivers".[61] So interpenetrating do these passages become that it seems impossible to determine exactly where space stops and Kahnweiler's body begins in a "psychic state" of becoming. In *L'évolution créatrice,* Bergson surmized this process:

> I find, first of all, that I pass from state to state. I am warm or cold, I am happy or sad, I work or I do nothing, I look at what is around me or I think about something else. Sensations, feelings, desires, ideas – these are the changes into which my existence is divided and which "colour" it.... a slight effort of attention would reveal to me that there is no feeling, no idea, no desire which is not undergoing change every moment.... This is especially so with psychic states that are more deeply internal, such as sensations, feelings, desires ... The truth is that we change without ceasing, and that the state itself is nothing but change.[62]

When viewed from a Bergsonian perspective, the interpenetration of human form, clothing, objects, space and light in Picasso's portrait could then represent the incessantly changing vitalism of Kahnweiler's "psychic states". Instead of Bonnat's perception of the external world as a stable collection of physically contained objects separated in space, the artworks and artefacts in Picasso's studio, which this art dealer knew so well from all his senses, appear to become interpenetrated within duration in what Maurice Blanchot calls "the silent thunder which reverberates amidst images dissolving into one another".[63] As Bergson explains, "[W]e shall no longer consider states of consciousness in isolation from one another, but in their concrete multiplicity, in so far as they unfold themselves in pure duration".[64] As he adds, "[I]nner duration, perceived by consciousness, is nothing else but the melting of states of consciousness into one another". Once they "begin to permeate and melt into one another", each seems to become "tinged with the colouring of all the others".[65] At the same time, Bergson was also addressing the interrelationship of *"l'élan vital"* to *"la force vitale"*, as signified by his six public lectures at Columbia University in February 1913 entitled *Spirituality and Liberty,* followed by his inaugural lecture given three months later as President of the British Society for Psychical Research entitled *Fantômes des vivants et recherche psychique.*[66] As Gleizes and Metzinger perceptively surmized:

> The Cubists taught a new way of imagining light ... Cubism will transport the three-dimensional object into a space at once spiritual and plastic in nature ... Profound realism merges insensibly into luminous spirituality.[67]

Picasso and "La Force vitale": Auratic Energies, Magnetic Luminosities and Phantasmatic Doubles

> There are more things in Heaven and Earth, Horatio, than we believe communally. And our century, effectively to be the epoch whence the invisible, the occult, relegated to the level of chimeras by positivists, seems to be revealed to us.[68]

Due to the rapid development of photographic technologies, the invisible dimensions of the occult were being constantly revealed by parapsychologist, magnetist and spirit photography. While Baraduc's soul photography illuminated the vital radiation of the psyche as the "vital force" and *L'Âme Humaine*, Commandant Darget's photographs captured "fluido-magnetic" bodily emanations, such as Rayons V (Vitaux). Rochas' psychic photographs exposed the "exteriorisation of sensibility", while Hector Durville's magnetist photographs revealed phantasmatic doubles.[69] Through photography, as Keshavjee surmizes in the previous chapter, many of France's acclaimed scientists were able to visualize an invisible force, while scientific technologies seemingly corroborated many discoveries of Occultism.

Through the Bibliothèque Nationale, its Cabinet des Estampes and Adrienne Monnier's *La Maison des Amis des Livres,* hermetic writings became increasingly accessible alongside those of the new sciences. Articles by Symbolists, Neo-Symbolists and Decadents on Occultism and Theosophy were published in the biomonthly Symbolist journal, *Le Mercure de France,* as often as those on the new sciences, particularly Henri Poincaré's *La Science et l'hypothèse* and Le Bon's *L'Évolution de la matière.* Conversely, articles by Occultists on the interrelationship of the new sciences and Occultism were also published in *Le Mercure de France*, as demonstrated by Sar Péladan's *Le Radium et l'Hyper-physique* identifying x-rays and radium with supernatural phenomena.[70] Conterminously, five Nobel Prize winners, the physicist, Jean Baptiste Perrin, the physiologist, Charles Richet and the scientists, Marie and Pierre Curie, attended seances, as did Bergson.[71] Joining hands with others at séances conducted by Eusapia Palladino from 1905, Marie Curie recalled "visions of light or luminescent points, visions of hands or limbs, sometimes ... as phosphorescent", including the luminosity of her own glowing hair.[72] After attending these séances,

Pierre Curie reported: "The result is that these phenomena really exist and it is no longer possible for me to doubt it". He then concluded: "In my opinion, there is a whole domain of completely new facts and psychical states of space about which we have had no conception".[73] Buoyed by this uncanny fusion of x-rays, electromagnetic rays and radium within an occulture of hauntology in which reality was being reconceived through invisible sources of energy, Hector Durville, assisted by his sons, Henri and Gaston, embarked upon a series of what he called "scientific experiments" at their School of Magnetism to unleash "la force vitale".[74]

With the support of Baraduc, Madame Blavatsky, Gérard Encausse "Papus", Péladan, Rochas and other well-established Occultists, in 1887 Hector Durville launched the Société Magnétique de France. Following the success of the 1889 *Congrès Magnétique International* alongside the *Congrès Spirite et Spiritualiste international* and *Congrès International de Hypnotisme expérimental et thérapeutique*, Hector and Henri Durville opened the École pratique du Massage et de Magnétisme on Avenue Mozart in Paris, with a branch in Lyon.[75] They also launched the *Journal du magnétisme et psychisme experimental,* and established their own publishing house, *Librairie du Magnétisme,* which published Hector Durville's treatises addressing the relationship of luminous projections and the human soul to "la force vitale".[76]

In his *Treatise Expérimental du Magnetisme* and *Magnetisme Personnel ou Psychique,* Hector Durville demonstrated how the magnetic poles punctuated every part of the human body, as signified by the positive and negative signs inscribed on anatomy correlating to the attraction and repulsion of horse-shoe magnets.[77] Likened to a flickering flame that could project as far as five metres either horizontally or vertically, Durville considered magnetic energy constituted "la force vitale".[78] Conceiving of the human body as both a receptor and transmitter, Durville commissioned drawings to illuminate how magnetic energies were not just attracted by the body but generated from it. So powerful were these magnetic energies that Durville likened them to rays and electrical sparks able to energize the body both physically and psychically. Since these magnetic energies seemed to radiate as a relatively invisible vital force field perceptible to clairvoyants, spiritists, theosophists, mediums, magnetizers and radiographers, as well as to particular cameras, Durville likened them to phantoms. Just before Picasso began his portrait of Kahnweiler (Figure 5.1), photographs of their luminous emanations were published in Durville's book, *Le Fantôme des Vivants*.[79]

Conceiving of photography as having a unique receptivity to their energies, Durville and Rochas commissioned photographs from Paul Nadar onto magnesium plates to reveal how magnetism was able to mediate a phantasmatic doubling of the physical body. The radiating auras generated from magnetized subjects were likened to magnetic waves connected to the phantasmatic etheric and astral body double, as theosophized by Annie Besant in *Man and his Bodies*, its French translation being cited extensively by Durville in *Le Fantôme des Vivants*, as well as in *Magnétisme personnel ou psychique*.[80] That the etheric and astral bodies could form alongside the physical body as phantoms was initially captured by Paul Nadar. The son of Félix Nadar, he had achieved such renown for his experimental photography that he became Eastman Kodak's agent in France and the photographer most commissioned by Durville and Rochas to capture phantasmatic doubles.[81]

Aware of the phantasmatic ramifications of this phenomenon, Durville and Rochas explored, with the aid of Nadar, how these luminous emanations took the form of phantoms after magnetism. Following the phantom's powerful vibratory energy, which Durville also likened to "la force vitale", he stipulated that their appearance followed a choreography of distinct stages invariably reflecting the state and gender of the physical body. While Durville explored this phenomenon with a range of women – Marthe, Léontine, Madame François, Madame Vix and Mlle Thérèse – it appeared to be most clearly captured by Nadar's photographs of Mme Lambert's phantasmatic double from February 1908.[82] (Figure 5.6).

A medium who Durville and Rochas frequently magnetised, Lambert had long studied with Rochas but had rejected Spiritism and spurned the séances of Eusapia Palladino.[83] Once magnetized by Durville, Lambert found that throughout the day and night a phantom would visit her. Initially the phantom appeared as a fluid vaporous mass of light, accompanied by spots of light floating in darkness. Yet gradually Lambert found that it transformed into the form of a luminous, white pillar or column, taller and larger than herself (Figure 5.6). After further magnetism, she watched it become increasingly luminous until it condensed into a human form taking the shape of a woman, with a body and face uncannily like her own. By no means did it shy away but continually visited Lambert until 1913 as subsequent photos reveal (Figure 5.7).

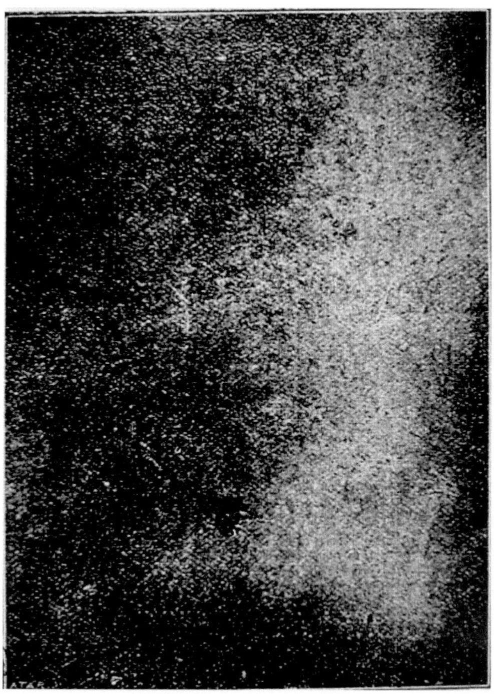

Figure 5.6 "Photographie prise dans l'état de dédoublement: Mme Lambert", Hector Durville, *Le Fantôme des Vivants: Anatomie et Physiologie de l'Âme. Recherches Expérimentales sur le Dédoublement des Corps de l'Homme* (Paris: Librairie de Magnétisme, 1909). (Author's collection and photograph.)

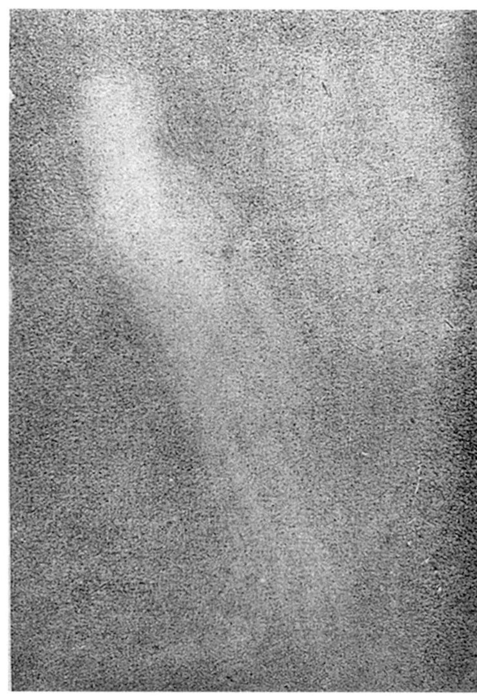

Figure 5.7 "Photographie prise dans l'état de dédoublement. Le sujet est invisible et à gauche se présente une vague forme humain, qui doit être le corps ètherique de celui-ci, Mme Lambert", Hector Durville, *Le Fantôme des Vivants: Anatomie et Physiologie de l'Âme. Recherches Expérimentales sur le Dédoublement des Corps de l'Homme* (Paris: Librairie de Magnétisme, 1909). (Author's collection and photograph.)

Situating itself about twenty centimetres from Lambert's left side, "it repeated", according to Durville, "like a shadow, all her movements and gestures".[84] Whenever she raised or lowered her arms, so did her phantom.[85] Rather than walking beside her, the phantom glided by her.[86] Animated by intense vibratory movements, Lambert found her phantasmatic double knocked loudly on tables, banged doors and created such intense energy and heat that it could be measured with a thermometer.[87] The power of the phantom to glide through walls and doors in front of the physical body or vertically behind it, covering a kilometre in fifteen seconds, appeared to be matched by its power to radiate white light in different forms.[88] Yet what appeared most striking to Durville was that "this invisible body carried with it the very principle of life, as well as will, intelligence, memory, consciousness, psychic sensibilities while the visible body does not possess any [of these] faculties".[89] In seeming to document how humans, once magnetized, could perpetually double with their etheric and astral bodies, these photographs appeared to demonstrate that human power was not confined to their physical body.

Amidst studies of these new energies, as indicated in Chapter 4, numerous French scientists exploring magnetism and what Albert Besnard and many others called *La force psychic,* conceptualized and visualized an invisible force as so dynamic and animating that it was able to transcend the material world. For Maurice Maeterlinck, this "prodigious force" was "the very fluid of life".[90] The possibility of its presence in Picasso's *Portrait of Kahnweiler* (Figure 5.1) alongside the seemingly supernatural radiance unleashed by x-rays, may be illuminated by a comparison of this portrait with the magnetic photography of *dédoublement* directed by Durville and the x-rays of Röntgen.

As if x-rayed by Röntgen, Kahnweiler's soft body tissue appears to have been penetrated by light in Picasso's painting while his bones seem to resist with such metallic paraphernalia as his fob-watch chain appearing as resilient to x-rays as the wedding ring on Röntgen's wife Anna's finger. Without any indication of a contour separating Kahnweiler's body from the surrounding space, like an x-ray there seems no longer an inside or outside, Kahnweiler's body seeming to penetrate Picasso's world as much as Picasso's world penetrates his body. The mysterious light that seems to flicker and vibrate within this world without any identifiable source may be identified with the radiance unleashed by x-rays and with *éther luminifère*, scientificized by Gustave Le Bon as able to penetrate and dematerialize all substances. This was also conceived by Madame Blavatsky as an "astral light" emanating from the "invisible universe".[91] Yet as these lights seem to glimmer in the space occupied by Kahnweiler's body, they may also pertain to phantasmatic energies and auratic luminosities rarely visible to humans but captured by photography, particularly paranormal photographic plates conceived by Rochas as having "a sensibility to capture, as one knows, thousands of stars invisible to our eyes".[92]

In Picasso's photograph of Kahnweiler (Figure 5.5), the light defines the contours of Kahnweiler's face, body and hand and locates his position in Picasso's studio. Yet in Picasso's portrait of his art dealer (Figure 5.1), there is no such clear distinction between Kahnweiler's anatomy and the space in which it was immersed. Instead a white inner light seems to weave its way through Kahnweiler's body while seeming to emanate from it and spiral around it in space, just as it does in Durville's photograph of Madame Lambert in the state of "dédoublement" (Figures 5.6 and 5.7). Since a halo of light seems to illuminate Kahnweiler's head and neck in Picasso's painting (Figure 5.1) while an oval of light mysteriously circumscribes his body, a model for Picasso's articulation of light and shade may have been drawn from the original negative photographic plate of Kahnweiler (Figure 5.8). As this arc of light in Picasso's painted portrait appears to sweep down Kahnweiler's head and flicker around the dark shaft where his charcoal suited torso seems to dissipate into darkness, Kahnweiler's clasped hands seem to become spot lit before this arc of light curves upward around the other side of his body to create the sensation of an aura.

The whites flecked across Picasso's picture plane and Durville's photographic planes may signify how hauntological light, through luminous vibrations, was able to shimmer and sparkle across space like Durville's description of the "jets of luminosity projected by the phantom", and the "force physique" illustrated in Chapter 4, captured by Besnard (Figure 4.6). For Durville and Rochas, these phantasmatic energies represented what Baraduc had termed in 1897, "la force vitale".[93] Yet with Kahnweiler appearing transparent, weightless, visible from all sides at once and auratically luminescent, Picasso seems to have also explored astral vision to capture him four dimensionally, with an aura, as defined by Annie Besant in *Man and His Bodies:*

> It is the man himself, manifest at once on the four planes of consciousness, and according to its development is his power of functioning on each; it is the aggregate of his bodies, of his vehicles of consciousness; in a phrase, it is the form-aspect of the man ... thus that we should regard it, and not as a mere ring or cloud surrounding him.[94]

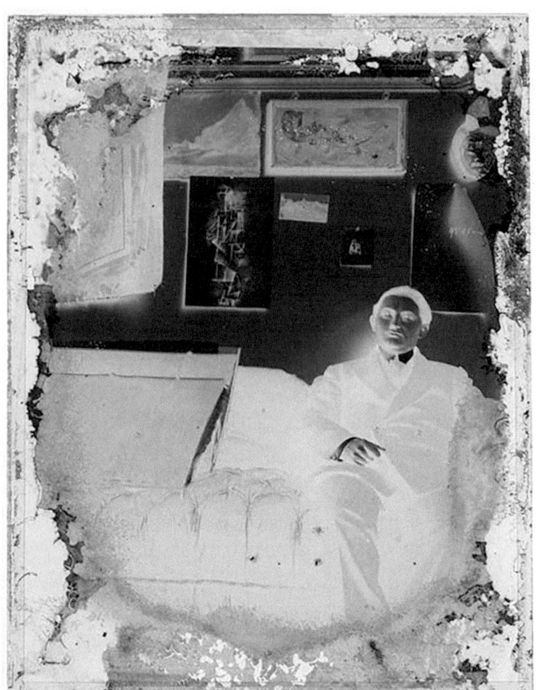

Figure 5.8 Pablo Picasso, *Daniel-Henry Kahnweiler in Pablo Picasso's Studio,* 11 Boulevard de Clichy, Paris, as photographed by Picasso, Autumn 1910; Succession Picasso – Gestion droits d'auteur; negative of Photograph Figure 5.5. (Musée Picasso de Paris ©RMN-Grand Palais. Public Domain.)

In French translation by 1902, Besant's auratic and astral vision theory was referenced by many, not just by Durville. Their relationship to *dédoublement* was also the subject of a three-part series in the Institut de Recherches Psychiques de France's first issues of *Le Monde Psychique*.[95] This concept also appeared integral to Bergson's *l'élan vital*. Life does not evolve mechanically and rationally, as Bergson stipulated in *L'Évolution créatice*. "Elle ne procède pas par association et addition d'éléments mais par dissociation et dédoublement".[96] Hence, as the body of Kahnweiler seems to dissolve into white light or uncanny luminescence, just as the body of Madame Lambert appeared to do in Durville's photographs (Figures 5.6 and 5.7), it may signify "la force vitale" of scientistic magnetists and occultists, as well as "l'élan vital" of Bergson within what he called a "métapsychique" reality.[97] As Bergson explained to the British Society for Psychical Research in his 1913 presidential speech, *Fantômes des vivants et recherche psychique:*

> There is, present and invisible, a certain metaphysic unconscious of itself – unconscious and therefore inconsistent, unconscious and therefore incapable of continually remodelling itself on observation and experience as every philosophy worthy of the name must do.[98]

Hence although Cubism has been historiographically located as an antithesis of Occultism, not only does it seem that Picasso was aware of this occultist revival but arguably became immersed within it as is also conveyed by his close relationship with Max Jacob and their collaboration on *Saint Matorel* and *Le Siège de Jérusalem*.

"An Art that Speaks a Secret Language": Max Jacob's and Picasso's Hermeticism, Mysticism and Occultist Cubism

No sooner had Picasso moved in with Jacob in 1901 on the Boulevard Voltaire than he had become absorbed in Jacob's occultist esotericism. During this time when Picasso was "cooped up with this part-time fortune-teller", as John Richardson puts it, Jacob taught him the rudiments of astrology, chiromancy and the Tarot.[99] From Jacob's appraisal of Picasso's creativity and palm-reading of his extraordinary lifeline, Jacob called Picasso "the Alchemist Prince", identified with Hermes Trismegistus, the creator of alchemy, hermetic and cabbalistic texts on magic and astrology, renowned for his priapic phallus.[100] Magic, mysticism and the occult held a fascination for Jacob, almost as much as they did for Picasso and the third member of their triumvirate: Apollinaire. They drew upon them as an hermetic *conjunctio oppositorum*, a means of reconciling opposites.[101] Despite Jacob's mystical apparition of Jesus in 1909 and conversion to Catholicism, his biographer, Jean Rousselot, maintains that "Jacob remained far more Occultist than Christian" while his poetry, like Picasso's painting, became increasingly hermetic.[102]

On returning to the Bateau Lavoir from the Bibliothèque Nationale on 28 October 1909, Jacob saw a vision of Jesus on one of his watercolours hanging on the wall of his tiny room situated below Picasso's studio at the Bateau Lavoir. Overwhelmed by its beauty, instantly he recalls falling on his knees.

> My eyes filled with tears … An ineffable feeling of well-being overcame me. I remained transfixed without understanding what was happening. Instantly I felt that I was becoming a man … a free man. Instantly also, as my eyes encountered the Ineffable Being, I felt divested of my human flesh, and two words simply chimed within me: To die, to be born.[103]

Six years later, he was baptized as a Catholic, Cyprien-Max Jacob, at the Couvent de Sien with Picasso as his god-father. Ten years later he became a mystic monk.[104] It was this mystical experience that proved the catalyst for his prose-poems, *Saint Matorel* and *Le Siège de Jérusalum*.

For Picasso it appears to have been the experience of the magic-making African masks and carvings in the Musée d'Ethnographie du Trocadéro that had affected him like an encounter with supernatural spirits.[105] The result was what Picasso purportedly called his first exorcism painting, *Les Demoiselles d'Avignon*, and his shift from painting harlequins and saltimbanques to Cubistic motifs. Yet after Jacob's apparition, Picasso's Cubist artworks became, according to Read, infused with "half hidden magic and a sense of impenetrable mystery".[106] For Jacob, this "impenetrable mystery" was integral to his immersion in Occultism and Hermeticism as a science of hidden wisdom that integrated the visible with the invisible forces of nature and the micro with macrocosmic as surmized by the Hermetic image and saying, "As above, so below". This Hermeticism was compounded by its association with the negative aesthetics of Neo-Symbolism and Néo-Mallarméism, as manifest by the literary journal, *La Phalange*, in which Picasso's Hermetic Cubism was referred to as *l'Art négatif*, as mentioned earlier. This art of negation entailed situating the subject within the visible and invisible experienced by all the senses over time and space but only with oblique rather than direct references to it through a Néo-Mallarméan process of connotation and evocation. "The work of art must be distant from the subject", Jacob explained, "That's why it must be situated".[107] This is illuminated by the first Cubist *livre d'artiste* commissioned and published by Kahnweiler and illustrated by Picasso: *Saint Matorel*.[108]

Although Derain had been initially chosen to illustrate *Saint Matorel*, he refused in no uncertain terms explaining "les préoccupations qui font l'objet de mon travail sont complètement absentes de ce livre".[109] Two years later he relented, illustrating Jacob's *Burlesque and Mystical Work of Saint Matorel*. Picasso's response was the opposite. To commemorate their hermetic relationship, Jacob added a telling dedication to Picasso when *Saint Matorel* was published in February 1911: "To Picasso, for all I know he knows, for all he knows I know".[110]

Consistent with Jacob's and Picasso's aesthetic of negation, there is no illustration of the main character. An employee of the Paris Metro, Victor Matorel straddles the character of a Hamlet "homunculus" and Jacob's alter ego, struggling between the forces of Hermeticism and Christianism. Opening with Matorel's confessions "revealing", in his words, "the secrets that God alone knows", the narrator of this Cubistic novel is Matorel's friend, Émile Cordier. A sweeper at the company, Cheiret et Cie, in the faubourg Saint-Antoine, who Matorel eventually converts, Cordier discloses that he shall reveal Matorel's adventures, both celestial and terrestrial.[111] "He speaks to me of the importance of earthly love, the cube and the circle", Cordier writes, "of armies of demons and angels ... He also speaks of his overwhelming hallucinations".[112] Yet by no means are these experiences consecutively sequenced in chronological order, as Jacob explains in his Prologue:

> Mais les formes sont immobiles et mobiles éternellement dans le ciel et il n'y a pas d'ordre chronologique pour Dieu ... Et nous sommes dans nos œuvres comme Jehovah dans les siennes. Il n'y a pas d'ordre chronologique pour nous.[113]

Instead, Jacob deploys Bergson's concept of time as duration fluidly oscillating between time present, past and future alongside spatial leaps between Matorel's adventures on earth and his celestial experiences narrated in mystical dialogues.[114] "Nothing functions as it is meant to", surmizes Antonio Rodriguez. "In this increasingly complexified plot, a continuous time frame is abandoned; logic transgressed, serious concerns laced with irony and the mystical juggled with the quotidian".[115]

To undertake the journey of his soul to God, Matorel has to pass through nine stages of purification. Like the nine concentric circles of Danté's *Inferno,* each contains their own earthly hells with Satanic temptations. In each stage, Matorel has then to negotiate the antithetical forces of God and Satan: Angels and demons; heaven and earth; hermetic knowledge and ardent sodomy. He also has to negotiate heterosexual temptation at the Moulin Rouge and most of all "the love of the infidel, Mademoiselle Léonie".[116] So formidable a temptress does she appear to be that two of Picasso's four illustrations are devoted to Mademoiselle Léonie. (Figure 5.9). Picasso had drawn her identity from, Jacob discerned, the acrobat performer at the Medrano Circus, who Picasso had encountered at the Taverne de l'Hermitage close to his new apartment, 11 boulevard de Clichy. After she became his model, Picasso had produced an Hermetic Cubist portrait of her in Spring 1910 entitled *Mademoiselle Léonie.*[117]

In *Saint Matorel*, Cordier narrates his encounter with Mademoiselle Léonie on the steps to the Metro Dauphine where she discusses her lovers while stipulating that Victor was the one she loved the most. Although Jacob never discloses how this apprentice laundress looks, nor does Picasso. In Picasso's illustration, Mademoiselle Léonie appears in fragments experienced over felt time and in different positions in

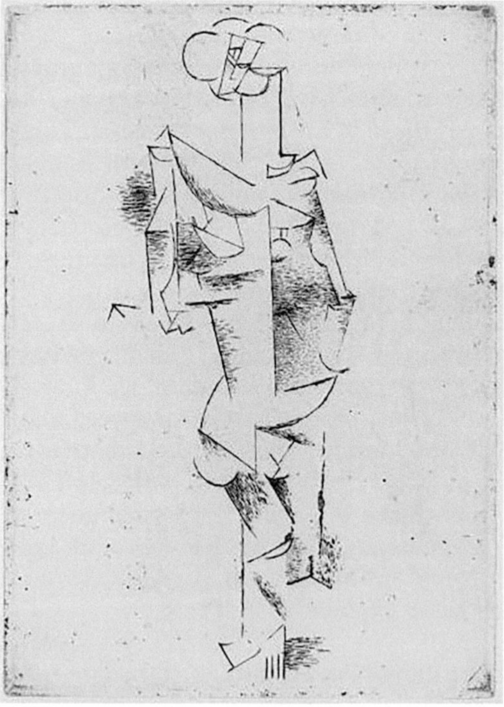

Figure 5.9 Pablo Picasso, *Mademoiselle Léonie*, 1910, etching; 7 15/16 × 5 9/16 inches: illustration, Figure 4; Max Jacob, *Saint Matorel* (Paris: Henry Kahnweiler Édition, 1911). (Author's collection and photograph.)

space as known through sensory memory.[118] Rounded forms in the upper part of her body may indicate the feel of her head with curling hair while the two parallel lines below may signify the cervical vertebrae of her long neck (Figure 5.9). The diagonal line shooting across the body may signify her collar bone above curvular lines to provide a hint of breasts while the downward diagonal meeting curvular lines may signify the feel of her torso billowing out at the pelvis with her flesh sweeping around her buttocks. Yet in keeping with Jacob's and Picasso's Hermeticism, the identity of these parts of Mademoiselle Léonie's body are only hinted at and conjectured relatively, not absolutely, through the interrelationship of Picasso's indexical signifiers. So unspecific are these indexical signifiers that this acrobat's anatomy, like that of Kahnweiler in Picasso's portrait of him, seems to become a body without organs. With Picasso's lines rarely meeting and left open in 'passages', Mademoiselle Léonie's corporeality seems to dissolve into phantasmatic luminescence as signified by the whiteness of Picasso's paper illustrated in Figure 5.9.

In the next chapter, Matorel begs the angel to purge him of all of his wicked and salacious memories of Mademoiselle Léonie. "Remove all obsessions with carnal love", he cries, "a room with a tiny iron bed for 100 francs on the Boulevard Barbès".[119] Although Jacob does not mention the chaise-longue that is meant to feature in Picasso's second etching of Mademoiselle Léonie, there are far more facets than in his previous image of Matorel's lover. While dark hatching rendered in multiple directions seems to suggest an abundance of body and pubic hair, it may also evoke the lovemaking of both Léonie and Victor on the iron bed that lingers in Matorel's fantasies. It may also invoke the homosexual experiences that Matorel, like Jacob, had pursued. "Must I admit that I have been a sodomite", Matorel reluctantly confesses, "without joy, it is true, but with ardour!"[120] In a pitiful state of remorse that Jacob calls "mystery psychosis", Matorel flees Paris for the Lazarist Convent in Barcelona, the subject of Picasso's final etching.[121]

At the convent, Matorel's immersion in Catholicism and mysticism is likened to Paul Verlaine's imprisonment after shooting his lover, Arthur Rimbaud, followed by Verlaine's reconversion to Catholicism and his prison writings as a "poète maudit" that eventually earned him the accolade, "prince of poets".[122] These poems Matorel "had read like a confessional", Verlaine being the first French poet that Jacob had introduced to Picasso when Jacob had envisioned himself as a 'poète maudit'.[123] In the convent Matorel experiences hallucinations during influenza in which he argues with the planetary angels and demons of Mars, Venus and Saturne, as well as Satan and the demon, Zazel, to signify his negotiation of the spheres of Heaven and attainment of Spirituality as Frère Manassé.[124] It is also in the convent that Matorel appears to have written spiritualist poetry and produced hallucinatory drawings although, in this burlesque mixture of clown and saint, how this happens is never elucidated. Instead the reader is told that on Matorel's death, his "truly hallucinating" drawings were sent to Apollinaire while his poetry went to Mademoiselle Léonie who passed it onto a publisher, who had it read at the Comédie Française. "How and why these things happen", Jacob's omniscient narrator concludes: "Ah! ... mystère ... mystère!"[125]

No less mysterious is Jacob's play, *Le Siège de Jerusalem: Grande tentation céleste de Saint Matorel* with its cast of forty including the Wolf of Saint Philip, Unicorn of Saint Thomas, Ram of Minerva, four angels and a choir plus Saint Matorel called Blanquetbleu, who is identified as dead.[126] Far from this mystery being clarified by the etchings commissioned from Picasso, it appears exacerbated by their highly tenuous relationship to the tests which Saint Matorel has to undertake in order to return to heaven – the greatest one entailing his entry into the besieged city of Jerusalem. Despite the few women in this almost male cast, two out of Picasso's three etchings are of women, possibly to evoke the Courtesan and the Lady who appears at the Adrianople Camp. The Courtesan, who constitutes one of the first tests, plays the temptress who endeavours to lure Saint Matorel repeatedly saying: "I shall do anything you want if you come with me".[127] That Picasso's nude woman with the guitar appears to be this alluring courtesan is signified by his relatively erotogenic signifiers discernible in Figure 5.10.

Following the darkness in which Picasso had shrouded his memories of the felt experience of Kahnweiler (Figure 5.1) and its correlation to the blackness of Picasso's photographic negative of his art dealer (Figure 5.8), Picasso appears to have heavily inked his hatching and cross-hatching in this etching to evoke the sensation in which this prostitute's naked body seemed to loom out of a nocturnal ambience like Durville's photographs of phantoms. Consistent with Picasso's other Cubist representations of women, the guitar indicated in the title is seemingly evoked by the bold sweeping curves on the upper left and lower right of the etching plate, fused with her flesh to signify how this prostitute's body may be played like a guitar and pierced through its sound hole. The 'demoiselles d'Avignon' gaze of her eye directly at Saint Matorel and the beholder, together with evocations of her rounded abdomen, bulging buttocks, voluptuous breasts with clearly delineated nipples and vulva, seem to position

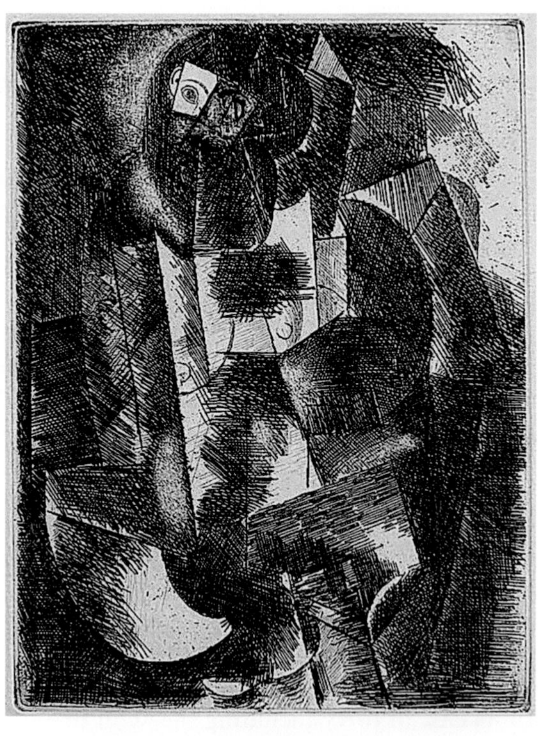

Figure 5.10 Pablo Picasso, *Femme nue à la guitare,* 1913, etching and drypoint, 15.7 × 11.6 cm; Max Jacob, *Le Siège de Jérusaleum*: *Grande tentation céleste de Saint Matorel* (Paris: Henry Kahnweiler Écition, 1914). (Author's collection and photograph.)

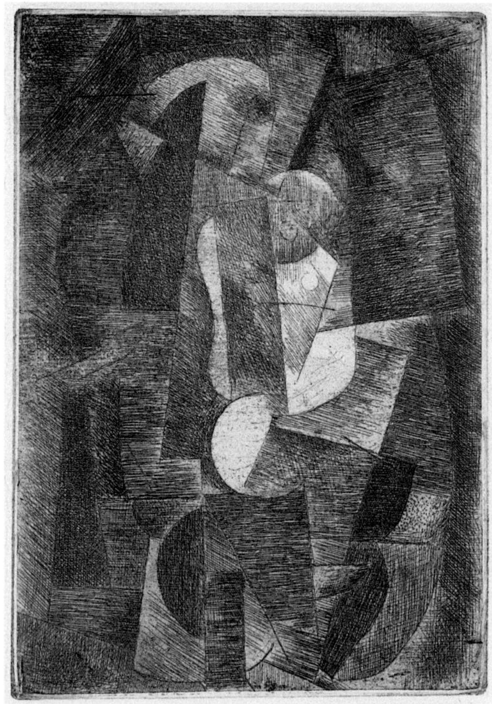

Figure 5.11 Pablo Picasso, *Femme,* 1913, etching and drypoint, 15.7 × 11.6 cm; Max Jacob, *Le Siège de Jérusalem: Grande tentation céleste de Saint Matorel* (Paris: Henry Kahnweiler Édition, 1914). (Author's collection and photograph.)

this Cubistic figure as a prostitute ready for business. When located within Jacob's text, this appears to be the moment when this seductress lures Saint Matorel before dramatically immolating: Literally bursting into flames.

By contrast, the "Lady" introduced in the second act with bishops and cardinals is the one whom the prince of Adrianople wishes to seduce due to the whiteness of her neck. This may be signified in Figure 5.11 by Picasso's light planes and white circular form around the centre of what appears to be a female body with faintly circumscribed breasts and a hint of nipples. Played off against the relatively explicit erotogenic signifiers of the prostitute looming out of darkness, it is the prostitute's body that appears to be pierced with vitalist rays of phantasmatic luminescence. As the materiality of Picasso's prostitute appears to dissolve into bursts and flickers of light and darkness, just like those to be found in the magnetism photographs by Durville (Figures 5.6 and 5.7), she appears to emanate out of a fusion of Bergsonian felt time within phantasmatic luminescent space: Both *l'élan vital* and *la force vitale*. Capturing the mystical experience of Jacob's visions, Picasso's Cubistic etchings then appear as celestial and as mysterious as Jacob's main character's spiritual conversion.

Hence while Cubism has been posited as an art that drew upon modernist experiences of the quotidian, by no means did Picasso's explorations of Hermetic Cubism exclude Bergsonism, hermeticism, mysticism or occultism. At this potent Neo-Symbolist, Occultist and Modernist juncture before the ruptures wrought by the unprecedented destructions and decimations of the First World War and the publication of Einstein's Theory of Relativity thereafter, it seemed possible to be an Anarchist, Bergsonist, hermeticist, mysticist, occultist, parapsychologist and Cubist in painting, photography and etching, as much as in prose. In the midst of Anarchisms, Bergsonisms, hermeticism, magnetism, mysticism, Neo-Symbolism and Occultism, the imposition of World Standard Time, Taylorism and the ergonomic reconception of the body as *"le moteur humain"* was spurned by Picasso and other Cubists. The positivist empiricism and scientistic realism inherent in Bonnat's *Portrait of the President of France, Armand Fallières* (Figure 5.2) and even Blanche's *Portrait of Henri Bergson* (Figure 5.3) were also rebuffed. Within the network of interrelationships spawned by *la bande à Picasso,* particularly their relationship to Bergsonisms, it was possible to conceive of a very different kind of portrait: This was a durational one that Picasso was able to draw from his sensory memories of his art dealer over 'felt time', not World Standard Time. Rather than depict Kahnweiler's surface appearance, Picasso seemed to capture Kahnweiler's "psychic states" and the "inwardness of his life", amidst the perpetual flow of becoming within Bergson's *l'élan vital* (Figure 5.1). Drawing upon photography, not just Picasso's explorations of capturing Cubistic architecture and Kahnweiler with his own camera but the scientistic phantasmatic photography developed by Baradac, Durville and Rochas, amongst others mentioned in Chapter 4, Picasso seemed to be able to fuse Bergson's *l'élan vital* with Baraduc's, Durville's and Rochas' concepts of *la force vitale.* Following their revelations of extensive experimentation with scientistic phantasmatic photography and Picasso's exploration of photographic negatives (Figure 5.8), Picasso was also able to capture auratic luminosities in his portrait of Kahnweiler and the dissolution of Kahnweiler's body without organs within nocturnal space. Finally, in portraying the mystical experience of Jacob's visions in *Saint Matorel* and *Le Siège de Jérusalum*, Picasso's Cubistic etchings appear as celestial and mysterious as Jacob's main character's spiritual conversion (Figures 5.9–5.11). As their phantasmatic luminescence seems to flicker uncannily through fragments of bodies, shards of objects and facets of unidentifiable forms, Picasso's etchings become endowed with a mysterious vitalism that can also be discerned in his Hermetic Cubist paintings, epitomized by his *Portrait of Kahnweiler* (Figure 5.1). Since this vitalism appears comparable to both *l'élan vital* and *la force vitale* identified in new scientistic research and apparent in their photography, it is then possible to deduce that this artwork and his Hermetic Cubist etchings constituted an Occultist Cubism created by Vitalist Picasso.

Notes

1 Gilles Deleuze, *Negotiations, 1972-1990,* trans. Martin Joughin (New York, NY: Columbia University Press, 1990; 1997) 143.

2 Fae Brauer, "Becoming Simian: Darwin, Picasso, Nolan and Creative Evolution", *Studies in History and Philosophy of Science,* 2023, forthcoming; Fae Brauer, "Becoming Simian: Devolution as Evolution in Transformist Modernism", *Picturing Evolution and Extinction: Regeneration and Degeneration in Modern Visual Culture,* eds. Fae Brauer and Serena Keshavjee (Lady Stephenson Library, Newcastle upon Tyne: Cambridge Scholars Publishing, 2015) 127–156.

3 Linda Dalrymple Henderson, "X Rays and the Quest for Invisible Reality in the Art of Kupka, Duchamp, and the Cubists", *Art Journal,* 47: 44 (1988): 323–340.

4 Gilles Deleuze, "The Logic of Sense," ed. Constantin V. Boundos; trans. Mark Lester with Charles Stivale (New York, NY: Columbia University Press, 1990) 87.

5 Akira Mizuta Lippit, "*From* Modes of Avisuality: Psychoanalysis – X-ray – Cinema", Chapter 15, *The Spectralities Reader: Ghosts and Haunting in Contemporary Cultural Theory,* eds. Maria del Pilar Blanco and Esther Peeren (London: Bloomsbury Publishing, 2013) 258.
6 Louis Vernéde, "Le Bergsonisme ou une Philosophie de la Mobilité", *La Phalange,* No. 72, 20 June 1912.
7 André Salmon, "Le Salon des Indépendants", *Paris Journal* (18 March 1910) pp. 4–5: *Ces cubistes ... se proclament volontiers anarchistes de l'art; venus après les journées d'émeute et de massacre, ils ne sont que des jacobins, ils académiseront à leur tour tyranniquement*; Paul Adam, "Avant la guerre sociale", *Paris-Journal* (2 May 1910) 1; Fay Brauer, *L'Art révolutionnaire : The Artist as Alien. The Discourses of Cubism, Modern Painting and Academicism in the Radical Republic,* PhD (University of London: The Courtauld Institute of Art, 1997) 236–316; also refer Patricia Leighten, *Reordering the Universive: Picasso and Anarchism, 1897-1914* (Princeton, NJ: Princeton University Press, 1989) 96–120.
8 *La Phalange,* No. 62 (20 August 1911).
9 *La Phalange,* No. 89 (15 November 1913).
10 Gaston Sauvebois, "Le Cubisme," *Le Gil Blas* (5 October 1911) 1.
11 André Salmon, *La jeune peinture française* (Paris: Société des Trente, Albert Messein, 1912) 27.
12 Fernande Olivier, *Picasso et ses amis* (1933: Librairie Stock, Delamain et Boutelleau, 1933) 161.
13 Peter Read, *Picasso and Apollinaire: The Persistence of Memory* (Berkeley/Los Angeles: University of California Press, 2008) 17–20.
14 Salmon, *La jeune peinture française*; also refer *André Salmon on French Modern Art,* trans.; annot. Beth Gersh-Nešić (Cambridge: Cambridge University Press, 2006).
15 Salmon, *La jeune peinture française,* 27.
16 Henri Bergson, *Essai sur les données immédiates de la conscience,* Thèse pour le Doctorat (Paris: Ancienne Librairie Germer-Ballière et Cie; Félix Alcan, Éditeur, 1889); *Time and Free Will: An Essay on the Immediate Data of Consciousness* trans. F. L. Pogson (London: George Allen and Unwin, 1910) 7–74; also refer Read, *Picasso and Apollinaire,* 56–60.
17 Pascal Rousseau, "The Magic of Images: Hallucination and Magnetic Reverie in the Work of Henri Rousseau," *Henri Rousseau: Jungles in Paris* (London: Tate Publishing, 2005) 192.
18 Michel Frizot, "The All-Powerful Eye: The Forms of the Invisible", *A New History of Photography,* ed. Michel Frizot; trans. Susan Bennett el al (Cologne: Könemann, 1998) 281.
19 Hector Durville, *Le Magnétisme considéré comme agent lumineux. Extrait du "Traité expérimental de magnétisme"* (Paris 1896); Hector Durville, *Le Fantôme des Vivants, anatomie et physiologie de l'âme. Recherches expérimentales sur le dédoublement des corps de l'homme* (Paris 1909).
20 Stephen Kern, *The Culture of Time and Space, 1880-1918* (Cambridge, MA; London, England: Harvard University Press, 1983; 2003) 10–11.
21 Bergson, *Essai sur les données immédiates de la conscience,* 1889.
22 Kern, *The Culture of Time and Space,* 11.
23 Fae Brauer, "Representing 'Le Moteur Humain': Chronometry, Chronophotography, 'The Art of Work' and the 'Taylored Body'", *Visual Resources: An International Journal of Documentation*, XIX: 2 (June 2003): 83–106.
24 Joseph Conrad, *The Secret Agent: A Simple Tale* (London: Methuen & Co., 1907) 17.
25 George Beard, *American Nervousness, Its Causes and Consequences. A Supplement to Nervous Exhaustion (Neurasthenia)*, (New York, NY: G. P. Putnam Sons, 1881).
26 Fae Brauer, "Capturing Unconsciousness: The New Psychology, Hypnosis and the Culture of Hysteria", *A Companion to Nineteenth Century Art,* ed. Michelle Facos (London et al: John Wiley & Sons, Inc., 2019) 243–262.
27 Bergson, *Essai sur les données immédiates de la conscience,* 485.

28 Henri Bergson, "The Intensity of Psychic States", *Time and Free Will: An Essay on the Immediate Data of Consciousness; Essai sur les données immédiates de la conscience,* 1889; trans. F. L. Pogson (London: George Allen & Unwin, 1910) 7-8. The translator acknowledges, p. x, that by this date, Bergson's *Essai* was in its seventh edition due to "the professional philosophers", as much as to "the ordinary cultivated public".
29 Henri Bergson, *The Creative Mind: An Introduction to Metaphysics*, as translated from Bergson's final published book, *La Pensée et le mouvant*, by Mabelle L. Andison (New York: Philosophical Library, 1946).
30 Bergson, "The Intensity of Psychic States", 21.
31 Bergson, "The Intensity of Psychic States", 17.
32 Bergson, "The Intensity of Psychic States", 15.
33 Bergson, "The Intensity of Psychic States", 8.
34 Henri Bergson, *L'évolution créatrice* (Paris: 1907; Presses Universitaires de France/ Quadrige, 1941) 272.
35 Alain Badiou, "Of Life as a Name of Being, or Deleuze's Vitalist Ontology", *Gilles Deleuze. Immanence et Vie,* ed. E. Alliez et al (Paris: Presse Universitaire de France, 1998).
36 Harry Paul, "The Debate over the Bankruptcy of Science in 1895", *French Historical Studies*, 5: 3, (Spring, 1968): 299–327, including Renan's vehement opposition to Occultism. For more on Bonnat's portraits, refer Fae Brauer, *Rivals and Conspirators: The Paris Salons and the Modern Art Centre* (Newcastle upon Tyne: Cambridge Scholars Publishing, 2013) 164–165; 218.
37 Mark Antliff, *Inventing Bergson: Cultural Politics and the Parisian Avant-Garde* (Princeton University Press, 1993).
38 Umberto Brunelleschi, *L'Assiette au beurre,* 1902.
39 Musée d'Orsay Documentation: Bonnat 1833-1922, *Gazette des Beaux-Arts* (1922): 12.
40 Paul Atkinson, *Henri Bergson and Visual Culture: A Philosophy for a New Aesthetic* (London, etc.: Bloomsbury Academic, 2021) 39.
41 A. Ferrier: 'Jacques-Emile Blanche, peintre et mémorialiste', *L'Oeil*, viii (1962): 108.
42 Brauer, *Rivals and Conspirators: The Paris Salons,* 107–138.
43 Brauer, *Rivals and Conspirators: The Paris Salons,* 30–31, 164, 306.
44 Pierre Daix and Joan Rosselet, *Picasso, The Cubist Years, 1907-1916: A Catalogue Raisonée of the Paintings and Related Works* (London: Thames and Hudson, 1979) 253, Plate 338.
45 Gary Tinterow, "Vollard and Picasso", *Cézanne to Picasso: Ambroise Vollard, Patron of the Avante Garde* (New York: Metropolitan Museum of Art, 2006) 111–112: *On the eve of the artist's departure for Barcelona and Horta de Ebro, Vollard gave Picasso 2,200 francs; following his return in September, he gave him 1,000 francs in October and another 1,000 in November.*
46 Tinterow, "Vollard and Picasso", 111.
47 Tinterow, "Vollard and Picasso", 111.
48 Tinterow, "Vollard and Picasso", 111.
49 Brauer, *Rivals and Conspirators,* 333–345.
50 Fae Brauer, "Dealing with Cubism: Daniel-Henry Kahnweiler's Perilous Internationalism", *Dealing Art on Both Sides of the Atlantic, 1860-1940,* ed. Lynn Catterton (Leiden/Boston/Tokyo: Brill International Publishing, 2017) 133–134.
51 François Garnier, *Correspondance de Max Jacob,* Tome I, dirigée par André Brissaud (Paris: Éditions de Paris, 1953), 40: "I would like to publish books illustrated by my painters. Apollinaire will be the first, Derain will illustrate *L'Enchanteur Pourissant.* Have you something?"
52 Roland Penrose, "Picasso's Portrait of Kahnweiler", *The Burlington Magazine,* "Modern Art (1908-25)", 116: 852 (March 1974): 129.
53 M. Merleau-Ponty, *Phénoménologie de la perception* (Paris: Librairie Gallimard, 1945).
54 Henri Bergson, *Matière et mémoire : Essai sur la relation du corps à l'esprit* (1896); *Matter and Memory: Essay on the Relation of Body and Spirit,* trans. Nancy Margaret Paul and W. Scott Palmer (London: George Allen and Unwin, 1911) 33.
55 Bergson, "Of the Survival of Images, Memory and Mind", *Matter and Memory,* 170.
56 Gilles Deleuze and Félix Guattari, *Anti-Oedipus,* trans. Robert Hurley, Mark Seem and Helen R. Lane (London and New York: Continuum, 2004).

57 Penrose, "Picasso's Portrait of Kahnweiler", 129.
58 Albert Gleizes and Jean Metzinger, *Du Cubisme* (Paris: Eugène Figuière, Éditeurs, 1912).
59 Marcel Duchamp, letter to Louise and Walter Arensberg, 28 January 1951; Archives of the Francis Bacon Foundation, Claremont, California, quoted by Arturo Schwarz, *Marcel Duchamp* (New York: Harry No. Abrams, Inc., Publishers, 1975) iii.
60 Patrick Jones, "Strange Distance: Bergson and Symbolism", 'L'anomalie en question', *TRANS – Revue de littérature génerale et compare,* No. 26 (2021): 3.
61 Bergson, *Matter and Memory,* 208-209.
62 Bergson, *Creative Evolution,* Chapter 1.
63 Maurice Blanchot, "Bergson and Symbolism", *Yale French Studies,* No. 4 (1949) 63.
64 Bergson, *Time and Free Will,* 73.
65 Bergson, *Time and Free Will,* 164.
66 Henri Bergson, « Le rêve », « "Fantômes de vivants" et "recherche psychique" » (Paris: Presses Universitaires de France, 2013).
67 Albert Gleizes and Jean Metzinger, *Cubism* (1912); Edward F. Fry, *Cubism* (London: Thames and Hudson, 1978) 108.
68 *Paris-Journal,* 19 May 1905, as quoted by Arthur L. Miller, *Einstein, Picasso: Space, Time and the Beauty That Causes Havoc* (London: Hachette, 2008) 40.
69 For an exploration of these relationships at this time to Duchamp, refer Fae Brauer, "Scientistic Magnetism and Hauntological Metarealism: The Phantasmatic Doubles of Duchamp and Durville", *Realisms of the Avant-Garde,* eds. David Ayers et al (Berlin/Boston: Walter de Gruyter, 2020) 42–75.
70 Joséphin Péladan, "Le Radium et les miracles, *La Nouvelle revue* (XXVI, 1 January 1904) 35-47; "Le Radium et l'hyperphysique", *Le Mercure de France* (50, Bo. 174, June 1904) 608–637.
71 Jason Ã. Josephson Storm, *The Myth of Disenchantment: Magic, Modernity and the Birth of the Human Sciences* (Chicago and London: The University of Chicago Press, 2017) 2.
72 Josephson Storm, *The Myth of Disenchantment,* 1.
73 Susan Quinn, *Marie Curie: A Life* (Plunkett Lake Press, 2019) 66.
74 Hector Durville, *Traité expérimental de magnétisme. Cours professé à l'École pratique de magnétisme et de massage* (Paris 1895); Hector Durville, *Le Magnétisme humain considéré comme agent physique, mémoire lu au congrès magnétique international* (Paris 1890); Hector Durville, *Lois physiques du magnétisme. Polarité humaine. Traité expérimental et thérapeutique de magnétisme. Cours professé à la clinique du magnétisme en 1885–1886* (Paris 1896).
75 *Congrès Magnétique International pour l'étude des applications du magnétisme humain, Séance du lundi, du 21 octobre 1889* (Paris: 1889).
76 Durville, *Le Magnétisme considéré comme agent*, 1896.
77 Durville, *Traité Expérimental du Magnétisme* (1890); *Magnétisme Personnel ou Psychique* (Paris: Librairie du Magnétisme, 1890; 1895; 1896); for more on Durville's publications and practices, refer Fae Brauer, "Magnetic Modernism. František Kupka's Mesmeric Abstraction and Anarcho-Cosmic Utopia", *Utopia. The Avant-Garde, Modernism and (Im)possible Life*, eds. David Ayers et al (Berlin/Boston: De Gruyter, 2015) 135–163.
78 Hector Durville, *Magnétisme personnel. Éducation de la pensée. Développement de la Volonté. Pour être Heureuse, Fort, Bien Portant et Réussir en tout* (Paris: Librairie du Magnétisme, 1899); refer illustrations: "Le Rayonnement psychique d'échange".
79 Durville, *Le Fantôme des Vivants*, 275.
80 Annie Besant, *L'Homme et ses Corps*, translation de l'anglais par F. B. (Paris: Publications théosophiques, 1902); Durville, *Le Fantôme des Vivants, 27–36.*
81 The youngest son of Félix Gaspard Tournachon, aka Nadar, Paul Nadar (1856–1939) successfully ran his father's third studio, 52 rue d'Anjou, and achieved renown for his aerial photography from hot-air balloons, use of artificial lighting, animation of still pictures and photo reportage.
82 Durville, *Le Fantôme des Vivants*, 167–169. Wilfried-René Chettéoui also points out Charles Lancelin's participation in these experiments and how "diverses phases d'extériorisation du 'double' ont été décrites par les sujets eux-mêmes", *La nouvelle parapsychologie. Une expérience métaphysique* (Paris 1993) 50.

83 Durville, *Le Fantôme des Vivants*, 167–169.
84 Durville, *Le Fantôme des Vivants*, 180.
85 Durville, *Le Fantôme des Vivants*, 183.
86 Durville, *Le Fantôme des Vivants*, 184.
87 Durville, *Le Fantôme des Vivants*, 195.
88 Léon Lefranc, "Le Corps Astral du Vivant. Forme – Matière – Couleur", in: *Le Monde Psychique,* 1911, 70–81.
89 Durville, *Le Fantôme des Vivants*, 353.
90 Maurice Maeterlinck, "Luck", *The Buried Temple: Works of Maeterlinck* (1910; 2015) 10.
91 Helena Petrovna Blavatsky, *Isis Unveiled: A Master-Key to the Mysteries of Ancient and Modern Science and Theology* (New York: J. W. Bouton, 1877); Gustave Le Bon, *L'Évolution de la matière* (Paris: Flammarion, 1905); Linda Dalrymple Henderson, "L'éther de l'espace, medium de l'art, de la science et de l'occultisme", *Repenser le medium* (Dijon: Les Presses de Réel, 2021) 71–103.
92 Lt.-Colonel de Rochas d'Aiglun, *Les Sentiments, La Musique et Le Geste* (Grenoble: Librairie Dauphinoise, 1900) 269; Brauer, *Scientistic Magnetism and Hauntological Metarealism,* 52.
93 Hippolyte Baraduc, *La Force vitale. Extrait de la Chronique Médicale,* 15 avril et 1er mai 1897 (Clermont (Oise): Imprimerie Daix Frère, 1897); *La Force vitale: Notre corps vital fluidique, sa formule biométrique* (Paris: P. Ollendorff, 1897).
94 Annie Besant, *Man and His Bodies* (London: Theosophical Publishing Society, 1896) 85.
95 Charles Lancelin, "Méthode de Dédoublement personnel", *Le Monde Physique : Organe Mensuel de "l'Institut de Recherches Psychiques de France" pour l'étude expérimentale des Phénomènes spirites, magnétiques, hypnotiques et occultists,* 1 (1 January 1911): 14–19; also refer Léon Lefranc, "Recherches Expérimentales sur le deuxième corps invisible du l'homme vivant (Le corps astral) à l'aide de magnétisme", 3 (1 March 1911): 8–13; Léon Lefranc, "Le Corps Astral du Vivant", Deuxième Partie, Chapître III, Essai du physiologie du corps Astral du vivant. FORME – MATIERE – COULEUR," 3 (1 March 1911): 72–80.
96 Bergson, *L'Évolution créatrice,* 1907; 1941, 90.
97 The entry for this painting on the Centre Pompidou website for their 2015 exhibition, *Marcel Duchamp, La Peinture, même*, refers to these white shapes as "des halos blancs et ocres [qui] évoquent un autre monde".
98 Henri Bergson, "Phantasms of the Living" and "Psychical Research": Presidential Address to the Society for Psychical Research, London, 23 May 1913; *Mind-Energy,* trans. H. Wildon Carr, ed. Keith Ansell Pearson and Michael Kolkman (Basingstoke: Palgrave Macmillan, 2007) 77.
99 John Richardson, with Marilyn McCully, *A Life of Picasso: Volume I: 1881-1906,* (New York, NY: Random House, 1991) 270.
100 Richardson, *A Life of Picasso,* 274. Roland Penrose likened Picasso's studio to an alchemist's den.
101 Read, *Picasso and Apollinaire*, 28, points out that Apollinaire's library contained numerous books on occultism and demonology.
102 Jean Rousselot, *Max Jacob au sérieux essai* (Charlieu: La Bartavelle, 1994).
103 Max Jacob, *Récit de ma conversion,* 1951; Garnier, *Correspondance de Max Jacob,* 37–39.
104 Jacob moved to the monastery, Saint-Benoît-sur-Loire, where his writing combined Heaven and Hell, Angels and Demons.
105 André Malraux, *La tête d'obsidienne* (Paris: Gallimard, 1974), cited in *Propos sur l'art : Pablo Picasso* (Paris: Gallimard, 1998) 138.
106 Read, *Picasso and Apollinaire*, 20.
107 Annette Thau, "The Esthetic Reflections of Max Jacob", *The French Review,* XLV: 4 (March 1972): 803.
108 Max Jacob, *Saint Matorel; illustré d'eau-fortes par Pablo Picasso* (Paris: Henry Kahnweiler, Éditeur, 1911).

109 Letter, Derain to Kahnweiler, 1909; Garnier, *Correspondance de Max Jacob,* 40.
110 Jacob, *Saint Matorel,* unpaginated.
111 Jacob, *Saint Matorel:* Cordier also promises to reveal "the sad beauties of Matorel's soul".
112 Jacob, *Saint Matorel.*
113 Jacob, *Saint Matorel: Prologue, Le Cycle Matorel.*
114 These dialogues are on wisdom, sexual passion and the epistemological question of man as animal, magician or saint.
115 *Max Jacob : Oeuvres,* Antonio Rodriguez (Paris: Quarto, Gallimard, 2012).
116 Jacob, *Saint Matorel,* Chapître II: An apprentice washer woman, Léonie lives with her mother and sister.
117 Joan Rosselet, "Mademoiselle Léonie", plate 340: Daix, *Picasso: The Cubist Years 1907-1916,* 254.
118 Jacob, *Saint Matorel:* Amidst non-chronological time and discontinuities of space and place, the narrative leaps from this encounter to Matorel's life in the seminary by which time Léonie had married, divorced, remarried and became a dancer at the Eldorado.
119 Jacob, *Saint Matorel.*
120 Jacob, *Saint Matorel: ... j'ai sodomite sans joie mais avec ardeur.*
121 Jacob, *Saint Matorel: ... psychose mystère.*
122 Verlaine is thought to have written some of his most powerful poetry in prison, as exemplified by his *Romances sans paroles;* refer *Paul Verlaine: A Bilingual Selection of his Verse,* trans. Samuel N. Rosenberg; ed. Nicolas Valazza (Penn State University Press, 2019; 2021).
123 Jacob, *Saint Matorel*; Judith Morganroth Schneider, "Max Jacob on Poetry", *The Modern Language Review,* 69: 2 (April 1974): 290–296.
124 Jacob, *Saint Matorel: Je crois que notre Eternité est en raison de notre Spiritualité.*
125 Jacob, *Saint Matorel: Comment et pourquoi ces choses arrivent? Ah mystère ... mystère!*
126 Max Jacob, *Le Siège de Jérusalem: Grande tentation céleste de Saint Matorel* (Paris: Eugène Delatre; Henry Kahnweiler, Éditeur, 1913); unpaginated.
127 Jacob, *Le Siège de Jérusalem: Je ferai tout ce que tu veux si tu viens avec moi.*

6 Chromatic Futurism

Vitalizing Painting, Sculpture, Music and Life's Energies

David S. Mather

> It is necessary to respond seriously [to Cubism] and to return, enriched as we are by French contact, to our great strength: The wild joy of color!
>
> – Umberto Boccioni, July 1913[1]

From its initial public reception, Futurist painting was haunted by the specter of cinema. For an artistic movement typically associated with modern technological and scientific advances, it may be surprising for some readers to learn that the Futurist painter and main theoretician of Futurism, Umberto Boccioni, strongly resisted the encroachment by mechanical reproduction into traditional artistic practices. Primarily, his resistance came in response to art critics who likened the physical motion depicted in early Futurist paintings to the moving pictures of film.[2] While this characterization may sound mild or even complimentary to contemporary ears, it amounted to a harsh condemnation at the time that connoted their having a lifeless appearance. As a key instance of this prejudice, Maxim Gorky noted in 1896, after viewing a film for the first time, that the colorless, silent images presented "a life devoid of words and shorn of the living spectrum of colors, a grey, silent, bleak and dismal life."[3] To Gorky, the filmic apparatus turned the natural world into grey ashes, effectively pulverizing it "into fragments and into dust."[4]

In a different context, the French vitalist philosopher Henri Bergson portrayed film with a similarly unflattering view that disparaged the literal and metaphorical effects of a mechanized visual perception.[5] He worked from the premise that mechanistic thinking and scientific analysis – and, by extension, the precise visual analysis of cinema – extracted and discarded the vital essences from any living processes. When reduced to analytical instances, or "stills," the continuous movement of life was destroyed.[6] For Bergson, vibrant coloration, by contrast, signaled irreducible qualities of natural experience, which photographic and film cameras readily disregarded when creating their ghostly, grey imagery. In undated notes (written before World War I), Boccioni carefully transcribed some of the main points of Bergson's critique of mechanistic thinking, including important references to the presumed dichotomy between mechanistic processes and the indivisible movement of life.[7] By applying Bergson's ideas to his artistic theories and practices, Boccioni was contesting the negative reception of Futurist visual works and simultaneously proposing they could counter the supposed diminishment of life's vitality by mechanical reproduction.[8] One of the main points and purposes of his complicated written and artistic

DOI: 10.4324/9781003045595-9

responses to this particular technical and discursive challenge would be to assert Futurist vitality vis-à-vis color.

In a text from 1914, Boccioni strongly dismissed any association of Futurist painting with the automatic processes of film and photography: "We, who are accused of seeing things outwardly, of cinematography, are the only ones working our way toward a definitive [construction,] which is an intuitive evolving creation."[9] Their futurist artworks did not simply trace the paths of figures or objects in motion, he claimed, but rather they revealed more emotional and intuitive trajectories. Boccioni's refutation of this negative association with film also supported his underlying argument in that text: It was Cubism, not Futurism, that embraced analytical and mechanical principles. As distinct from Brauer's exploration in Chapter 5 of the multiple dimensions of Bergsonian Vitalism captured by Picasso's Cubism, Boccioni claimed, more specifically, that Picasso's pictures were "the result of an impassive scientific calibration."[10] According to this logic, Boccioni rejected Cubism and, by extension, cinema, for the same reason that some critics mocked Futurism, which was due to its perceived lifelessness. He noted, "The analysis of an object is always made at the expense of the object: That is, by *killing* it."[11] Not merely analytical, Cubism seemed to this Futurist to be practically criminal: "Picasso, ... by putting a stop to the life in the object, kills the *emotion*."[12]

A few months earlier, the poet and critic Guillaume Apollinaire had praised Picasso for having the skill of a surgeon dissecting a corpse, whereas Boccioni seized on this concept of dispassionate visual analysis to portray Cubist works as cadaverous and drained of life.[13] In addition, Picasso's muted palette during what Brauer calls in Chapter 5 his *Occultist Cubism* provided additional grounds for Boccioni to discount the Cubists' contributions, since they "have emphasized pure chiaroscuro, seasoning it with French grays and cold tones ... lacking all vitality."[14] Fortunately, alongside his criticism of Picasso's formal reduction, Boccioni argued affirmatively in favor of Futurism's chromatic vitality, which he regarded as mirroring aspects of Bergson's philosophy. Initiated in response to the double threat of cinema and Cubism, and then developing in various direction over nearly a decade, Futurist chromatic principles and practices covered a wide spectrum of connotations, which will be discussed in this chapter. These include associations with emotional and intuitive intensities, sexual virility, real and imagined trajectories, mass urban populations, musical and cinematic analogies, an interest in prismatic variations, connotations of patriotism, and even vitalist affirmations of death and destruction.

From the start, Futurist chromatism represented the continuation of an historical shift away from Divisionist theories of color in painting and the application of adjacent brushstrokes with contrasting, or complementary, colors.[15] In France during the 1880s, Michel Chevreul's influential principles of chromatic harmony, which entailed carefully calculated juxtapositions of complementary colors, help sparked a chromatic reconceptualization of painting among many French Postimpressionists.[16] While the Italian school of Divisionism adopted many features of this stringent French chromatic system, the Futurists, like the Fauvists, pursued alternatives to what was perceived as the stifling precision of both naturalistic shading and the Divisionist system of diffuse, contrasting strokes.[17] For artists as different as Henri Matisse and Boccioni, a concern with maximizing chromatic intensity led each artist to emphasize (in very distinct ways) the unmitigated intensity of pigments by expanding the picture area of colors beyond single adjacent strokes.[18] In "The Technical Manifesto of Futurist Painting," published in March 1910 before any Futurist visual works were exhibited, the five signatories declared their love for a polychromatic world that avoided muddy hues and revealed a chromatic fount of human passion:

> It will be readily admitted that brown tints have never coursed beneath our skin; it will be discovered that yellow shines forth in our flesh, that red blazes, and that green, blue and violet dance upon it with untold charms, voluptuous and caressing.[19]

Their credo of vitalistic color attributed evanescent qualities to the human visage as well:

> How is it possible still to see the human face pink, now that our life ... has multiplied our perceptions as colorists? The human face is yellow, red, green, blue, violet. The pallor of a woman gazing in a jeweller's window is more intensely iridescent than the prismatic fires of the jewels that fascinate her like a lark.[20]

No longer composed of studied variations on a single dominant hue, a woman's face could gleam with the bright colors of radiant jewels, whereas her passionate gaze implicitly links her material desires with the objects on display.

As if to illustrate their manifesto's poetic image of strong chromatic impulses, Boccioni's 1911 painting *Modern Idol* (Figure 6.1) depicts a woman looking into a

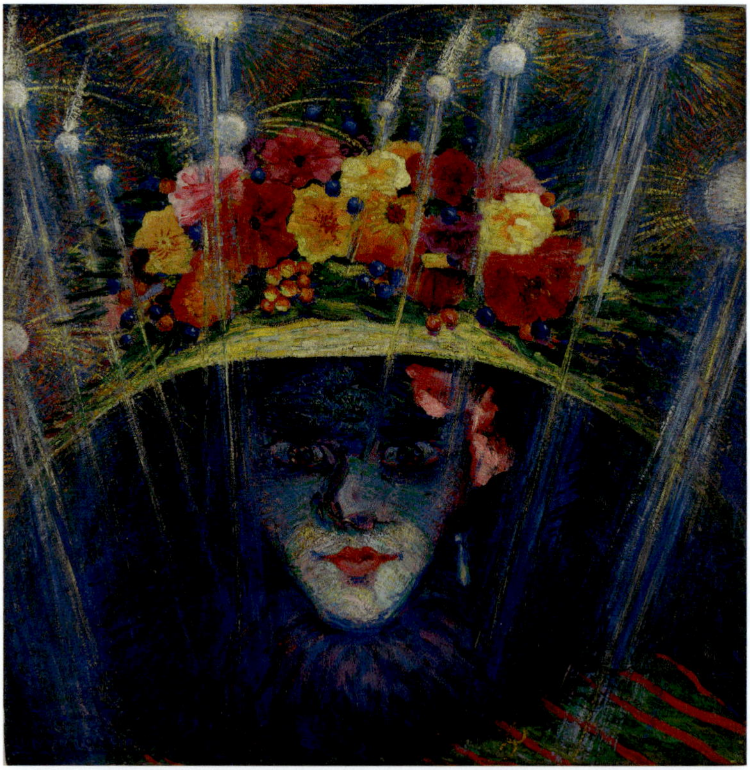

Figure 6.1 Umberto Boccioni, *Modern Idol*, 1911, 59.7 × 58.4 cm. (© Estorick Collection, London/Bridgeman Images.)

jewelry shop window. If, in the manifesto, the colorfully shining gems are directly equated with the mesmerized figure, it is notable that the vibrant colors in the painting do not correspond with the rather pale features of the woman's face (mentioned in the text as "pallor"), but rather they emphasize her sumptuous hat. In lieu of chromatic similitude, the analogy between jewels and the woman in this artwork pivots around their shared status as objects of visual and commercial pleasure. The artist pays attention to her, a figure identified as a prostitute, in an analogous manner to how she looks at the jewelry.[21] Yet something has gone awry with this correlation between the woman's material desire and the artist's fascination with her. Along with her ghostly pallor, Boccioni undermines visual pleasure by suppressing the signs of feminine beauty and delight, which had been evident in his preparatory sketches for this work.[22] With sharply delineated eye lids and pupils, the young woman no longer seems mirthful or enchanted but rather appears obsessively fixated on the jewels in the window.

As an image of consumerist obsession doubled – in terms of her acquisitive impulse for the jewels and libidinal responses to her – the painting comes across as intentionally repulsive, especially when contrasted with Boccioni's more whimsical painting of commercial pleasure from the same year, that paralleled Bergson's theory of laughter, entitled *The Laugh*.[23] Boccioni's painting subtly debases this object of sexual and visual desire – a debasement that, if one accepts Sigmund Freud's analysis, serves to circumvent a psychological prohibition against taboo sexual relations or else to counteract a fear of sexual impotence.[24] Because the prostitute's deathly white complexion sharply contrasts with the chromatic intensity of the hat, the Futurists' vitalistic theory of color appears to have been inhibited, at least temporarily, by the artist's resistance to the supposedly life-affirming qualities of his theme.[25] In this instance, Boccioni's rendition of dysfunctional commercial and bodily desires serves to acknowledge one of the disturbing pictorial implications of vitalistic color: Vibrant surrounding colors can make individuals or objects seem less colorful or even drained of life by comparison. Although the 1910 "Technical Manifesto" advised artists to use a bright palette to express a young woman's brimming, gem-like vitality, Boccioni's rendition of this transfixed figure at the shop-window offers a complex critique of material desires, which effectively satirizes worldly and fleshly pursuits when placed in the context of aesthetic contemplation.

In the same 1910 "Technical Manifesto," the Futurists also espoused a more dematerialized kind of chromatic intensity that was not directly associated with pictorial representations of human anatomy and consumerist reflections:

> Your eyes, accustomed to semi-darkness, will soon open to more radiant visions of light. The shadows which we paint shall be more luminous than the highlights of our predecessors, and our pictures, next to those of the museums, will shine like blinding daylight compared with deepest night.[26]

Triggered by radiant light, this version of Futurist chromatic vitalism seemed to arrive for the viewer in a revelatory flash, through which even painted shadows would appear more vivid than all previously painted highlights.[27] Crucially, the Futurists claimed these chromatic effects derived from the principle of "innate complementarity," which clearly references, but also cleverly reformulates the Divisionist doctrine for applying complementary colors.[28] While the Futurists' version of complementarity in those early years directly acknowledged various Italian Divisionist painters, such as Giovanni Segantini, Geatano Previati, and Vittorio Grubicy de Dragon, it also signaled an unmistakable shift toward "more luminous" coloration.[29] In 1911, Boccioni succinctly summarized the prevailing attitude among the Futurists: "The strict application of Divisionist ideas about complementary colors leads to a decrease in intensity down to gray."[30] Apart from its frequent use of bright, unmixed pigments, Divisionism advised applying only small quantities of the contrasting hues, which, according to Boccioni, gave an overall gray appearance. In his view, the vibrant colors in Futurist paintings could dispel all manner of "lifeless" representation – whether cinematic, Cubist, or Divisionist. Along with a desire for bright coloration came a deepening faith in various types of bold contrast, as Boccioni stated: "We believe complementarity to be an attitude of the spirit."[31] In this way, the Divisionist idea of contrasting brushstrokes was refashioned by Boccioni into a doctrine of unrestrained emotive intensities, which presupposed a similar lack of restraint when representing various kinds of perceptual or referential phenomena. When recalibrating Futurist painterly attitudes and techniques via complementarity, Boccioni espoused what he described as "the wild joy of color" (as cited in the epigraph). This system of beliefs and formal principles was predicated on highlighting chromatic intensities to express the vitalistic forces inherent in any phenomena depicted – and this chromatic vitality assumed multiple forms and connotations across the Futurist visual arts.

Adopted by Boccioni as far back as 1903, Divisionist-inspired painting techniques had continued during his initial involvement with Futurism and contradicted Futurist claims to the contrary. In his 1910–1911 painting *The City Rises*, the long, contiguous brushstrokes depict old and new features of a modern and modernizing Milan in a backlit scene of a construction site teaming with activity. The complementary colors make the human and equine figures stand apart from the urban background to draw the spectator's attention to the sheer physical labor needed to transform the city into a metropolis.[32]

Boccioni's exuberant palette seems to choreograph this vast expenditure of kinetic forces. Soon after completing this large work, Boccioni experimented with painting techniques in another image depicting a construction site, overseen by his mother standing on an upper-floor apartment balcony. In his 1911 work *The Street Enters the House* (Figure 6.2), Boccioni applied frenzied dabs of pigment next to more well-defined strokes, abutting delicately blended areas – all in the space of the blouse worn by the woman. His overall idea to commingle far and near elements in the visual space is translated into the distant (and diminutive) horses becoming entangled in the nearby railing of the balcony.[33] Along with his abrupt chromatic and tonal shifts, the pictorial space grows noticeably congested, as the surrounding buildings lean toward the construction pit, like the hypervigilant figure of the woman. The beholder's capacity to decipher Boccioni's jampacked surface seems to demand the perceptual equivalent of the labor being depicted. If the elongated strokes and color contrasts in *The City Rises* emphasized an energetic circulation of physical forces through the visual field, *The Street Enters the House* reveals an artist stretching beyond his Divisionist roots to explore other painting techniques. Boccioni's eventual breakthrough, which informed the chromatic practices of the other Futurists, arrived the following year, but it was initially (and perhaps unintentionally) signaled by an errant citation.

The term "interpenetration of planes" first appeared in Futurist writings in Boccioni's "Technical Manifesto of Futurist Sculpture" (April 1912), referring to what the author characterized as a recent achievement of Futurist painting, one

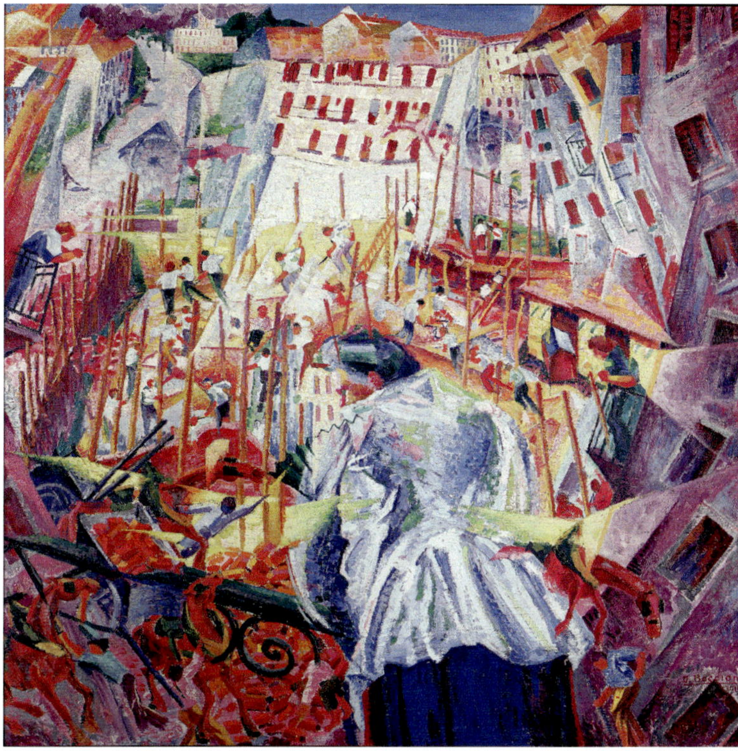

Figure 6.2 Umberto Boccioni, *The Street Enters the House*, 1911, 100 × 100.5 cm. (© Sprengel Museum, Hanover/Bridgeman Images.)

purportedly proposed a few years earlier in "The Technical Manifesto of Futurist Painting."[34] Yet, this phrase was not mentioned in the earlier text that, by contrast, specified a variation on Divisionist chromatism: "innate complementarity."[35] Whether intentionally or not, Boccioni had substituted one term for the other when describing how Futurist ideas can be applied to various mediums (i.e., painting and sculpture), as well as to diverse materials (i.e., wood, metal, and glass).[36] No longer specific to colors and brushstrokes, this new concept encompassed diverse materials that included, but were not limited to oil painting. Apparently, the Divisionist-inspired principle of "innate complementarity" was assimilated by Boccioni into the more inclusive concept of "planar interpenetration." Given the similarity between the phrases in the original Italian, *complementarismo congenito* and *compenetrazione dei piani*, the erroneous citation may have simply resulted from an inadvertent lexical substitution rather than from sleight of hand.[37] Yet, this change from a painting-specific concept to the more generalized formal principle that fueled his budding sculptural practice also coincided with a shift in Boccioni's use of color *within* his paintings – from interwoven brushstrokes in 1910–1911 to interlocking planes by 1912 – a shift that led him to another artistic innovation, but that also introduced other pictorial challenges.[38]

Expressing chromatic vitality in terms of planarity presented Boccioni with a specific pictorial problem concerning a lack of cohesion among adjacent planes, which

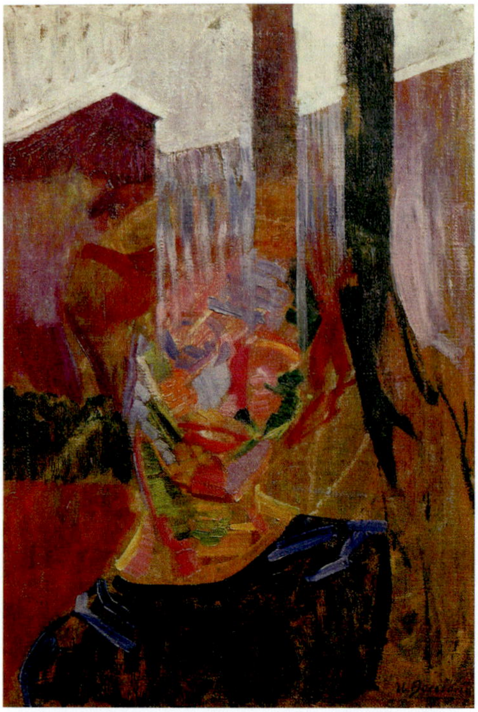

Figure 6.3 Umberto Boccioni, *Decomposition of the Head of a Woman*, 1911, 61 × 46 cm. (Private Collection, Turin; Author's Photograph.)

became apparent only after his elongated brushstrokes were dispatched. In his 1911 painting, *Decomposition of the Head of a Woman* (Figure 6.3), a young woman named Ines, with whom the artist had been involved, stands at an open window with the luminous backlighting visually penetrating and complicating her contours. The bright contiguous facets of her face appear to approach the gem-like qualities announced in the earlier manifesto. Unlike his dysfunctional *Modern Idol*, this buoyant figure – with a bare neck, relaxed, smiling mouth, and soft eyes – more closely aligns his use of chromatic intensity with visual desire and with his own sense of virility. In a lecture that same year, Boccioni mused: "If objects appear colored more or less according to the emotion that invests them, why not paint the *sensation* these variations arouse?"[39]

In *Decomposition of the Head of a Woman*, an oscillating pattern of red and green brushstrokes creates interlocking planes on the figure's chin, cheeks, and neck, while the patches of red-orange on her cheeks, off-white left of her nose, and lavender and light blue on her chin remain unblended and retain their formal distinctness. While these effusive colors might satisfy Boccioni's search for emotional content, the distinct, contrasting patches remain disconnected to a greater extent than the interwoven strokes in his previous works. These adjacent planes float freely on the loose background washes of color, unmoored by the need for greater pictorial cohesion. This lack of cohesive blending among adjacent patches likely offered the artist a provisional solution to the demand for heightened emotional intensity, which the artist explicitly formulated in terms of luminosity: "To put mixed colors on the canvas means to lose 75 percent in luminosity. Now, an artist cannot be indifferent to that loss, feeling the imperious need within himself to make his own work come alive in perfect response to his own time."[40] If Boccioni's bright pigments sought to project vitalistic aliveness and to resist indifference, then the prospect of constraining or coordinating those patches of color may have been perceived by him as undermining his premise of emotional intensity. At this point, it seems his idea of chromatic vitalism – which he associated with visual desire and sexual virility – could accommodate these disconnected facets as an acceptable consequence of overflowing emotion. Before returning to Boccioni's unexpected chromatic solution to this same problem of cohesion later in this chapter, it can be informative to review some of the most interesting developments with color by the other Futurists, including Boccioni's former teacher, the painter Giacomo Balla.

By the middle of 1912, Boccioni and the other Futurists convinced Balla of the need to update his Divisionist-inspired techniques, which he had been applying and teaching in Rome for decades. Amidst his reevaluation of painting techniques, Balla adopted a variation on the chronophotographic method introduced twenty-five years earlier by the French physiologist Étienne-Jules Marey and his long-time assistant, Georges Demenÿ.[41] The first paintings in this vein include Balla's late 1912 paintings *Dynamism of a Dog on a Leash* and *The Hands of the Violinist*, both of which imitated chronophotography's multiple, contiguous traces of kinetic motion that have been registered on a single picture surface.[42] From Boccioni's perspective, Balla's approach to painting was going in the right direction by addressing the themes of kinetic motion. However, it was still too literal and carried unfortunate associations with mechanically reproduced motion, a criticism that Boccioni was still vigorously refuting.[43]

Soon thereafter, Balla undertook a daring stylistic maneuver in his 1913 work *Automobile Velocity*. For this work, the horizontal repetition of fragments of a car moving from right to left across his picture plane became more abstracted, though they were still heavily indebted to Marey's indexical procedure. Gradually, Balla's interest in tracking kinetic motion from one instant to the next led him, at first, to focus on the perceptual effects caused by motion and, then, to compose abstract (viz., non-naturalistic) patterns of contiguous colored tesserae.[44] In *Girl Running on a Balcony*, a work from 1912 (Figure 6.4), Balla took inspiration from his daughter playing with a ball, but he distributed the figural elements across the entire image surface. The unblended areas of color follow a grid-like structure when tracing the temporal fractions of her left-to-right movement, which are strongly correlated visually with the evenly spaced, vertical supports of the railing. The resulting pattern of chromatic traces downplays the effects of contour and perspectival depth in order to convert her kinetic path into a semi-abstract visual language. With its interlocking chromatic planes imitating the hue and tonality of the running girl, his 1913 work *Iridescent Interpenetration No. 4* goes further by suppressing any observable, referential phenomena (even in its title).[45] No longer retaining any discernable link to chronophotography or kinetic forces, this flat, mosaic-like image systematically explores color relationships as a pictorial end in itself. While Balla's chromatic vitality initially emerged from figural imagery and moved to abstract patterning, he subsequently applied these patterns to the human figure – in the form of textile and clothing designs that were meant to connote the new rhythms of modern life.

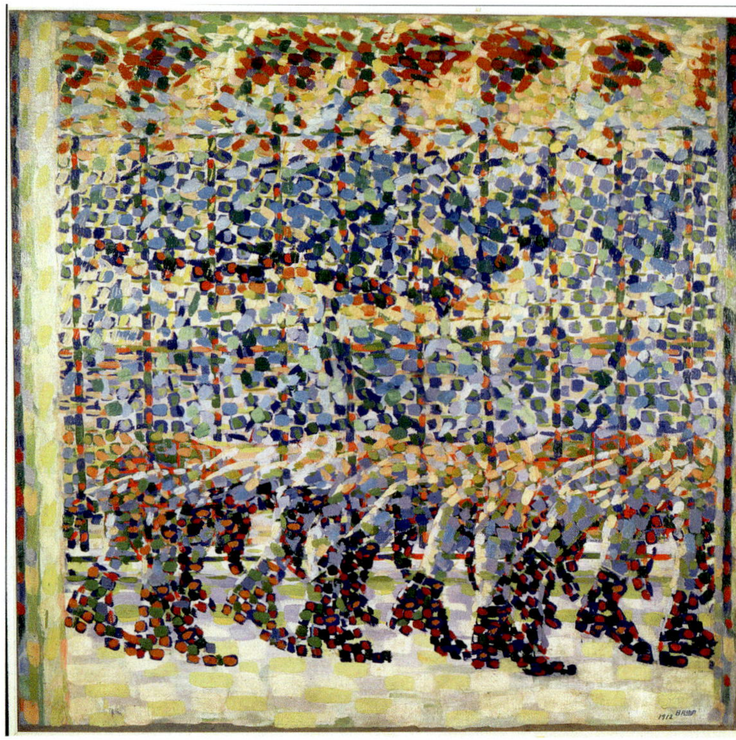

Figure 6.4 Giacomo Balla, *Girl Running on a Balcony*, 1912, 125 × 125 cm. (© Galleria d'Arte Moderna, Villa Reale, Milan/Bridgeman Images.)

In the 1913 text, "Futurist Manifesto of Men's Clothing," Balla decried the prevalence of "colorless, funereal" clothes of the past and demanded instead the creation of "daring clothes with brilliant colors and dynamic lines."[46] Returning to the jewelry shop window, Balla imagined a city street invaded by athletic refractions of colored light: "Everything will begin to sparkle like the glorious prism of a jeweler's gigantic glass-front, and all around us we shall find acrobatic blocks of color."[47] In the 1914 version of this same text entitled "The Anti-Neutral Suit," his idea of bold attire came to symbolize an invigorated national spirit: "We want to color Italy with Futurist audacity and danger, and at long last, give the Italians aggressive and cheerful clothes."[48] If the explosion of luminosity in Balla's visual practices overflowed the boundaries of figuration, it also encompassed kinetic forces extending into the domains of mass politics.

In May 1915, Balla documented the densely packed crowds that gathered regularly in Rome to agitate both for and against Italian military intervention in World War 1. In a series of semi-abstracted paintings of these large political demonstrations, curvilinear swathes of vibrant color function as an all-purpose, adaptable motif to connote the mass, agitated populace of the metropolis, while the hues of the Italian flag explicitly mark adjacent chromatic swathes in the 1915 painting *Waving–Patriotic Demonstration* (Figure 6.5). As patriotic emblems parading through the streets, his abstracted kinetic-crowd motifs appear to mimic a phrase from his 1914 manifesto that visualized those wearing his clothing to be "living Futurist flags."[49]

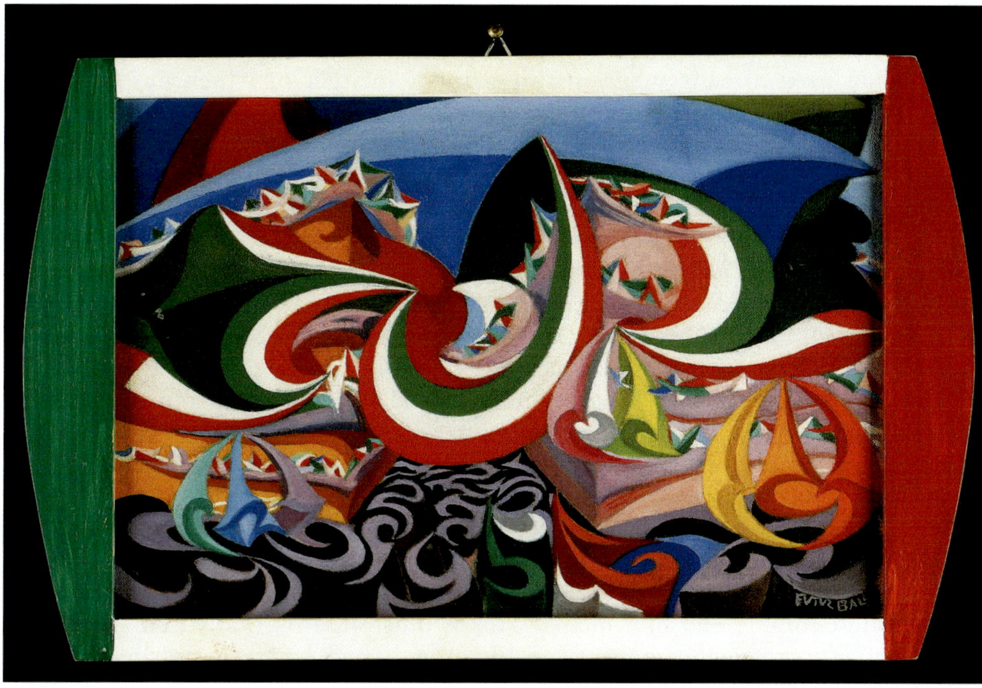

Figure 6.5 Giacomo Balla, *Flag Waving*, 1915, 20.4 × 30 cm. (© De Agostini Picture Library/G. Cigolini/Bridgeman Images/Artists Right Society [ARS], New York/SIAE, Rome.)

After suppressing the bodily and physical traces of motion in his earlier works, as demonstrated by the technical shift from *Girl Running on a Balcony* to *Iridescent Interpenetration No. 4*, Balla's imagery had become less directly referential. Through this technical and conceptual reevaluation, he discovered that motion-inspired chromatic effects could vitalize a wide range of social and political themes.

Alongside the attempts by Boccioni and Balla to correlate color with vitality, other Futurists explored some of the complicated linkages among color, music, painting, and film. This is illuminated by the Ravenna-based artist Bruno Corra's 1912 manifesto "Abstract Cinema – Chromatic Music," which chronicled his and his collaborators' efforts prior to their entry into the main Futurist group (based in Milan).[50] At first, Corra refers admiringly to the chromatic effects of Futurist painting, noting the pleasurable effects of "chromatic harmony."[51] Then, he describes experiments undertaken by him and his associates that explored the analogy between color and music. In the vein of Wassily Kandinsky's concept of musical color, Corra proposes techniques for combining nonrepresentational colors into chords and creating visual symphonies.[52] One experiment involved making a "chromatic piano," which was an actual keyboard, now lost or destroyed, that controlled a set of painted electric bulbs. However, the result was considered unsatisfactory, because it produced only a limited amount of luminosity.

Their desire for "stupefying intensity" precipitated an interest in cinema, according to Corra, their musical analogy for color leading them to experiment with film imagery.[53]

Yet instead of using a film camera with its photosensitive medium, their group ran hand-painted film strips in front of a projector bulb.[54] Corra described their two films having been already made in this manner (neither of which has been preserved) while outlining ideas for three additional "color symphonies," which were also to be made without a camera.[55] For certain screenings, they assembled an audience, wearing only white clothing, in an otherwise empty, white projection room. For these film projects, the projector had, in effect, replaced the painter, while a blank canvas served as the projection screen on at least one occasion.[56] Convinced that chromatic intensity could inspire new social configurations, Corra and his Ravenna-based cohorts arrived at a mechanical mode of chromatic abstraction by way of hand-made film sequences. However, when officially joining Futurism in 1913, they abandoned this abstract form of color cinema, possibly due to the threatening implication that the projectors would be replacing painters.[57] For the recalcitrant Boccioni, mechanical visual reproduction offered a travesty of vitalistic activities that could not be permitted to usurp the role and status of the Futurist painters.

At the same time that Boccioni resisted the encroachment by machines into the domain of chromatic vitality, significant scientific and technical attention before World War 1 had been paid to the longstanding problem of reproducing naturalistic color in photographic and film media. This is illustrated by Louis Lumière's longstanding ambition to replicate the colors of nature, which had been only temporarily diverted by the invention of black-and-white moving pictures in the 1890s.[58] After his family's company sold the Cinematographe patents to Pathé in 1902, Lumière returned to the problem of reproducing naturalistic colors: His success culminated in 1903 with the invention of the Lumière Autochrome, a color glass photographic plate that became commercially available in 1907.[59] Along with its scientific ingenuity, his color photographic medium demonstrated an expressive capacity akin to traditional painting, with one writer noting: "What the artist's brush achieves, the Autochrome plate can do automatically."[60] Another critic even referred to the Italian Divisionists when extending this same analogy between Lumière's invention and painting.[61] Yet by 1910, when the Futurist painters were forming as a group, the complex technical and commercial aim of reproducing naturalistic color had been propelling photographic and film research for decades.

This quest for chromatic vitality fueled a common perception throughout Europe, as well as the Futurist fear that photography and film would eventually replace painters in the task of faithfully copying nature.[62] Given their practical interest in realistic coloration, the enthusiasts of panchromatic film would likely have agreed with the Futurists about the importance of color to modern experience, which Boccioni succinctly summarized in 1911: "Everything in modern consciousness aspires to luminosity."[63] However, for the Futurist painters, the search for greater luminosity entailed finding ways to convey chromatic vitalism outside or beyond the narrow confines of naturalism.

In Futurist painter Enrico Prampolini's 1913 manifesto, "Chromophony – the Color of Sounds," color was a specialized feature of optical perception, which amounted to a medium of universal translation – from a wide range of sonic and environmental data into the visible spectrum.[64] While artists could perceive diverse types of stimuli through colorized data, those without this capacity were labelled "living corpses, cold souls, those beings dedicated to competence and hard work."[65] In effect, the colorless world of the factory represented the enemy of chromatic vitality. By contrast, the chromatic imagination seemed to liberate people from the conditions of mechanized labor. Almost apologetically, Prampolini noted that music was a useful analogy for color, but it was not for him a literal, lyrical source of inspiration as it had been for Corra and the Futurist artist Luigi Russolo.[66] That is, music and particularly the concept of musical intervals provided an elegant model for structuring differences, which the artistic use of nonrepresentational color might emulate. Concluding with a riddle of sorts, his manifesto claims to have been written for those who could not yet perceive chromatically, "for those suffering from blindness of the mind and dumbness of the eyes."[67] In effect, by reading Prampolini's text, his readers were initiated into a chromatic system for making or interpreting abstract color images based on diverse kinds of phenomena, not only sonic or musical sources.

When likewise seeking alternatives to naturalistic shading and chiaroscuro lighting, the Futurists painters, Gino Severini and Carlo Carrà each separately advocated for a similar theory of chromatic analogy, in which various types of data could be transcribed with "*plastic analogies*" or "*color analogies*."[68] Severini called this method of registering sensations and non-sensory data as the "spherical expansion of light in space."[69] In his painting *Spherical Expansion of Light (Centrifugal)* (1914) (Figure 6.6), he constructs an abstract visual analog for the diverse sensations of "speed, heat, smell, noise, etc."[70] From its yellow center to its blue, violet, and black edges, the progression among contiguous hues manifests a thermal principle of chromatic

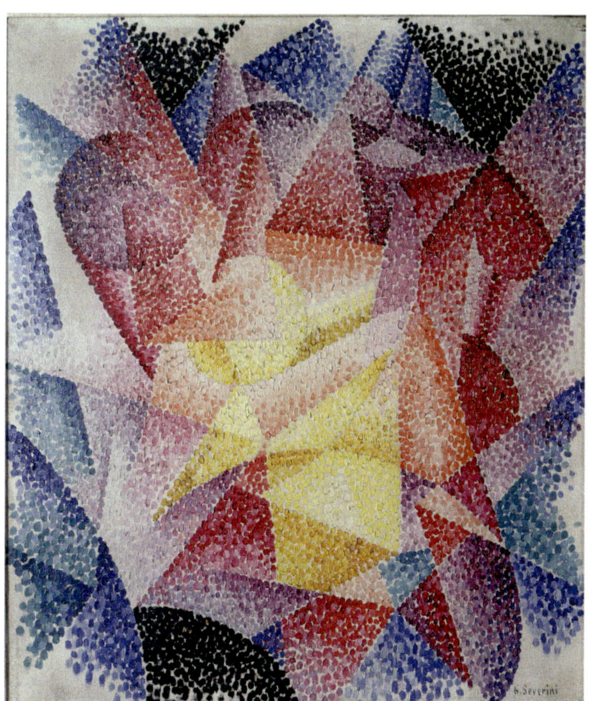

Figure 6.6 Gino Severini, *Spherical Expansion of Light (Centrifugal)*, 1914, 61 × 50.8 cm. (© Luisa Ricciarini Collection/Bridgeman Images.)

intensities that configured a strong polarity between hot-light and cool-dark.[71] Alongside the several cosmic and religious connotations in his theory and imagery, Severini mentioned centripetal and centrifugal nerves, which are now referred to as *afferent* and *efferent* nerves, and his *Spherical Expansion* artworks bear a striking resemblance to anatomical diagrams of the human nervous system from a period textbook (Figure 6.7).[72]

Closely aligned with Severini's approach, Carrà applied a similarly expansive concept of color analogies to his abstract and semi-abstract works from this time, outlining his version of "spherical expansion" in his manifestos "Plastic Planes as Spherical Expansions in Space" (published in March 1913) and "Painting of Sounds, Noises, and Smells" (from August 1913).[73] The Futurist idea to represent various phenomena with energetic, radiating bands of color appears to respond to a difficult problem faced by Modernist painters: Once the traditional rules of perspective and chiaroscuro were superseded, what were colors meant to signify? Like Prampolini, Severini and Carrà would vastly expand the discourses and practices of analogical color, which could purportedly be used to visualize all manner of internal and external forces – thermal, olfactory, musical, temporal, physiological, astronomical, etc. While these artists approached color as an *optical* means to translate various phenomena non-naturalistically, Boccioni sought to preserve an inherently *embodied* dimension of chromatic

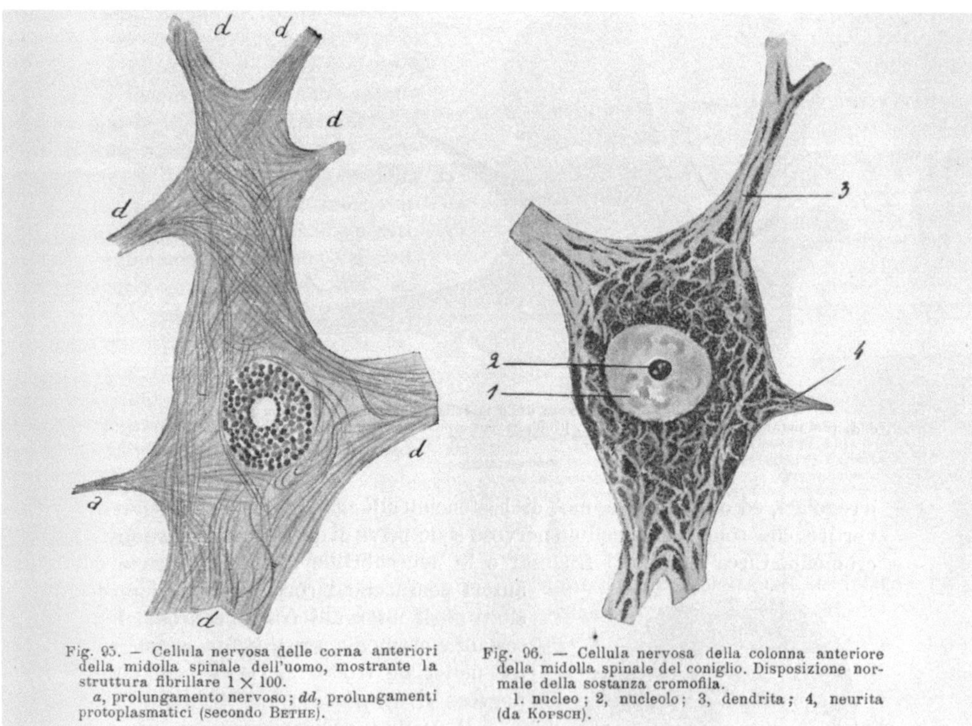

Figure 6.7 Anatomical diagrams of nerve cells in the spinal column from Giulielmo Romiti's "Anatomia generale" in *Trattato di Anatomia Umana*, Vol. 1 (Milan: F. Vallardi, 1912), 105. (Courtesy of the New York Academy of Medicine Library.)

intensities, which were rooted in emotional investments, physiological exertions, and other experiential sources.

As the culmination of Boccioni's approach to color, his painting from between 1913 and 1914 entitled *Dynamism of a Football Player* images a complex interpenetration of internal sensations, external data, and physical forces in different ways to that conceived by Delaunay in his *L'Equipe de Cardiff* (Figure 3.1). The vigorous athletic activities depicted resonate thematically with the Futurist ideology of action, while the visual style of this image illustrates how vitalistic bodily forces radiate into the surrounding environment.[74] The bright yellow and white areas of the figure's chest reveal a luminescent core, while bright red and orange contrast with the more subdued bands of blue and purple. With its forces extending beyond the anatomical limits of the figure, this painting offers a succinct visual analogy for the intertwining forces of Bergsonian intuition, internal passion, physiological exertions, and environmental conditions. For Boccioni, this broad network of relations could be revealed by the unifying field of color, as he explained: "We want physical forces to be diffused into the environment and to overlap and to flood one over the other like vibrations, caught in the vortex of those vibrations that together intensify the overall light in a painting."[75]

Along with preserving heat and intensity through prismatic luminosity, Boccioni's *Dynamism of a Football Player* also resolved the earlier issue of cohesion among adjacent patches of color by filling the spaces between brighter areas with gray, beige, and brown strokes. No longer isolated, fragmentary, or diffuse, his chromatic intensities appear to move through a continuous field, within which the emotional investments of a figure spill forth and merge with its surroundings.[76] Also, unlike the optical and mechanical principles of color espoused by the other Futurists, Boccioni's use of color suggests an embodied medium that can register an expanded range of experiential content. Rather than transcribing any sort of disembodied data into color, he deployed vibrant color to relay how vitalistic forces well up within a body and extend into the environment. By considering his paintings in the context of his writings, especially when expressing his resistance to Cubism and cinema, it becomes possible to recognize how his concept of vibrant coloration effectively translated ideas of vitalism, particularly modern vitality and virility, into a set of non-naturalistic and non-mechanical visual strategies.

Alongside the various life-enhancing connotations of Futurist chromatic vitalism, a destructive dimension was also evident from the start. In the early "Technical Manifesto," the immaterial forces of light and motion were conceived as disrupting and dissolving the boundaries of material objects and people: "Movement and light destroy the materiality of bodies."[77] Later, certain visual qualities associated with destruction were directly correlated with the Futurists' understanding of chromatic vitality. This is illustrated by Severini's description of the concept of spherical expansion as "a complex form of realism which *totally* destroys the integrity of the subject-matter – henceforth taken by us only *at its greatest vitality*."[78] For Severini, bright, non-naturalistic color led to a lack of visual recognition in painted imagery that appeared to move beyond the limits of human understanding. Likewise, Boccioni embraced vitalistic destruction when adding this rather jarring formulation into his 1911 lecture on aesthetic rejuvenation: "Everything moves towards catastrophe! And one must have the courage to surpass oneself until death."[79] To innovate artistically, for him, required assimilating new experiences and habits that, in turn, seemed to alter the known world. Also, a violent dimension of luminescent experimentation emerges in Bruno Corra's description of the now-lost film, *The Rainbow,* which culminated with "ever-increasing intensity until it finally explodes with dazzling violence."[80] Elsewhere, when concluding his "Chromophony" text, Prampolini identified a program of creative experimentation that relied on a similar tendency toward radical negation: "Destroy, destroy, in order to rebuild consciousness and opinion, culture and the genius of art."[81] Such straightforward acknowledgment of the role of destruction in Futurist chromatic vitalism supplemented the life-enhancing and socially regenerative dimensions of Futurists' philosophies, thereby opening it to a wider spectrum that could encompass both creative and destructive effects.

Ultimately, Boccioni's passion for capturing life's energies alongside his disparaging of Picasso's Cubism, dismissal of Cubism's relationship to Bergsonism and his antipathy to the lifelessness of film, propelled him to develop chromatism as an aesthetic strategy integral to *Vitalist Modernism.* From the 1910 "Futurist Painting: Technical Manifesto" onward, chromatism was explored in diverse ways, as this chapter has demonstrated by its exploration of Futurist art by Balla, Carrà, Corra, Prampolini and Russolo. When what Boccioni called "the wild joy of colour" was synchronized with dynamic compositions in painting, sculpture, abstract cinema, chromatic music, film imagery and electric bulbs, it was designed to invoke intuitive intensities, the energy of football, the passion of nationalism, the sexual urges of virility and the destructive compulsion of the death drive to counter any diminishment of life's vitality. Hence, these chromatic principles largely revolved around revealing qualities of human experience that, according to these Futurists, could not be adequately captured by any other artistic style or by any mechanical form of visual reproduction.

Notes

1 Umberto Boccioni's letter to Ardengo Soffici is dated between July 15 and 22, 1913. Umberto Boccioni, *Lettere Futuriste*, ed. Federica Rovati (Rovereto, Italy: Egon and Museo di Arte Moderno e Contemporanea di Trento e Rovereto, 2009) 77; 258; my translation.

2 Roger Allard, "Les Beaux Arts," *Revue indépendente*, no. 3 (Aug. 1911) 134. In an article on Boccioni's work, Henri des Pruraux, "Il sogetto nella pittura," *La Voce* 4, no. 44 (31 October, 1912), 13, decried the photographic medium that presented a type of mechanical deception.

3 Maxim Gorky, "The Lumière Cinematographe (Extracts)", 1896; reprinted eds. Richard Taylor and Ian Christie, *The Film Factory: Russian and Soviet Cinema in Documents* (Cambridge, MA: Harvard University Press, 1988) 25.

4 Gorky, 1896, 25: Other film critics from this period likewise condemned this seeming reduction of vitality to "a series of ugly and barren sights" and to "a world of all things black and white!"; see Piero Antonio Gariazzo, *Il teatro muto* (Turin: S. Lattes, 1919) 100; and Terry Ramsaye, "Color Photography and the Motion Picture," *Photoplay* (15, no. 4, March 1919): 84–86.

5 Bergson, *Creative Evolution*, 2005, 331–333, applies the terms *cinematograph* and *cinematographical* to a mechanistic view of time and experience, in keeping with his notion of mechanical thinking when the Lumière brothers invention was made public in 1895. Despite this reference to a specific technology, Bergson employs the term in an elaborate, ongoing analogy for determinism and mechanistic philosophy; for the relationship of Bergsonism to early twentieth-century artists in France, see Mark Antliff, *Inventing Bergson: Cultural Politics and the Parisian Avant-Garde* (Princeton, NJ: Princeton University Press, 1993).

6 A recurring example of this principle in Bergson writings is Zeno's Paradox, an ancient Greek analogy that explained life's indivisibility by illustrating the opposite – the folly of over-analysis; see Henri Bergson, *Time and Free Will: An Essay on the Immediate Data of Consciousness*, trans. F. L. Pogson (Mineola, NY: Dover, 2001) 112–115; Bergson, *Matter and Memory*, trans. Nancy Margaret Paul and W. Scottt Palmer (Mineola, NY: Dover, 2004) 250–253; Bergson, *Creative Evolution* (Paris: Alcan, 1907; New York, NY: Cosimo Classics, 2005) 335–340.

7 Umberto Boccioni, Notes on Henri Bergson's *Matière et memoire*, Getty Research Institute, Special Collections, Boccioni papers, 1899–1986, Accession no. 880380, Box 3, folder 29.

8 For additional discussion of Bergson's influence on Boccioni, refer Mark Antliff, "The Fourth Dimension and Futurism: A Politicized Space," *Art Bulletin* (82/4, December 2000): 722–733; see also Antliff, *Inventing Bergson*, 163–164.

9 Umberto Boccioni, "What Divides Us from Cubism"; Ester Coen, *Umberto Boccioni: A Retrospective* (New York, NY: Harry Abrams and the Museum of Modern Art, 1988) 248. For clarity, I added "construction" to reflect a recurring phrase from his writings during this period, as in his letter to A. Nino Barbantini, February 12, 1912; see Umberto Boccioni, *Gli scritti editi e inediti*, ed. Zeno Birolli (Milan: Feltrinelli, 1971) 347.

10 Boccioni, "What Divides Us from Cubism"; 243.

11 Boccioni, "What Divides Us from Cubism", 243; on 247, Boccioni says "Picasso, when he dissects a figure, slices it into bits and pieces, breaks it down into its elements, kills it."

12 Boccioni, "What Divides Us from Cubism", 244.

13 Guillaume Apollinaire, *The Cubist Painters* (Berkeley: University of California Press, 2004) 13: "Picasso studies an object like a surgeon dissecting a corpse".

14 Umberto Boccioni, "What Divides Us from Cubism"; 246. In Chapter 5 of this volume, Fae Brauer explores Picasso's fragmentation of forms, multiple viewpoints and monochromatic palette with magnetism photography of phantasmatic energies, occultist concepts of "la force psychique" and Bergson's "psychic states".

15 Influential treatises on color include Michel Eugène Chevreul, *De la loi du contraste simultané des couleurs et de l'assortiment des objets colorés* (1839); Ogden Rood, *Modern Chromatics, with Applications to Art and Industry* (1879); Charles Henry, *Introduction à une esthétique scientifique* (1885) and *Cercle cromatique* (1888); Paul Signac, *D'Eugène Delacroix au néo-impressionisme* (1899 and 1911).

16 Michel Eugène Chevreul, *The Laws of Contrast of Color*, trans. John Spanton (London: Routledge, 1857) 62: "In the harmony of contrasts, the complementary assortment is superior to every other."

17 Yve-Alain Bois, "On Matisse: The Blinding; for Leo Steinberg," trans. Greg Sims, *October* 68 (Spring 1994) 93, describes Matisse's desire to escape the "suffocating" Divisionist system and cites Matisse's description of Divisionism as an impersonal system in which "everything is treated in the same way." The Italian school of Divisionism interpreted the French idea of treating complementary hues equally as being well-suited to its commitment to social equality; see Vivien Greene, "Painted Measles: The Contagion of Divisionism in Italy", *Divisionism/Neo-Impressionism: Anarchy and Arcadia*, ed. Vivien Greene (New York, NY: Guggenheim Museum, 2007) 22–23; see also Vivien Greene, "*Pittura ideista*: The Spiritual in Divisionist Painting", *California Italian Studies* (5/1, 2014): 1–16.

18 Henri Matisse, "Statement to Teriade: On Fauvism and Color" (1929), *Matisse on Art*, ed. Jack Flam (Berkeley: University of California Press, 1995) 84: "[L]ight is not suppressed, but is expressed by a harmony of intensely colored surfaces." An alternate account of Boccioni's long process of reevaluating his own Divisionist-inspired techniques can be found in Vivien Greene, "The Path to Universal Synthesis: Boccioni's Development from Divisionism to Futurism," *Boccioni's* Materia*: A Futurist Masterpiece and the Avant-garde in Milan and Paris*, ed. Laura Mattioli Rossi (New York, NY: Solomon R. Guggenheim Foundation, 2003) 23–33.

19 Boccioni, Carrà, Russolo, Balla, and Severini, "Futurist Painting: Technical Manifesto" (1910), *Futurist Manifestos*, ed. Apollonio, 29.

20 "Futurist Painting: Technical Manifesto" (1910).

21 Christine Poggi *Inventing Futurism* (Princeton: Princeton University Press, 2009) 203 and 206–213 describes the features of Boccioni's "cocotte" as representing degeneracy and sexual promiscuity. In the same chapter, she discusses the figures of prostitutes in other works from this year. As a recurring motif in Boccioni's work and texts from this period, the consumer-prostitute figure personified for him a complex and dysfunctional mode of sexual attraction and materialistic exuberance. Examples in his paintings from this period of dysfunctional materialism include two 1910 paintings titled *The Riot*, both of which depict crowd violence stemming from an altercation betwen two prostitutes in a popular commercial district in Milan; see Christine Poggi, "*Folla/Follia*: Futurism and the Crowd," *Critical Inquiry* 28 (Spring 2002): 709–748. Also, Poggi, *Inventing Futurism*, 199–213, offers an interpretation of *Modern Idol*, which identifies a sense of psychic dysfunction in the artist's depictions of women. A later example of this same general motif can be found in his free-word poem "Small Dress Shoes + Urine" from 1913, in which Boccioni loosely narrates an encounter with a prostitute in a London commercial district; see David S. Mather, "Mobilizing Desire in Umberto Boccioni's 'Small Dress Shoe + Urine'", *Getty Research Journal* 6 (Los Angeles: Getty Research Institute) 185–194.

22 Coen, *Umberto Boccioni*, 114: One catalog entry for this painting, which illustrates and mentions two of his preparatory works, describes "a picture fraught with mysterious force and cruel aggression" and her face "staring, [with] almost mad eyes." Elsewhere, Poggi, *Inventing Futurism*, 197–203, notes the possibly pathological dimension to this figure's intense gaze.

23 A spectator may observe that the mesmerized object of male (viz., the artist's) desire in *Modern Idol* hardly represents an artistic muse in the traditional sense, and this short-circuited mode of visual pleasure may impact the empathetic responses of the spectator. Italian neuroscientist Vittorio Gallese has recently argued that human visual perception entails cognitive patterns that can trigger empathetic responses in observers; see Vittorio Gallese, "Embodied Simulation: From Neurons to Phenomenal Experience," *Phenomenology and the Cognitive Sciences* 4, no. 1 (2005): 23–48; Vittorio Gallese, "The Roots of Empathy: The Shared Manifold Hypothesis and the Neural Basis of Intersubjectivity", *Psychopathology* 36, no. 4 (2003): 171–180 also refer Henri Bergson, Le Rire: Essai sur la signification du comic (Paris: Félix Alcan, Éditeur, 1900); Laughter: *An Essay on the Meaning of the Comic*,1st English translation, 1924.

24 Sigmund Freud, "On the Universal Tendency to Debasement in the Sphere of Love" (1912); Sigmund Freud, *Standard Edition of the Complete Psychological Works of Sigmund Freud*, Vol. 11 (London: Hogarth and the Institute of Psychoanalysis, 1957) 179–190.

25 By simultaneously appeasing and frustrating a spectator's expectation of visual pleasure, based on its varied intensities of color, the painting adeptly transfers qualities of (male) sexual anxiety onto the (female) object of desire signaling, perhaps unexpectedly, a sense of Boccioni's erotic impulse.

26 Boccioni, Carrà, Russolo, Balla, and Severini, "Futurist Painting: Technical Manifesto" (1910); *Futurist Manifestos*, ed. Apollonio, 29.

27 In its blinding, sensorial intensity, this Futurist description of colored light closely resembles statements by Matisse about his own shift from Divisionism to increased luminosity. In "On Matisse: The Binding", 1994, 85, Yve-Alain Bois and Greg Sims recalled the effects of the midday sun in Corsica as "frightening" and as having a "brilliance, which is intolerable", and noted: "The search for color did not come from studying paintings, but from the outside – that is, from the revelation of light in nature".

28 I have rendered the Italian term *complementarismo* in English as *complementarity* in order to reflect its wider application to the English translations of various color theories of the nineteenth century; see John Gage, *Color and Culture: Practice and Meaning from Antiquity to Abstraction* (London: Thames and Hudson, 1993) 171; 205.

29 Boccioni, Carrà, Russolo, Balla, and Severini, "Futurist Painting: Technical Manifesto" (1910); *Futurist Manifestos*, ed. Apollonio, 29. The Futurist reformulation of the Italian Divisionist principle of complementarity expanded the idea of contrasting strokes to render local, observed colors by using many colors to depict individual emotive and intuitive forces, which defied and unsettled their Divisionist colleagues.

30 Umberto Boccioni, "Notes from Rome lecture in May 1911"; Coen, *Umberto Boccioni*, 239.
31 Boccioni, "Notes from Rome lecture in May 1911"; Coen, *Umberto Boccioni*, 235 (translation altered per note 27).
32 John C. Welchman, "Color, light, and labor: Futurism and the dissolution of work", *Work and the Image, Vol II: Work In Modern Times – Visual Mediations and Social Processes,* eds. Valerie Mainz and Griselda Pollock, (Aldershot: Ashgate, 2000) 61–90.
33 For more on the subject of Boccioni's painting of his mother, see Virginia Spate, "Mother and Son: Boccioni's Painting and Sculpture," *In Visible Touch: Modernism and Masculinity,* ed. Terry Smith (Chicago: University of Chicago, 1997) 107–138; also see, Laura Mattioli Rossi, "Boccioni Between Painting and Sculpture," *Boccioni's Materia,* ed. Laura Mattioli Rossi, 2004, 35–45.
34 Boccioni, "Technical Manifesto of Futurist Sculpture" (April 1912), *Futurist Manifestos*, ed. Apollonio, 52. Bergson employs the term *interpenetration* to describe a continuous, indivisible duration. Unlike Bergson, the Futurists sought to explore the material and aesthetic consequences of interpenetration, while maintaining the same basic metaphysical premises as Bergson – contrasting color to mechanistic analysis; see also Fae Brauer's Chapter Five in this volume for additional scrutiny of planar interpenetration in Cubism.
35 *Futurist Manifestos*, ed. Apollonio, 29.
36 *Futurist Manifestos*, ed. Apollonio, 64–65.
37 Given Boccioni's careful citations in this and other published texts, this substitution seems to have been unintentional. Often inadvertent lexical substitutions are triggered by similar words with otherwise unrelated meaning. However, Boccioni's change was not simply a random substitution error given that the two terms are also closely related conceptually. On lexical substitution errors, see Merrill F. Garrett, "Processes in Language Production", *Linguistics: The Cambridge Survey: Vol. 3, Language: Psychological and Biological Aspects*, ed. Frederick J. Newmeyer (Cambridge: Cambridge University Press, 1988) 69–96.
38 Two Boccioni paintings from 1912, *Horizontal Construction* and *Materia*, illustrate how his painterly application of "planar interpenetration" relied on a recurring formal structure that disrupted the contours of his figures, especially with distant visual elements overlapping spatially nearer ones; see *Boccioni's Materia,* ed. Rossi.
39 Boccioni, "Notes from Rome lecture"; Coen, *Umberto Boccioni*, 239.
40 Boccioni, "Notes from Rome lecture"; Coen, *Umberto Boccioni*, 235.
41 Among various scholarly texts discussing the influence of Marey on Balla, see Giovanni Lista, *Balla* (Modena: Galleria Fonte d'Abisso, 1982) 45–47; Giovanni Lista, *Balla: La modernità Futurista* (Milan: Skira, 2009) 42–67; Poggi, *Inventing Futurism*, 27–29; 110–146. In early 1912, Balla met Anton Giulio Bragaglia, who lived near Rome and who, along with his brother Arturo, was experimenting with photographic techniques inspired by Marey's chronophotography. The Bragaglia brothers briefly become associated with Futurism in 1913, which ended when Boccioni had them expelled. On the dispute within Futurism relating to the Bragaglias' research, see Poggi, *Inventing Futurism*, 141–142; also see David S. Mather, "The Bragaglias' Unreality", Chapter 1, *Futurist Conditions: Imagining Time in Italian Futurism* (London: Bloomsbury Academic, 2020) 47–94.
42 Discussion of the chronophotographic lineage of Balla's paintings can be found, among other sources, in Poggi, "Photogenic Abstraction: Giacomo Balla's *Iridescent Interpenetrations*", *Inventing Futurism*, 109–149.
43 In January 1913, Boccioni wrote to Severini that Balla's works were "too photographic and episodic"; Boccioni, *Lettere Futuriste*, 60–61; 241–244.
44 One interpretation of his abstracted imagery of velocity correlates the diverse sensory stimuli with the specific behaviors of his lines and shapes; see Lista, *Balla: La modernità Futurista, 58–59*; Poggi, *Inventing Futurism*, 27–29; Fabio Benzi, "Giacomo Balla: The Conquest of Speed," *Italian Futurism, 1909–1944: Reconstructing the Universe,* ed. Vivien Greene (New York, NY: Solomon R. Guggenheim Museum, 2014) 104; 106 n. 12.

45 A detailed account of this shift in the visual works of Balla can be found in Poggi's *Inventing Futurism,* Chapter 4.

46 Balla, "Futurist Manifesto of Men's Clothing" (December 1913); in *Futurist Manifestos*, ed. Apollonio, 132.

47 Balla, "Futurist Manifesto of Men's Clothing", 1913; for discussion of how these brightly colored outfits related to a novel approach to commercial industry, see Mather, *Futurist Conditions,* 112–116.

48 Giacomo Balla, "The Anti-Neutral Suit" (September 1914), Virginia Dortch Dorazio, *Giacomo Balla: An Album of His Life and Work* (New York, NY: Wittenborn, 1969). Berghaus suggests that prior to its publication, Marinetti rewrote Balla's original text to reflect a military interventionist attitude; this later version differs in important ways from an earlier draft: "Futurist Manifesto of Men's Clothing", Günter Berghaus, *Futurism and Politics: Between Anarchist Rebellion and Fascist Reaction, 1909-1944* (New York, NY: Berghahn Books: 1996) 76.

49 Giacomo Balla, "The Anti-Neutral Suit" (September 1914); Balla, *Scritti*, 32. Elsewhere in this text, the patriotic message has been made clear by his references to the *tricolore*, literally meaning "three colors", often used to designate the Italian flag.

50 Corra's text appeared in a volume of essays written by him and his brother Arnaldo Ginna, entitled *Il pastore, il gregge e la zampogna* (Bologna: Libreria Beltrami, 1912). These citations, however, come from a translated version, reprinted in *Futurist Manifestos*, ed. Apollonio, 66–70. Bruno Corra and Arnaldo Ginna were pseudonyms used by the brothers Bruno and Arnaldo Ginanni-Corradini; see Giovanni Lista, *Cinema e fotografia Futurista*, 19.

51 Bruno Corra, "Abstract Cinema – Chromatic Music" (ca. 1912) in *Futurist Manifestos*, ed. Apollonio, 66.

52 In *Concerning the Spiritual in Art* (1911; Boston, MA: MFA Publications, 2006) 41; 52, Wassily Kandinsky proposed music as a useful analogy for the abstract language of painting.

53 On the aim of producing grand chromatic effects for large audiences, see Bruno Corra, "Abstract Cinema – Chromatic Music," *Futurist Manifestos*, ed. Apollonio, 67.

54 Before deciding to project hand-painted films, they altered the projector, which created "a cataclysm of incomprehensible colors"; however, they repaired the device and continued with their search for increased (and comprehensible) luminosity; Corra, "Abstract Cinema – Chromatic Music," *Futurist Manifestos*, 68.

55 Corra, "Abstract Cinema – Chromatic Music," *Futurist Manifestos*, 70.

56 Corra, "Abstract Cinema – Chromatic Music," *Futurist Manifestos*, 68. Notably, Lista, *Cinema e fotografia Futurista*, 26, calls these chromatic experiments *cinepittura* ("cine-painting"), capturing a sense of the threat posed to painters by these technological experiments.

57 Lista, *Cinema e fotografia Futurista*, 24–32.

58 Henri Kubnick, *Les Frères Lumière* (Paris: Librarie Plon, 1938), 79, describes the invention of cinema as "a short episode" ["un court épisode"] in Lumière's career.

59 According to Maurice Trarieux-Lumière, the inventor's grandson, Louis Lumière is reported to have said that color photography, not cinema, was his greatest achievement; see Maurice Trarieux-Lumière's preface to Nathalie Boulouch's *Les Autochromes Lumière: la couleur inventée*, 1995 (unpaginated).

60 A. Personnaz, "L'Esthetique de la plaque Autochrome", presented at the 5ième Congres International de Photographie, 1910; Boulouch, *Les Autochromes Lumière* (unpaginated).

61 L. Pellerano, "L'Autochrome et ses applications artistiques", *La Fotografia Artistica* 8 (1909), 128; cited in Boulouch, *Les Autochromes Lumière,* unpaginated.

62 Despite the astonishing successes by researchers in early cinema and in color photography, neither Lumière nor anyone else could solve the complex technical and financial issues associated with manufacturing and distributing color film stock for widespread use – at least prior to Technicolor's public debut in 1922; see Scott Higgins, *Harnessing the Technicolor Rainbow*, 2007, 4. Even after creating a market-ready color film medium (February 1909), with the primary research done by George Albert Smith,

Charles Urban's company found financing difficult to secure in a film industry dominated by black-and-white production and distribution; see Charles Urban, *A Yank in Britain* (East Sussex, England: Projection Box, 1999) 78.

63 Boccioni, "Notes from Rome lecture"; Coen, *Umberto Boccioni*, 235.

64 Enrico Prampolini, "Chromophony – the Color of Sounds," *Futurist Manifestos*, ed. Apollonio, 115, noted the human capacity "to express in chromatic terms the sound waves and the vibrations of all movements within the atmosphere." Prampolini, 118, proposed the external and internal realms were filled with invisible information that could be transcribed into color through an "intuitive chromatic stimulus" that resided in the artistic imagination.

65 Prampolini, "Chromophony", 117.

66 Luigi Russolo's 1911 painting entitled *Music* offers a simplified version of the analogy between sound and image, in which a pianist plays a keyboard and colorful, concentric circles radiate from the figure's blurred silhouette to reveal an array of Greek masks of comedy and tragedy, symbolizing the varying moods of a musical composition.

67 Prampolini, "Chromophony", 118.

68 Gino Severini, "The Plastic Analogies of Dynamism" (1913), *Futurist Manifestos*, ed. Apollonio, 122. Severini, 123, states: "*Using color analogies one can obtain the greatest luminous intensity, heat, musicality, optical and constructional dynamism.*"

69 Severini, 124.

70 Severini, 121.

71 The use of color to chart a range of expanded tonalities manifested a similar premise to Kandinsky's theory of color, seemingly drawing upon Goethe's notion of polarity in the color spectrum in which hues overlay the tonal spectrum from white to black; see John Gage, *Color and Culture*, 202; 207.

72 The chromatic and tonal qualities in Severini's *Spherical Expansion of Light*, as well as his description of "centrifugal" and "centripetal" nerves in his 1917 manifesto "La peinture d'avant-garde," closely resembles anatomical diagrams in Giulielmo Romiti's "Anatomia generale", *Trattato di Anatomia Umana,* Vol. 1 (Milan: F. Vallardi, 1912); the illustrations of these nerve cells in the spinal column demonstrate a similarly strong tonal contrast to create a binary visual relationship for depicting divergent biological concepts; also see Gino Severini, "La peinture d'avant-garde", *Écrits sur l'art* (Paris: Editions Cercle d'Art, 1987) 83.

73 Carrà's texts appear in translation in *Futurist Manifestos*, ed. Apollonio, 91–92; 111–115. In 1912, Carrà entitled two works *Centrifugal Forces* and *Spherical Expansion of Prisms: Centers of Force of a Boxer*, the latter being listed as *Force Centres of a Boxer*, Trask and Laurvik, *Catalogue de Luxe*, 1915, and as simply *Boxer*, Taylor, *Futurism*, 1961, 79.

74 The Futurist ideology of action is summarized in F. T. Marinetti's texts "The Necessity and Beauty of Violence" (1910) and the first section of "Destruction of Syntax – Untrammeled Imagination – Words-in-Freedom" (1913); see F. T. Marinetti, *Critical Writings*, ed. Günter Berghaus (New York, NY: Farrar, Straus and Giroux, 2006) 60–72; 120–123. Elsewhere, Marinetti called their desire to combine art-making with sociopolitical action as "art-action" (*arte-azione*); see Marinetti, *Guerra sola igiene del mundo* (Milan: Ed. Futuriste di "Poesia," 1915) 6.

75 Boccioni, "Notes from Rome lecture"; Coen, *Umberto Boccioni*, 237. This translation has been slightly altered to preserve its original sense. The artist characterized his figures in motion as vortexes of "light's colored vibrations."

76 Boccioni, "What Divides Us from Cubism," Coen, *Umberto Boccioni*, 243, argues that a more seamless progression among adjacent hues would restore the qualities of visual continuity, whose loss he blamed on the Cubists' "impassive scientific calibration that destroys all dynamic heat, all violence, and all incidental variety of forms." Umberto Boccioni, *Pittura e scultura Futuriste*, ed. Zeno Birolli (Milan: Abscondita, 2006) 64, stated: "The psychological objective analysis of figures has killed *unity*, *heat*, *action*, which are the fundamentals of creating an artwork... But precisely this dynamic heat, violence, and incidental variety make the forms have a life outside of intelligence and project them into the infinite."

77 "Futurist Painting: Technical Manifesto" (1910); *Futurist Manifestos*, ed. Apollonio, 30.
78 Severini, "The Plastic Analogies of Dynamism"; *Futurist Manifestos*, ed. Apollonio, 122.
79 Boccioni, "What Divides Us from Cubism", Coen, *Umberto Boccioni*, 244.
80 Bruno Corra, "Abstract Cinema – Chromatic Music," *Futurist Manifestos*, ed. Apollonio, 69.
81 Enrico Prampolini, "Chromophony – the Colors of Sounds" (1913), *Futurist Manifestos*, ed. Apollonio, 118. In Robert Delaunay and Sonia Delaunay, *The New Art of Color*, ed. Arthur Cohen (New York, NY: Viking Press, 1978) 78, notes unpublished in his lifetime, Robert Delaunay, the subject of Chapter Three in this volume, spelled out his prediction of a pessimistic end to the Futurist visual and social experiments: "You are really marching toward death."

Part III NEO-VITALISM

Absurdity, Dysfunctionality, Inversion and Socialism

7 Was Dada Vitalistic?

Brandon Taylor

One clue to an answer is to notice how implacably hostile Dada artists were to mechanism and to the properties of the machine. This was not only a reaction to the alarming mechanization of war itself. "Life is confined and shackled", the German poet Hugo Ball wrote about the days before the First World War. "Economic fatalism prevails; individuals without being asked are assigned their interests, roles and character; the church is no more than a redemption factory; literature is a safety valve". What was needed, Ball wrote, was "a league of those who want to escape from the mechanical world, a way of life opposed to mere utility. Orgiastic devotion to the opposite of everything that is serviceable and useful".[1]

In neutral Zurich in the spring of 1916, where Ball and others were by now assembled, a decision was made to open a cabaret. It would be named "Voltaire": in Ball's words "the anti-poet, the king of jackanapes, prince of the superficial, anti-artist". Such at least was how Ball hoped the Cabaret would be: semi-serious, mocking – and for the most part a performance. Those qualities would signify a new kind of life; and for him, nothing would better provide an index of the living and the vital than a group of human voices enunciating rhythmically and expressively against a background of other random noises. As Ball wrote in his diary for 30 March 1916, "the human voice represents the soul, the individuality in its wanderings with its demonic companions. The noises represent the background – the inarticulate, the disastrous, the decisive". The *poème simultané* [simultaneous poem] "tries to elucidate the fact that man is swallowed up in the mechanistic process. In a typically compressed way it shows the conflict of the *vox humana* with a world that threatens, ensnares, and destroys it".[2] Some three weeks later, Ball would perform on stage by himself, this time in a manner that parodied the dreary mechanism of the liturgical voice while suggesting the life-enhancing energy of the flow of utterance when left to its own patterns of growth and decay. "I made myself a special costume" noted Ball of his performance of the 23rd June. "My legs were in a cylinder of shiny blue cardboard, which came up to my hips so that I looked like an obelisk" (Figure 7.1). "Over it I wore a huge coat collar cut out of cardboard, scarlet inside and gold outside. It was fastened at the neck in such a way that I could give the impression of wing-like movement by raising and lowering my elbows. I wore a high, blue-and-white striped witch-doctor's hat … I could not walk inside the cylinder so I was carried onto the stage in the dark and began, slowly and solemnly:

> gadji beri bimba
> glandridi lauli lonni cadori
> gadjama bim beri glassala
> glandridi glassala tuffm i zimbrabim
> blassa galassasa tuffm i zimbrabim …"

DOI: 10.4324/9781003045595-11

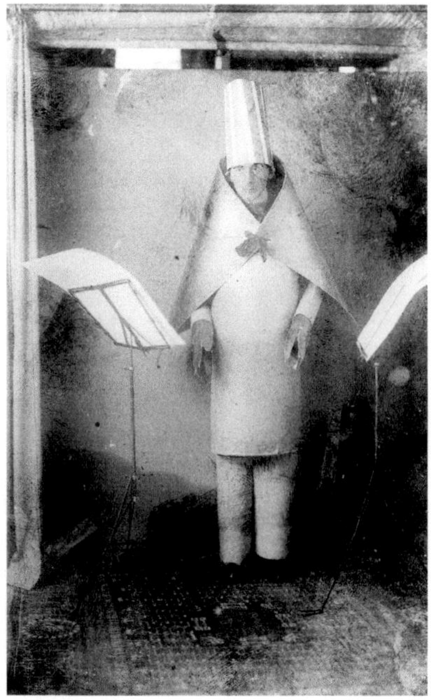

Figure 7.1 Hugo Ball performing at the Cabaret Voltaire, 23 June 1916, photograph; author unknown. (Courtesy of Kunsthaus Zurich, Dada-Sammlung Inv. VI:5.)

Ball's *lautgedichte* [sound-poem] was not only a bitter parody of a priest – intoning stale formulae while dressed in a robotic suit, arms flapping senselessly – but a vehicle for the release of other kinds of forces once conventions of sense-making had been stripped away. "The Dadaist loves the extraordinary and the absurd", Ball had written in his diary a couple of weeks before. "In the midst of the enormous unnaturalness" – the war, and technical culture – "the direct and the primitive seem incredible to him".[3]

The question is whether these energies and forces amount to an alliance with what others of his generation understood as "vitalism" – and if so, how that alliance worked. Ever since Aristotle, vitalism and vitalist thinking had functioned to preserve a sense of free volition in a world increasingly understood as law-like and predictable. Amid the planning and regulation of the First World War, the noise and randomness of the Cabaret signified that freedom well. Here is the Alsatian artist Hans Arp, like Ball in Zurich in refuge from the war: "On the stage of a gaudy, overcrowded tavern there are weird and peculiar figures. Total pandemonium. The people around us are shouting, laughing, and gesticulating. Our responses are sighs of love, bursts of hiccupping, poems, the mooing, and miaowing of medieval Bruitists".[4] In part this was a reference to Luigi Rusollo's *L'Arte dei Rumori* [The Art of Noises] manifesto of 1913, cited in Chapter 6 of this book, in which the Italian Futurist movement took to itself the observation that in previous centuries "life went by in silence, or ... in muted tones".[5] The invention of sounds as against mere sound inaugurated the long history of Western music, whose main obsession became one of temporal sequence, eventually of rhythm, pattern and harmonic progression. We Futurists preferred familiar noise, Russolo had suggested, even the whistles, buzzes, and scrapings of the modern factory. What he called Bruitism "has the power to conjure up life itself", to become a source of "innumerable surprise", of "unexpected sensual pleasure".[6]

Directness and the appearance of randomness are certainly qualities of Arp's ink drawings of 1915 and 1916, the earliest done before the Zurich Cabaret began. It reminds us that Arp was already in thrall to the idea of "natural becoming" such as could be found in the German Romantic poetry of Schiller and Novalis. In 1915, living in Ascona to the south of Zurich, he had delighted in the watery detritus Lake Maggiore would wash up randomly on the shore, and made several ink drawings at the time. Unplanned and unpredictable transformation are the principal visual qualities of these drawings, whose squirming and mobile forms appear perpetually "on their way" to something without ever seeming in any way complete. Arp's scissor-cut collages of 1916–17, several of them done with Sophie Taeuber, the artist and dancer affiliated to the experimental dance school of Rudolf von Laban and with whom he formed a relationship at the end of 1915, participate in another kind of withdrawal from planned or determinate facture. A small group of torn-paper collages from 1915–19 and a far larger group from after 1930 are also relevant to the vitalist orientation that Arp, as an artist, would never entirely betray. By setting aside the scissors entirely he could allow the paper-edges to configure as they may, their ragged, miniscule fibres randomly disposed and visible in what the viewer sees (Figure 7.2). As he later described these works, they were made "without will ... according to the law of chance", a principle that "embraces all laws and is unfathomable, like the first cause from which all life arises".[7] The sequence of *papiers déchirés* made – or that made themselves – after about 1930 have the look and feel not of purposeful design but of an event of nature seemingly close to that of botanical life itself. Such events were *sinnlos* [sense-less], according to Arp, in contrast to how a sentence with its customary internal syntax comes to be sense-full.

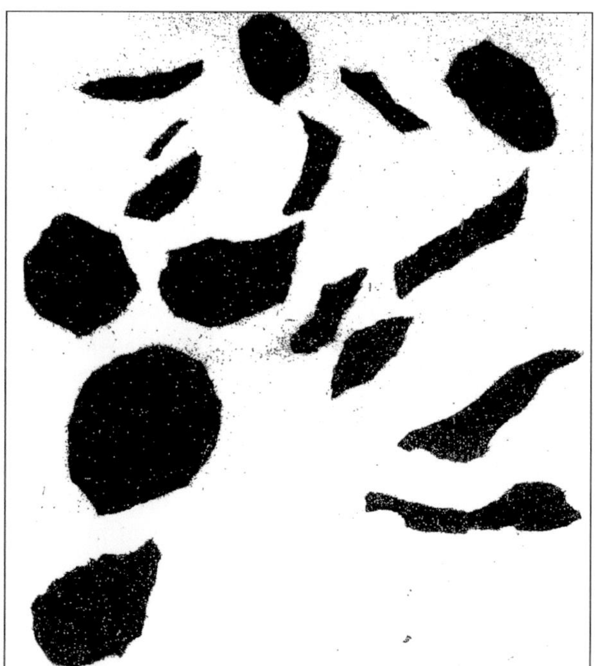

Figure 7.2 Hans Arp, *Premier papier déchiré*, c. 1932, collage, 28 × 22 cm. (Fondation Arp-Hagenbach, Locarno; gift of Marguerite Arp-Hagenbach; ©DACS London 2021.)

To the Dada artists of Zurich, "sense-lacking", "sense-evading", and "sense-emptying" were more or less equivalent terms. And there are first-hand accounts – by Tristan Tzara, Huelsenbeck, and others – of how the avoidance of sense-*making* in the work of art was at the same time virtually a sufficient condition of that aesthetic freshening occurring. In his later career Arp wrote of the stone and plaster sculptures that he also began to make around 1930. They too were to be looked upon as an unintended natural becoming; a sequence of germination, growth and maturation that can best be described in the terminology of biological or zoological life. "A small fragment of one of my plastic works presenting a curve or contrast that moves me is often the germ of a new work", he recorded. "I intensify the curve or the contrast, and this determines new forms. Among the new forms two grow with special intensity. I let these two continue to grow until the original forms have become secondary and almost expressionless, [then] I suppress one of the secondary forms so that the other becomes more apparent ... I work [like this] until enough of my life has flowed into its body".[8]

It was an attitude to art-making that under the impulse of Dada became widespread. Arp's approach to materiality, both in 1915–16 and later, can be put aside that of the Russian artist Vladimir Tatlin in far-away Moscow in the same years, as also indicated in Chapter 10 of this book. In an eye-witness report we hear of Tatlin sorting through piles of cast-off pieces of glass, wood, and metal while preparing to make reliefs during the years of the First World War. In Nicolai Tarabukin's words, the process was one of "letting material determine form for the artist [and] not the other way around".[9] Tatlin's celebrated "culture of materials", though seldom identified with Dada, was one in which the artist cedes a certain amount of agency to material itself. More widely in that generation, receptiveness on the part of an artist to something "just happening" with material was believed not only to allow material to retain its best qualities (the "truth to materials" doctrine that has thrived ever since) but also to bring to the fore a kind of negative-positive agency of the artist in which surprise and humour could thrive and where a novel kind of ontological density could begin to appear. These included effects without an obvious cause; the sympathy of things for each other, and an accumulation of meaning on its own. "A picture of the world takes place", the Dada film-maker and artist Hans Richter said of much of the art of that time, "in which, beside 'causal' experiences, others that were previously unknown and unmentioned find a place".[10]

For some, such rebalancing of the processes of artistic manufacture went by the term "creative indifference" – *schöpferische Indifferenz* in the words of the German writer and philosopher Salomo Friedländer (who invented the pen-name "Mynona" for himself, or "anonym" written backwards). He was an occasional participant in the Zurich activities and treated the Cabaret to his verses on 29th March 1916. For him and others, "creative indifference" at one level was an attitude of passive awareness towards events combined with a readiness to act expediently, having understood how material can often seem to fall into conjunction, self-arrange, or present an appearance that the artist him/herself had not manipulated. In another version, it could verge upon complete indifference to the affairs of the material world. Arp himself is known to have taken a close interest in medieval mystical sources in which comparable attitudes had been expressed. Writings by Meister Eckhart (1200–1327), Johannes Tauler (1300–61), Nicholas of Cusa (1401–64), and Jakob Boehme (1575–1624) were also popular in the Monte Verità community at Ascona – as well as widely advertised by contemporaries like Rudolf Steiner of whose writings, along with those of Kandinsky, most artists either read or knew something.[11] A not untypical passage in Eckhart's treatise "On Detachment", written in the early 14th century, reads "I find no other virtue better than a pure detachment from all things; because all other virtues have some regard for created things, whereas detachment is free from all of them". Eckhart offers the following image of how the spirit, God, can present itself to the individual: "It is far greater for me to compel God to come to me than compel myself to come to God: and that is because God is able ... to unite himself far better with me than I could unite myself with God"; moreover "there cannot be perfect detachment without perfect humility, because perfect humility proceeds from annihilation of self".[12] In such an outlook, attachment to earthly things could be serfdom, or bondage. In Eckhart's preaching, *eigenschaftigkeit* [attachment] must be mitigated by *gelassenheit* [release, or "letting be"], while the eventual gain in this tradition is nothing other than life itself.

There was much in such speculation that appealed to the ethos of the Zurich community. The energetic and pugnacious Tzara, who liked to express himself in paradoxes wherever he could, would assert in his Weimar lecture of 1922 that Dada never did want to be "modern" in the sense of making a formal progression or advance; rather it hoped to effect an escape from the forward march of style, in fact "a return to a quasi-Buddhist religion of indifference, to an even, calm state of mind in which everything is equal and unimportant". Then, and characteristically, Tzara added: "What's more I could, with the same air of conviction, maintain the opposite".[13] Ball too frequently referred to the bankruptcy of the idea of conviction – of adherence to party, philosophy, religion, country or race. And in Dada no less than in the medieval texts, such lack of commitment could now convert itself into a fully-fledged commitment to "life" itself. "The Dadaist no longer believes in the comprehension of things from *one* point of view", Ball affirmed in a diary entry for 12th June 1916, "and yet he is still convinced of the unity of all beings, of the totality of all things". Above all, the Dadaist "knows that life asserts itself in contradiction". Tzara, in the manifesto that helped launched Dada in Paris after its publication in 1918, goaded his readers with a similar paradox. "While we put on a show of being facile, we are actually searching for the central essence of things and are pleased if we can hide it ...". Tzara ends on an unmistakably vitalist note: "DADA DADA DADA ... the interweaving of contraries and of all contradictions, freaks and irrelevances: LIFE".[14]

In other Dada manifestations "creative indifference" merged effortlessly with the critique of mechanism already mentioned. Yet that critique was never direct, and in Dada could take the form of parody, even mimicry, of how mechanism had invaded the social world, threatening its spontaneity. "Dada" was not yet a word when Marcel Duchamp arrived in New York in June 1915 – like his friend Francis Picabia – in flight from the war in Europe. He had earlier been an outlier in the Paris Cubist community, from the outset taking more interest in machine imagery and arcane symbolism than in the Cubist exploration of memory and interplay of signifying systems, or the play of light, dark and planarity over time in the *Occultist Cubism* of Picasso explored in Chapter 5. Duchamp had done machine-like drawings and paintings that seemed to trace movement, exemplified by his *Nude Descending a Staircase* of 1912, or elaborate complex machine analogies, several of which came to dominate the organization of the *Large Glass* of 1915–23 with its tale of frustrated sex between mechanical devices and made entirely out of glass, metallic foil with a rigid frame. Duchamp's early so-called "ready-made" objects, the *Bicycle Wheel* of 1913, a tyre-less wheel attached to the top of a kitchen stool, or *Bottle Rack*, a piece of labour-saving café equipment purchased in a Paris store, equally speak of mechanical repetition but with the difference that here, human agency is withheld in order that the objects selected could not fit into any of the categories identified with art (Figure 7.3). The wheel was "just a distraction", Duchamp said disarmingly. "I didn't have any special reason to do it, or any intention of showing it, or describing anything. No nothing at all like that ...". It would sit in the corner of his studio and just rotate, nothing more – but also nothing less, given its self-declaration as an image of free energy only barely entropic, seemingly capable of living on its own terms. In the case of *Bottlerack*, "I just bought it", Duchamp said later, on account of its very lack of appeal aesthetically, either good or bad. "You have to approach something with an indifference, as if you had no aesthetic emotion ... visual indifference, and at the same time the total absence of good taste". He went on: The term "ready-made" "thrust itself upon me as perfect for these things that weren't works of art, that weren't sketches, and to which no art terms referred".[15] In rendering aesthetic choice itself quasi-mechanical, Duchamp had not only parodied the ordinary mechanical devices of his day but in doing so, had prompted other energies to invade the sanctuary of aesthetics itself.

Even before 1915, Duchamp's friend Francis Picabia had been at work on a series of drawings and paintings in which the machine is at once seemingly precise and at the same time lacking in actual function – as if the human and the personal stood at opposite experiential poles. In these we see car-engine parts that seem to stand for the human subject; wires, tubes and pulleys making up an allegory of function and mis-function; determinism and the shallow foundations of determinism all combined in a single complex image. Picabia's *Here: This Is Stieglitz Here* – to take an example from a series of "portraits" he completed in 1915 – presents in meticulous hand-drawn

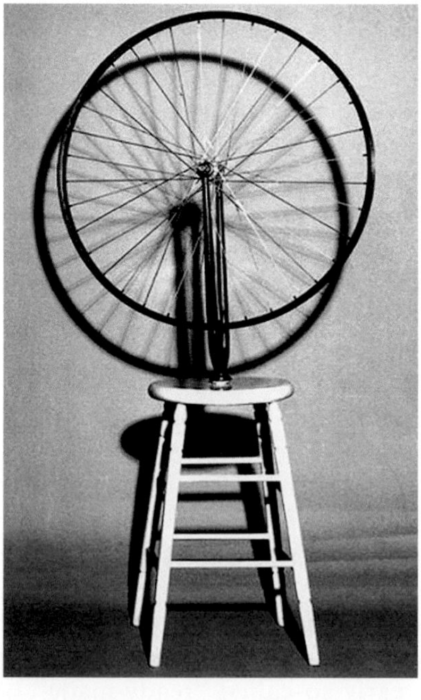

Figure 7.3 Marcel Duchamp, *Bicycle Wheel*, 1913; photograph of a lost work. (©Association Marcel Duchamp/ADAGP, Paris, and DACS, London 2021.)

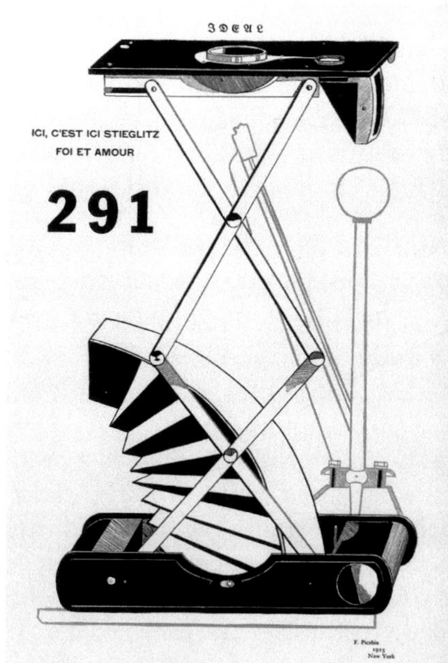

Figure 7.4 Francis Picabia, *Here, This Is Stieglitz Here,* 1915, ink, graphite and cut-and-pasted printed papers on paperboard. (Courtesy of The Metropolitan Museum New York; ©ADAGP, Paris and DACS, London 2021.)

detail the photographic impresario in a guise of a contraption that does not work (Figure 7.4). In fact, the contraption is a car handbrake and gearstick positioned in the rear-ground of a camera bellows that has become detached from the camera lens and collapsed downwards, implying perhaps the collapse of Stieglitz's *291 magazine* and gallery amidst the claim advanced by them that photography was the medium of a truly modern art.[16] Picabia was exploring elsewhere. An independently wealthy and frequent international traveller – between Barcelona, Zurich, Paris and New York – he soon began publication of his own magazine, *391*, in which literary fragments and simple dysfunctional diagrams would appear together. Alongside his machine imagery he was an experimental poet and writer, including verses in the form of lists or phrases as well as "automatic" and upside-down texts. A game with mechanism and indifference is evident in most of those writings too. In his literary work Picabia "hates professionalism", Tristan Tzara tells us approvingly. "He writes without *working* ... and doesn't control his feelings ... His poems have no ending, his prose works never start". Picabia is a writing machine possessed of a "non-zoological vitality" – Tzara's words – of a kind that belongs to him alone.[17]

Receptiveness combined with indifference in a mixture alive to the play of opposites: Such was an attitude especially important in Berlin – if we listen to Richard Huelsenbeck who, on his return there at the end of 1917, describes the city as "one of mounting hunger and hidden rage". With the Russian revolutions erupting to the east, with revolutionary socialism in the Berlin streets, and with Germany facing military defeat and the collapse of the Wilhelmine order, the affiliation of younger artists with new experiential and social values was probably inevitable. The noise and confusion of the city were essential to Huelsenbeck, to whom "Bruitism", the celebration of noise and disorder, "is life itself". Like Dada, he says, Bruitism forces one to decide: "There are only Bruitists, and others ... The screeching of a brake ... could at least give you a toothache ... Bruitism is a kind of return to nature".[18] No wonder that Huelsenbeck's attitude could look like Futurism by another name: A recognition of superhuman energies ready to animate the social fabric which was in a sense already latent within it. "Simultaneity is against what has become", Huelsenbeck says, sounding like Marinetti or Boccioni, "and for what is becoming"; the Dada group that formed in Berlin "had nothing to do with the childlike stammering to which some have tried to link it". "What is this Dadaism that I am advocating this evening?", he asked at the beginning of his *First Dada Lecture* on 23rd January 1918. He gave his own answer. It consists "of individuals who know life first-hand; it includes real characters, people of destiny with a capacity for living. People with a honed intellect, who understand they are facing a turning-point in history".[19] Huelsenbeck's sound-picture from *En Avant Dada* gives a sense of the energies just then impinging from the turmoil of the street. "The screeching of a streetcar brake and the crash of a brick falling off the roof next door reach my ear simultaneously and my (outward or inward) eye rouses itself to seize ... a swift meaning of life ... I obtain an impulse towards direct action ... I feel the form-giving force behind the bustling of the clerks in the Dresdner Bank and the simple-minded erectness of the policemen". And then, in a remark that locates vitality in the clash and clamour of the revolutionary city, he adds: "I become aware that I am alive".[20]

For all Huelsenbeck's recognition of Futurist energies, his "First Dada Manifesto", read before an audience in Berlin on 12 April 1918, made clear in noisy terms that Dada transcended all cliques and movements of art – even life itself. "Life appears as a simultaneous confusion of noises, colours and spiritual rhythms", he reiterated, "and is thus incorporated – with all the sensational screams and feverish excitements of its audacious everyday psyche and the entirety of its brutal reality – unwaveringly into Dadaist art. This is the dividing line that separates Dada from all previous artistic directions, most particularly from FUTURISM".[21]

It is in its claim to an "audacious everyday psyche" that Dada makes its commitment to a "life" contained in ordinary matter, and in different kinds of ordinary matter in combination. It is a claim that anticipates Huelsenbeck's meeting with the Hanover artist, Kurt Schwitters. "Dada demands the use of new materials in painting", Huelsenbeck says in the "First Manifesto": "a piece of cloth rips between one's fingers, one says "yes" to a life wishing to elevate itself by negation. Affirmation-negation: the gigantic hocus-pocus of Being fires the nerves of the true Dadaist".[22] Schwitters had first encountered the Berlin Dada group through his contacts with the Der Sturm gallery network where he had exhibited as an expressionist painter. Whether or not he was following Huselsenbeck's words in the "First Manifesto", Schwitters' method from the late part of 1918 – coinciding with the Armistice, the abdication of the Kaiser and the beginnings of the new Republic – consisted in scouring the streets and waste-heaps of Hanover to collect scraps of material and gluing them in complex spatial patterns on an often-miniscule scale. In one sense, Schwitters' use of pieces of rubbish was a bricolage posture that echoed those practices of "creative indifference" that had already marked Dada in Zurich and New York. In another, it harked back to the rag-picker motif in Charles Baudelaire, to the modern man and woman as celebrant of "the ephemeral, the fugitive and the contingent" in modernity.[23] For all that, Schwitters was turned down when he applied to join Berlin Club Dada late in 1918, perhaps because of his former links with Der Sturm. Yet it may have prompted him to introduce his own term, "Merz" – it could be a noun or a verb – which to him meant "essentially the totality of all imaginable materials that can be used for artistic purposes, and technically the principle that all of these individual materials have equal value ... the wheel off a pram, wire mesh, string and cotton balls – these are of equal value to paints".[24]

Though such a de-hierarchisation is similar to that which Ball had applied to words and sounds on the stage, that Arp had applied to torn and cut paper, and that Huelsenbeck had applied to the screech of a brake and a brick crashing from the roof, Schwitters' own vitalist intuition – if such it be – was to release a sort of ontological energy from bits of anonymous matter in a procedure that would have consequences for all of modern art. Assembling a group of small components and intricately interweaving them, their assembled power far exceeded their potentialities taken one by one. It was in the nature of a discovery for Schwitters that in that assembly process even the least promising material scraps became dynamic. "The artist creates by choosing, distributing and reshaping materials", he said modestly. "In Merz art, the crate cover, the playing card, the newspaper clipping become surfaces. Strings, brushstrokes or pencil strokes become lines. A stuck-on sandwich-wrap becomes glazing" (Figure 7.5).[25]

The release of perceptual and material energies locked up in unassuming things was not very different from the method at that moment being tested with destruction

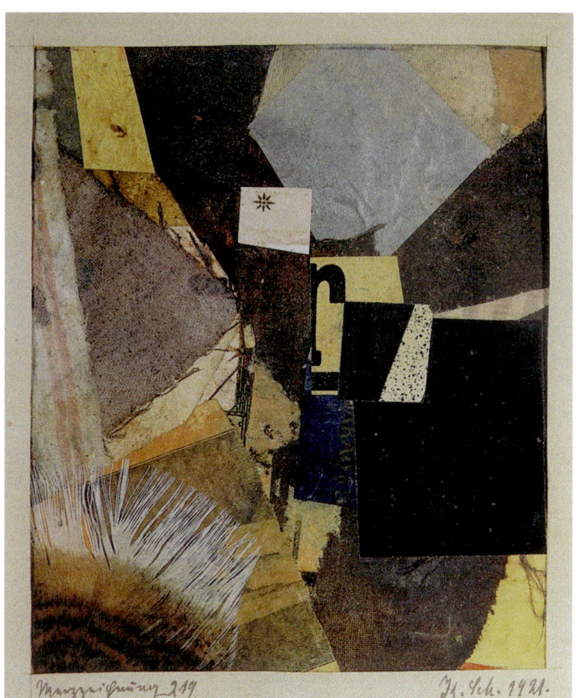

Figure 7.5 Kurt Schwitters, *Merzzeichnung 219,* 1921, 17.3 × 14 cm, collage, and gouache. (Courtesy of Kunsthandel Achenbach, Dusseldorf, ©DACS London 2021.)

by the Dutch artist Theo van Doesburg, with whom Schwitters, perhaps surprisingly, embarked on a Dada Tour of the Netherlands in the early months of 1923. Van Doesburg's own position in relation to Dada's vitalism was no less subtle than Schwitters' own. In the early part of his artistic career, his main allegiances were to dialectical systems based on the philosophies of Georg Hegel and the Dutch metaphysician Mathieu Schoenmaekers. Yet as an artist, even then he was moving to a position in which his abiding allegiance was to the principle of no system at all – other than the fundamental one that for every assertion or commitment an equal and opposite one must immediately follow. By the time of the Holland tour with Schwitters, Van Doesburg was already aligning himself with an "elementarist" position in which a limited number of basic visual and graphic elements (line, mass, direction) were subject to modification by other such elements in a dynamic and generative manner. Van Doesburg's series of paintings known as *Counter-Compositions* from 1923 would be such that vertical and horizontal dimensions were first contradicted by an intruding diagonal before each such diagonal generated a further diagonal departure – from which the process of inner contradiction would continue (Figure 7.6). The implication was that no existing philosophical system was adequate to encompass the never-to-be-completed work of art.

Even by 1921, van Doesburg had become impatient with every kind of metaphysical and aesthetic theory. This predicament suggested an interesting conceptual turn – namely a recognition that his own identity was material for internal contradiction too. With the help of an alter-ego, "Aldo Camini", a name selected to sound like an Italian Futurist, he was ready to articulate a posture of anti-philosophy in which, as van Doesburg-Camini now says, "every systematic hypothesis is impractical". "Philosophy is either charlatanism or childishness ... the only truth that interests me is *life*".[26] Scientific and philosophical systems are "outside of us", and function as a bait with which 'life' can only be caught and tamed. "I replace the scheme of birth, flourishing and decay with continuous evolution", he says with a nod to the

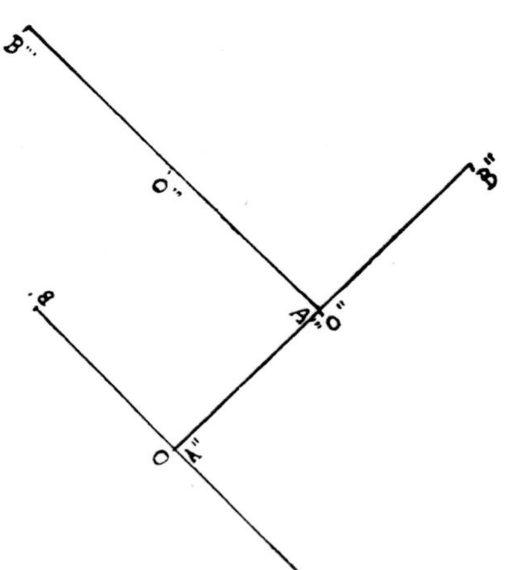

Figure 7.6 Theo van Doesburg, *Diagram showing changeable equilibrium,* "Painting: From Composition to Counter-Composition", *De Stijl,* series XIII, 1926.

influential philosophy of Bergson. Development and decay go together. "Nowhere do we find an end. Everything is in a state of development".[27]

Such a dynamic understanding of how subjectivity and the world must be comprehended was an essential aspect of van Doesburg's allegiance to what to him was the Dada idea. Generative paradox lay at its very core. A certain "I. K. Bonset", another moniker and the one under which van Doesburg conducted the tour of Holland with Schwitters, prepared an extraordinary pamphlet for the perusal of anyone attending their rumbustious and disruptive evenings. Dada "was not made but came into being", van Doesberg-Bonset exclaimed there. Dada "creates a 'point of indifference' beyond man's understanding of time and space; Dada is yes-no, a bird on four legs, a square without corners … We neo-vitalists, Dadaists, destructive constructivists have laid bare the abcesses which hide the world's body by crying 'look look look here here here [but then] nothing nothing nothing'". Van Doesburg-Bonset ends his pamphlet with a remark that has three further and unusual vitalist allegiances in mind. "The most outstanding and intelligent people nowadays", he asserts, "like [Albert] Einstein, [Charlie] Chaplin and [Henri] Bergson, want to be counted amongst the Dadaists – who declare in almost all their manifestos that they do not want anything, know anything, are anything". Each new Dada recruit must have seemed significant in his own way. Einstein was the Dada who revealed the paradoxes of curved space, unevenly distributed rates of time, and relative rather than absolute gravity. Chaplin qualified due to his screen account of the social universe as essentially comic – having performed in Paris and Berlin during his Europe tour beginning in September 1921 and having caused a sensation in the European press. Van Doesburg's pamphlet ends by returning to the vitalist principle. "To think that Dada means only destruction is to misunderstand life", as he again puts it, "of which Dada is the expression".[28]

And Bergson? "We were very interested in Bergson", Hugo Ball tells us in his diary of the Zurich Cabaret.[29] It seems clear that Bergson was now a Dada not only on account of his dynamic conception of sensation (and also therefore of cognition), but also for his assault on the phenomenology of negation as a component of experience at its most fundamental level. Perhaps van Doesburg had noticed that part of *L'Évolution créatrice* in which Bergson had tackled the conundrums surrounding "The Idea of Nothing". In that section, Bergson had argued strenuously against the illusion that nothing is the lack of anything – that the void is rendered intelligible by imagining the removal of everything formerly within it, that nothing or nought [*le néant*] is merely the absence of every something [*l'existence*]. On the contrary, Bergson had said in a remark that pre-dates the Dada rebellion but seems to rehearse it, "the image of nothing is an image full of things, an image that includes at once that of the subject and that of the object and, besides, a perpetual leaping from one to the other and the refusal ever to come to rest finally on either".[30]

The image serves well to detach Dada from its reputation as merely destructive and to underline once more its commitment to a continual interplay and rebalancing of contraries. Bergson's insistence upon a "perpetual leaping" between contraries resonates well with Dada's more general adherence to a dialectic of extremes – construction and destruction, assertion and denial, receptiveness and indifference – as well as a further duality that for obvious reasons has exerted a special appeal in the visual arts. This is the dialogue of presence and absence in visual form – of colour, articulation, texture, mass, indeed any category, empty or full, in the experience of the visual world. Could absence itself – emptiness, nullity, negation, non-occurrence – function as a prior condition of fullness of form? That way, so several Dadas believed, semantic vacancy or indeterminacy could also function as the beginning of meaning itself: A larger, higher meaning that accumulates by virtue of being ungraspable, beyond mundane comprehension and beyond fixed bourgeois categories of "the empty" or "the full".

A blank piece of paper or an empty page. Silence, a gap between sounds. Darkness. Sensory nothingness. We find several exemplifications of the full-empty dialectic in European art between 1915 and about 1923. Fulness and nullity together are the single most prominent visual and material attributes of Kazimir Malevich's *Black Square* paintings and drawings (as witnessed directly and not just in poor photographic reproduction). First shown in the notorious Last Futurist Exhibition of Paintings 0.10 in St Petersburg in 1915, they became pervasive in the manner that Malevich called Suprematism (his journal *Supremus* was originally to be called *Nul*). Across the corner at 0.10 there hung a black near-square carefully positioned within – never on – a white ground; tonal complementaries are thick with presence and immediacy, with their own declarative *faktura*. Malevich called their co-existence "the zero of form": A black-square-on-white-ground as both presence and absence simultaneously in the sight and sensory envelope of the viewer. And that duality was not in fact new. Did Malevich know the black square diagram in Robert Fludd's magnum opus *History of the Two Worlds*, published in Germany in 1617–21? Itself a richly cross-hatched copper-plate engraving – "inked to oblivion" in the words of a recent exhibition of the book – it was designed to picture the emptiness of deep space before the creation of the heavenly bodies and the appearance of animal and eventually human life (Figure 7.7). In Fludd's treatise the black square is bordered on each side by the words *Et sic in infinitum* – "and thus to infinity".[31]

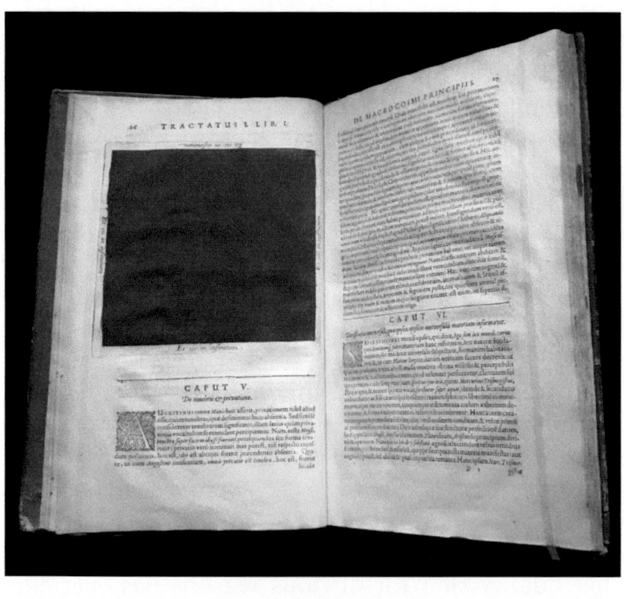

Figure 7.7 Robert Fludd, *Utriusque Cosmi Mairoris Scilicet et Minoris Metaphysica, Physica atque Technica Historia*, 5 volumes, Oppenheim 1617–21, p. 26. St John's College Oxford.

Whether or not Malevich was inspired by the texts of medieval mystical speculation, the interplay of presence and absence was foundational to several artists of the Russian avant-garde who functioned seemingly independently of Western Europe and its deeper metaphysical traditions, even as they often seemed cognizant of them. Those connections and correspondences are becoming increasingly well documented.[32] Ivan Puni, one of the organizers of 0.10, arrived in Berlin in the autumn of 1920 as himself, apparently an exponent of the full-empty work. Puni's *Untitled (The Hunger Plate)* of c.1918 asks to be read as a full signifier of extreme hunger, even starvation, at the time of Russia's post-Revolutionary civil war, at the same time as it accomplishes this signification by means of an empty plate – a geometrical circle positioned at the centre of a bare wooden table (Figure 7.8). It should be remembered too that, aside from the machine, one of the founding targets of the Russian avant-garde had been rational language, especially the confinements of bureaucratic language and established political power. Elaborate wordplay had distinguished Malevich's poet colleagues, Aleksei Krychenykh and Velimir Khlebnikov in the pre-Revolutionary period. In the chaos of the post-Revolution, a resort to semantic negation remained a powerful and potentially dialectical tool, as is elaborated in Chapter 10 of this book. And yet the Nothingist (*Nichevochestvo*) group that flourished briefly in Rostov-on-Don may not have heard the word "Dada" by the time of their trans-rational [*zaumnyi*] activities between 1920 and 1922.

The remainder of this chapter will briefly narrate the passage of Dada into neo-Dada – a passage in which several of the vitalist sympathies of original Dada remained more or less intact. In 1935, the young musician, artist and composer John Cage became acquainted with two innovations that would bring his study of Schoenberg's musical system to a close and help spark a revival of Dada-like attitudes and activities across the visual and performing arts throughout the West. When Cage saw the early works

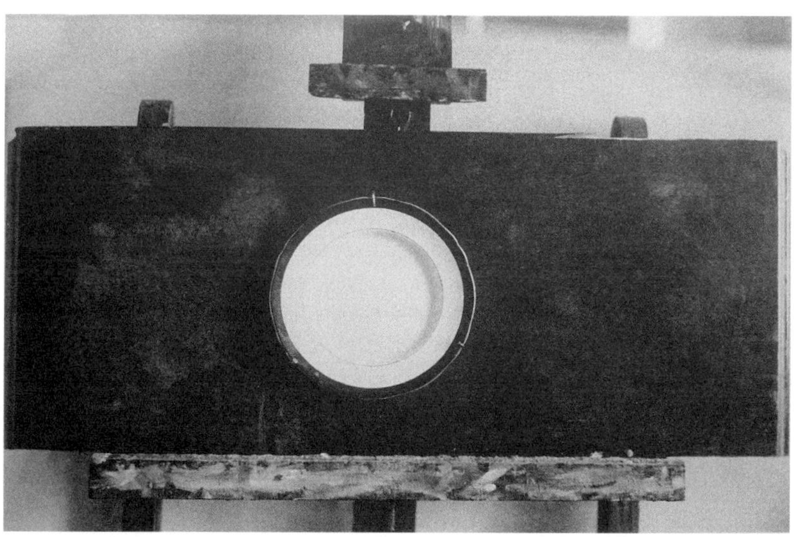

Figure 7.8 Ivan Puni, *Untitled (Hunger Plate),* c.1918 (lost work), photo: gelatin silver print. (Courtesy of State Mayakovsky Museum, Moscow, HB-1813, ©ADAGP, Paris and DACS, London 2021.)

of Duchamp at the home of Walter and Louise Ahrensberg in California, he learnt for the first time what the term "ready-made" approximately meant.[33] On reading Luigi Russolo's *Art of Noises* manifesto of 1913, he quickly concluded that the future of music lay not in conventional orchestral or instrumental sound, but in noise. Noise "has the power to pull us into life itself", Russolo had written. Russolo's noise categories, including human and animal ones such as mumbles, hisses, murmurs, wheezes and sobs – to Cage, could be ready-mades too.[34] In Seattle in the late 1930s, when Cage came under the tutelage of the painter Mark Tobey while both were teaching at the Cornish School of Art, he began to understand Tobey's conviction that anything can be interesting, that no hierarchy of value existed between one sensory experience and any other: From that moment Cage's neo-Dada manner in music and painting was born.[35] Tobey proved foundational to Cage's interest in the cross-overs between music, performance, and the visual, and in the process of deciding temporal structure with the help of randomized rules. By non-deterministic means such as consulting the *I Ching* or the tossing of coins, Cage discovered that visual absence and sonic absence – silence – could be more or less aligned. Armed with a growing fascination for Zen Buddhism as well as for the medieval mystics, Cage was soon integrated into the artistic community in New York and became a regular attendee at their meetings and events. A statement from his "Lecture on Nothing", read before the Artists' Club on Eighth Street, New York in 1949, demonstrates the quietism that by then marked Cage's distance from the attitudes of his much earlier Dada peers.

It was an important moment in the formation of a late-Dada community in New York, a community whose detailed knowledge of the Zurich Cabaret would remain sketchy until the publication of Robert Motherwell's anthology in 1951, *The Dada Painters and Poets*. A characteristic passage from Cage's "Lecture on Nothing" reads: "More and more ... I have the feeling ... that we are getting nowhere ... Slowly, ... as the talk goes on ... we are getting nowhere ... and that is a pleasure ... It is not irritating ... to be where one is .... It is only irritating ... to think one would like to be somewhere else".[36] Having met the young Robert Rauschenberg at Black Mountain College soon after Rauschenberg's enrolment there in late 1949, Cage would soon write about the artist's work that it established the possibility of "looking anywhere, not just where someone arranged that you should ... [in Rauschenberg] there is the same acceptance of what happens and no tendency towards gesture or arrangement". And it was with Rauschenberg and Duchamp in mind that he cited Bergson in support of a credo that in art as in life, "something and nothing need each other".[37] By now, interest in European Dada was becoming palpable. Cage would shortly write another lecture, probably in February 1951, titled "Lecture on Something", ostensibly on the music of the composer Morton Feldman. In the meantime he had become absorbed in Zen Buddhism through the writings (and later the Columbia lectures) of Daisetz Suzuki, and in Meister Eckhart, with whose words Cage ended his lecture. They are words that the Zurich community had surely known: "Earth [that is something] has no escape from heaven [that is nothing]; flee she up or flee she down, heaven still invades her, energizing her, fructifying her, whether for her weal or for her woe".[38]

Whether Dada can be called vitalistic should remain, in the end, undecidable. Dada's dislike of final principles and fixed solutions is part of the reason. Its mobility and inscrutability is another. The following anecdote concerning Cage in 1949 supports it. In Cage's version of Dada, Meister Eckhart's proposal that the Nothing "energises" and "fructifies" the Something, and Bergson's insistence that the Something and the Nothing "need each other", are paradoxes that cannot be proven but must be understood within the human task of wresting meaning from matter's entropic collapse, its decay into non-meaning. Accordingly, for the end of his "Lecture on Nothing" Cage prepared some answers to any questions that audience members might ask – to be offered irrespective of what the actual questions were. Cage's first prepared answer can function as a concluding thought on whether Dada in any of its forms was vitalistic. "It's a very good question", Cage was ready to say. "I should not want to spoil it with an answer".[39]

Notes

1 Both passages are from H. Ball, *Flight Out of Time: A Dada Diary*, edited and introduced by John Elderfield, (Berkeley: University of California Press, 1974) 3–4.
2 Ball, *Flight Out of Time*, entry for 30 March 1916, 57.
3 Ball, *Flight Out of Time*, entry for 12 June 1916, 65.
4 Jean Arp, "Dadaland" (1948), in M. Jean (ed), *Jean Arp: Collected French Writings* (London: Calder and Boyars, 1974) 234.
5 L. Russolo, *The Art of Noises* (1913), in U. Apollonio (ed), *Futurist Manifestos* (New York: The Viking Press 1973) 86.
6 Russolo, *The Art of Noises*, 86.
7 Hans Arp, "And So the Circle Closed", in *Arp: On My Way: Poetry and Essays 1912-1947*, (New York: Wittenborn Schulz, 1948) 77.
8 Hans Arp, "The Germ of a New Plastic Work", *Arp: On My Way*, 1948, 70.
9 N. Tarabukin, *Le Dernier Tableau* (written 1916); (Paris: Editions Champs Libres, 1972) 104.
10 H. Richter, *Dada: Art and Anti-Art* (1964); (London edition: Thames and Hudson, 2016) 60.
11 Rudolf Steiner, *Mystics of the Renaissance and Their Relation to Modern Thought* (Berlin 1901; New York and London: G. P. Putnam & Sons, 1911).
12 Meister Eckhart, "On Detachment", in *Meister Eckhart: The Essential Sermons, Commentaries, Treatises and Defense* (trans E. Colledge and B. McGinn), (New Jersey: Paulist Press, 1981) 286.
13 Tristan Tzara, "Lecture on Dada", Weimar, 23 September 1922, in *Seven Dada Manifestos and Lampisteries*, (Calder, London and Riverrun Press, New York, 1977); *Merz*, Jan 1924, 108.
14 Hugo Ball, *Flight Out of Time*, entry for 12th June 1916; and T. Tzara, *Manifesto 1918*, in R. Huelsenbeck, *Dada Almanac*, 1920, ed. M. Green (London: Atlas Press, 1998) 127–8, 131, 132.
15 Paul Cabanne, *Dialogues with Marcel Duchamp* (London: Thames and Hudson, 1971) 47, 48, 48, 48, respectively.
16 Francis Naumann, "Francis Picabia", *New York Dada 1915-23* (New York: Harry N. Abrams, 1994) 60–1. Above the machine itself we read the word "Ideal" presented in capitalized script, at once affirming a Tzara-like commitment to always changing one's mind.
17 Tristan Tzara, "Francis Picabia", *Dada 4 et 5*, Zurich, May 1919; *Seven Dada Manifestos and Lampisteries*, trans. Barbara Wright (John Calder Nov., 1977) 81.
18 Richard Huelsenbeck, *En Avant Dada* (1920), in R. Motherwell (ed), *The Dada Painters and Poets*, (New York: Wittenborn, Schulz Inc, 1951; Harvard University Press edition 1981), 25–6.
19 Richard Huelsenbeck, "First Dada Lecture", 23 January 1918; in Huelsenbeck, *Dada Almanac*, 1998, 113.
20 Richard Huelsenbeck, *En Avant Dada*, 1920, 39, 35–6.
21 Richard Huelsenbeck, "First Dada Manifesto", 12 April 1918, *Dada Almanac*, (London: Atlas Press, 1992), 44–9.
22 Richard Huelsenbeck, "First Dada Manifesto", 1918; 1992, 44–9.

23 Charles Baudelaire, “The Painter of Modern Life” (1863), in *Charles Baudelaire: The Painter of Modern Life and Other Essays* (London: Phaidon Press, 1995), 12.
24 Kurt Schwitters, “Merz Painting”, 1919, in *Kurt Schwitters: I Is Style*, Amsterdam, n. d. 91.
25 Schwitters, “Merz Painting”, 91.
26 Van Doesburg (Aldo Camini), “Caminoscopy: an anti-philosophical view of life without any thread or system”, *De Stijl*, Vol. IV, 5 (June 1921) 66–7; here J. Baljeu, *Theo van Doesburg* (London: Studio Vista, 1974) 113. My emphasis.
27 Van Doesburg, “The Will to Style”, *De Stijl*, Vol V, 4 (April 1922) 59–62; Baljeu, *Theo van Doesburg*, 117.
28 Van Doesburg, “What Is Dada ???????”, *De Stijl*, The Hague, 1923; Baljeu, *Theo Van Doesburg*, 133.
29 Hugo Ball, *Flight Out of Time*, entry for 9 September 1917.
30 Henri Bergson, *L’Évolution créatrice*, 1907; *Creative Evolution*, 1911, A. Mitchell translation (New York: Henry Holt and Co, 1911; New York reprint 1944) 304.
31 R. Fludd, *Utriusque Cosmi* ..., Oppenheim 1617–21, 26 (from the exhibition at the Weston Reading Room, Oxford University, 2019).
32 See for instance M. Tupitsyn (ed), *Russian Dada 1914-1924* (Madrid: Museo Nacional Centre de Arte Reina Sofia, and Cambridge, Mass., The MIT Press, 2018).
33 For this and other biographical information on Cage’s early life, see K. Larson, *Where the Heart Beats: John Cage, Zen Buddhism and the inner Life of Artists* (New York: Penguin Books, 2013).
34 Russolo, *The Art of Noises*, 87.
35 For an argument about the eruption of neo-Dada in American art of the 1940s and 1950s, see C. Craft, *An Audience of Artists: Dada, Neo-Dada, and the Emergence of Abstract Expressionism* (Chicago: University of Chicago Press, 2012).
36 Cage, “Lecture on Nothing” (c.1949), *Silence: Lectures and Writings*, 124.
37 Cage, “On Robert Rauschenberg, Artist, and His Work” (1961), *Silence: Lectures and Writings*, (London: Calder and Boyars, 1986) 100, 102; and “Experimental Music” (1957), *Silence: Lectures and Writings*, 12.
38 Cage, “Lecture on Something” (c.1951), *Silence: Lectures and Writings*, 145.
39 Cage, “Lecture on Nothing” (c.1949), *Silence: Lectures and Writings*, 126.

8 Henri Bergson and Surrealism

Art, The Vital Impetus and The Persistence of Memory

Donna Roberts

If we consider the elements of Henri Bergson's philosophy that focus upon the value of experience and inner life and the complexities of temporality, or upon the fluidity of being and the dynamic of a vital creative impetus, we might wonder why its influence has not been more discussed in relation to the Surrealist movement. In this chapter, I shall explore the relations of Surrealist thought and art with Bergsonism as a vitalist evolutionary philosophy and a philosophy of experience. The question that joins these two seemingly disparate intellectual entities is the problematic of *Life*, specifically, how life and knowledge cannot be strictly separated, and how Bergson and the Surrealists argued that different methods are required to better understand life beyond the intellectual habit of separating its inner dimensions from its outer dimensions.

In *Creative Evolution*, Bergson presented a theory of nature as inherently driven by the *élan vital* – the vital impetus – thereby offering a vivid organic model for human creativity that found particularly fertile ground in the visual arts. Although the influence of Bergson's ideas on modern art is well established, it seems that both the political position of Surrealism and its evidently Freudian concerns have worked to blind-side inquiry into the possibilities of a latent Bergsonian current within Surrealist art and thought.[1] Nevertheless, since Bergson's philosophy aimed to overcome, on the one hand, the reduction and misrepresentation of experience by the dualistic approaches of idealism and materialism and, on the other, the intellectual tendency to analyse the world according to fixed categories rather than open-ended flux, Surrealism can be aligned to it. Moreover, the Surrealists' critical scrutiny of prevailing ideological and philosophical concepts of nature and the place of the human subject within it closely corresponds with Bergson's critique of classical and Enlightenment thinking on nature and our complex experience of reality and time.

Bergson is barely mentioned in the annals of Surrealism, other than in a 1931 polemic as a philosopher *not* to read and one condemned by the leader of the Surrealist movement, André Breton, for his participation in French nationalistic propaganda during the war of 1914–18.[2] In locating the Bergsonian currents within Surrealism, it is then necessary to dig quite deeply into the subtleties of language, reference points, and the rich hinterland of ideas from the fields of philosophy, psychology and art which were densely interconnected in the early decades of the twentieth century. The argument in this chapter for connections between Surrealist ideas and the writings of Bergson is premised on the fact that Surrealism developed in France at a time when intellectual and creative life had been saturated with Bergson's ideas for the previous twenty years or more, as indicated in the *Introduction* to this book and the

DOI: 10.4324/9781003045595-12

chapters that follow. Additionally, the thinkers who influenced the Surrealists were either directly influenced by Bergson or developed their ideas according to a similar psycho-biological framework, as exemplified by the poetics of Pierre Reverdy and the psychoanalyses of Sigmund Freud.

Given Bergson's immense contribution to discourses on the experience of time, perception, the unconscious, memory, and dreams; non-ordinary states of consciousness and the philosophical critique of dualistic thinking, there seems much connective tissue with Surrealism. So many Surrealist artists explored the complexities of time and memory, the fluidity of material morphologies, and the power of art and poetry to tap into a vital vein of creative energy freed from the confines of intellect and attention to utilitarian life.

Drawing upon these vital interconnections, I shall first discuss Bergson's view on art and his evolutionary view of life as open-ended, self-creative, and interconnective in relation to Vitalism. In the subsequent part, I shall explore how life was conceived in relation to Vitalism and experimentation as a means to attain knowledge in André Breton's and Roger Caillois' Surrealist thinking on art and nature. In the final section, I shall focus on specific Surrealist artworks, reading Salvador Dalí's *The Persistence of Memory* (Figure 8.1), Yves Tanguy's *There! (The Evening Before)* (Figure 8.2) and his *Heredity of Acquired Characteristics* (Figure 8.3) through Bergson's ideas on the fluidity of time and the nature of memory.

Bergson's *élan vital*, temporality, and Vitalism

In *Creative Evolution*, Bergson presents his evolutionary concept of the *élan vital* as the "original impetus of life", a counter to both Neo-Darwinian mechanisms and Neo-Lamarckian finalism.[3] For Bergson, life develops from an original state of unity into the divergent directions of the natural kingdoms. What is important in this text with regards to Surrealism can be defined by three points. The first point concerns Bergson's rejection of mechanism as a closed system independent of temporality and his development of ideas concerning life as complex, inter-relational, and co-evolutionary. The second concerns Bergson's rejection of finalism as predicated upon a teleological principle by which life proceeds as if with some pre-determined aim or end. The third concerns Bergson's idea of life as a self-creating impetus, which finds a precedent in the romantic nature philosophy of Johann Wolfgang von Goethe and Friedrich Schelling. Maurice Merleau-Ponty made a connection between Bergson and romantic nature philosophy in terms of an organic history in which divergent multiplicity originated in an undivided state. As Merleau-Ponty states: "Like Schelling, Bergson conceives of nature as a 'primordial lost non-dividedness,' and the *élan vital* takes up the question of a natural productivity without the idea of a final end."[4] It is within such an understanding of the *élan vital* that we can locate Surrealist practice in general, in terms of poetry, art, and automatism, including the more ephemeral automatist praxis of wandering through city and countryside while receptive to affective encounters with objects, places, or beings.

Commenting on how, in *Creative Evolution*, Bergson develops his theory of temporality as duration into the *élan vital*, Richard A. Cohen identifies Bergson as having modified his theory of time according to his understanding of evolution as a "unitary notion of organic development, cumulative process [and] inner growth".[5] It is through Bergson's grasp of the "integral developing interrelationship of matter and spirit, mind and body", writes Cohen, that he "launches an *eco-logical* vision of the contemporary world" setting Bergson apart from previous thinkers of change.[6] As distinct from such philosophers as Heidegger who take account of the lost paradigm of change and transformation in Western thought, Bergson bases his philosophy in evolutionary theory. In so doing, he posits a theory of the bias of the intellect towards fixity and distinction as a product of evolution itself. "The intellect", writes Bergson, "is not made to think *evolution* in the proper sense of the word – that is to say, the continuity of a change that is purely mobile."[7] It is only through the method of intuition that Bergson believes we can overcome thinking in terms of fixity or conceiving of time and change in spatial terms. It is through this integral organic model that Bergson's philosophy seems to overcome apparent dualisms, as Cohen argues:

> [Bergson's] integral and dynamic unity of mind and body, matter and spirit, past and present – elaborated by Bergson in many writings through the notions of growth and creative evolution, borrowed from and reflecting the organic world – inaugurates a new epoch in the Western spirit.[8]

Similarly, Elizabeth Grosz has defined Bergson's *élan vital* with a particular sense of its historical and unifying nature as "an inner force that life shares with the forms of life that come before it, linking it to a vast chain of life that no living being, including Man, may be conscious of, yet which produces life interconnected in its every detail to all other living forms".[9] Grosz's emphasis on the historical dynamic of Bergson's vital impetus relates to the development into evolutionary terms of Bergson's theory of memory, where it is defined as the co-existence of an infinite plurality of durations. "Is it not plain", asks Bergson, "that life goes to work here exactly like consciousness, exactly like memory?"[10] This shift in *Creative Evolution* is described by Gilles Deleuze as the point in Bergson's philosophy "where life itself is compared to a memory, the genera or species corresponding to coexisting degrees of this vital memory".[11] This point is particularly significant in relation to reading Surrealism with attention to its poetic and visual strategies, which so often invoke both memory and the organic (and inorganic) processes within mind and nature in order to unbind the principle of identity, while blurring distinctions between natural forms or between species.[12] This kind of melding of memory and organicism is exemplified by the paintings of Tanguy, as shall be illuminated.

Nevertheless the question of what might be specifically *vitalist* in the connections between Bergson and Surrealism is by no means straightforward. Not least because Bergson's status as a vitalist philosopher is contentious but also because transposing a discussion of Surrealism into the complexities of what Vitalism was and was not, or indeed is and is not, would require unreasonable allowance for speculation. Nonetheless, the question of whether Bergson was in fact a vitalist is worth addressing because it opens out a comparative approach to thinking about Bergson and Surrealism to a shared historical context concerning the ongoing problematic of *life*. Both Grosz and Keith Ansell Pearson, for example, have argued that Bergson is not a vitalist, that is, according to the classical definition of the term:

> Bergson is careful to distinguish his position from Vitalism, the claim that there is a special substance, force or form that distinguishes life from non-life, although his concept of the *élan vital* is commonly regarded as a form of Vitalism. Vitalism

itself, of course, takes many forms. But in his sense, in the sense of a special force distinguishable from other natural or material forces, Bergson cannot be regarded as a vitalist.[13]

Grosz's view here can be further elucidated by Monica Greco's writing on Vitalism, which she has defined as a shifting concept that constantly requires a historical frame of reference.[14] Since its eighteenth-century formulation in opposition to the classical Newtonian reduction of all life to mechanical principles, according to Greco, the premises of Vitalism have changed and have even been reversed.[15] For Greco, Vitalism has generically been defined as the notion that "living organisms are fundamentally different to non-living entities because they contain some non-physical element or are governed by different principles than are inanimate things."[16] Greco claims that early twentieth-century philosophers such as Bergson and Alfred North Whitehead developed an anti-mechanistic view of life that has since fed, along with that of Georges Canguilhem through the work of Deleuze, into another historical frame of reference for understanding Vitalism.[17] This more recently developed frame of reference thus presents the Vitalism of such early-twentieth-century philosophers as not based on a fundamental difference between animate and inanimate entities but, rather, as a pervasive view of life as irreducible to the paradigm of the machine in which there is *no* fundamental distinction between the animate and the inanimate. As Greco writes: "the generic concept of 'life' understood as a being's capacity to be active, to be sensitive and to grow, no longer applies exclusively to biological beings but has rather become irrelevant beyond biology, across nature."[18] Such a view of Vitalism would be significant when considering the attention afforded by Surrealists to the impersonal effect of the inorganic, as exemplified by the attention to stones and minerals in the writings of Breton and Roger Caillois as ideal models for human thought and artistic aspiration. Breton's eulogy to the crystal in *Mad Love* (1937), for example, reveals an acute sensibility towards the creative forces of the inanimate: "I have never stopped advocating creation, spontaneous action, insofar as the crystal, non-perfectible by definition, is the perfect example of it."[19] Although Jane Bennett in her book on *Vibrant Matter* seems entirely unaware of the contributions of Surrealism to this discourse, Breton's creative paradigm finds a belated echo in Bennett's concepts of "vibrant matter" and "vital materialism" that she bases on "the vitality of matter and the lively powers of material formations."[20]

In order to develop a conceptual background for exploring shared connections to Vitalism in Bergson and Surrealism, we would therefore need to move away from the long-standing view of Vitalism as essentially distinguishing animate from inanimate to these recent formulations of Vitalism. One such view that would draw together Bergsonian and Surrealist thinking would be that which, according to Greco, would place Vitalism within an "implicit enthusiasm for the 'vital' as a signifier of contingency, potentiality and the possibility of change."[21] Another recent reformulation of Vitalism which would apply both to Surrealism and Bergson's thinking would be that proposed by Robert Mitchell, who sees in both artists and scientists drawn to neo-Vitalism a shared sense that "life is not a self-evident fact that can be taken for granted but rather a source of perplexity that demands new modes of conceptual and practical experimentation."[22] Surrealism would thus seem to align directly with Mitchell's concept of "experimental Vitalism", as opposed to theoretical Vitalism, as he explains: "Experimental vitalists begin with a *sense* that life cannot be fully explained by current scientific concepts and assumptions and then develop experiments in order to provoke new questions and concepts about life and living beings."[23] Surrealist practice, whether in the visual arts, poetry, or affective encounters, may then be perceived as exploring such experimentation with the problematic of life in all its complexities, which includes dreams and the irrational.[24]

Bergson, Breton, Caillois: Art and Nature

It is notable that in his 1948 book on Breton, Julien Gracq described Breton's identification of Surrealism with psychic automatism in the *Manifesto of Surrealism* (1924) as highly Bergsonian: "One cannot fail to see in Surrealism an irreproachably Bergsonian enterprise of inspiration, where the obsessive fear of abyssal dives into the forbidden domain, in the midst of Freudian monsters, seems to hold no place."[25] It was only some years later, Gracq proposes, that this properly Bergsonian model of inspiration was supplanted by one more resolutely Freudian. To appreciate Gracq's understanding of Surrealism, and automatism in particular, as following a Bergsonian model of inspiration, it may be helpful to refer to Bergson's numerous comments on the visionary character of the artist and poet. This was conveyed in his 1900 publication, *Laughter: An Essay on the Meaning of the Comic*, and Bergson's 1911 lecture, "The Perception of Change," as well as in his comments on the image in both *Laughter* and his 1903 essay "Introduction to Metaphysics". The latter hugely influenced Pierre Reverdy's notion of the image as a juxtaposition of disparate entities, which became key to Surrealist theory.[26] Yet it is in the former two texts that Bergson characterises the artist-poet as a revelatory figure, very much like Arthur Rimbaud's seer, which Breton adopted as a model for Surrealist insight. The artist-poet is for Bergson a "revealing agent",[27] capable of bypassing utilitarian concerns, entering into processes of interiority and duration, who contrives "to make us see something of what they have seen".[28] Breton's characterisation of the Surrealist artist or poet as delving deep into a visionary imagination or unconscious world thus has much in common with Bergson's view. Furthermore, Bergson's sense of the powers of the artist-poet to activate a recognition of such hidden depths is an idea that becomes vital to the Surrealists' claim for the transformative effect of art and poetics able to pass from aesthetics and literature into the realms of life and politics. Artists, for Bergson:

> "impel us to set in motion, in the depths of our being, some secret chord which was only waiting to thrill. So art [...] has no other object than to brush aside the utilitarian symbols, the conventional and socially accepted generalities, in short, everything that veils reality from us, in order to bring us face to face with reality itself."[29]

Had Breton been more receptive to acknowledging Bergson and not, in my view, disavowing his significance due to ideological imperatives, the genealogy of Bretonian Surrealism might have looked differently. Furthermore, if Breton had not implicitly rejected biologically influenced philosophy as reductive and Darwinistic, he may have found an ally in Bergson, especially the Bergson of *Creative Evolution*. Not only does Bergson's critique of the limits of intellect resonate with Breton's thinking but also with Bergson's notion of life as an open-ended, organic, and irrepressibly self-creative impetus that finds a parallel in the Surrealists understanding of creative life in the broadest sense. In Breton's thinking about environments, such as Paris, Tenerife, or Martinique, his highly reflexive and lyrical view of the complexities of experiencing an environment finds a parallel with Bergson's thinking on the necessarily relational character of life. According to Keith Ansell-Pearson, for Bergson, an organism – human included – does not simply adapt to environments, it *responds*, it *modifies* its relationship with its milieu.[30]

This insistence on an open system seems to relate receptivity to both internal impetus and external influence, which characterises the Surrealist position. Moreover, in an interview with Breton given in 1941, Breton – a keen observer of nature – clearly conveyed his own conviction in the open, reciprocal, interdependent dynamic of life, notably the *relations* between species being as equally important as their morphological structures. When speaking of the butterflies he had observed in the countryside around New York, Breton declared:

> Ought the description of a plant omit that of the caterpillar or the larva which, more or less electively, lives on it? The affinity of an animal organism with such a species – is it not as significant as its type of inflorescence, for instance? But the mania for classification tends to get the upper hand over all the real methods of knowledge. I'm really afraid that natural philosophy has not advanced a step since Hegel.[31]

Despite the prescience of his observation about species interaction, Breton's reliance on G.W.F. Hegel here appears to reveals his blindness to Bergson's writings on nature. Although Breton's observations might well sound akin to more recent ecological thinking, it was Bergson who made the break from Hegelian dialectics to the evolutionary model that has offered a bio-philosophical basis for ecological thinking ever since. As Ansell-Pearson has shown, Bergson's ideas find their place alongside ecologically focused biology for which complexity theory and co-evolution are at the core: The "new biology of complexity theory emphasises the fact that an environment cannot be separated from what organisms are and what they do".[32] In Bergson's open system, owing to original evolutionary origins, species exist in sympathy with one another in "reciprocal interpenetration".[33] In a text entitled *Bergson and the War Against Nature,* Peter Gunter also affirms Bergson's co-evolutionary thinking of presumably the kind that Breton had in mind when observing the symbiosis between plants and insects. Humans and all living things, Bergson asserts, "hold together", as he elaborates:

> They are parts of a whole, no members of which can be understood as existing in splendid isolation from the others. Man, from his own vantage point, is not alien to the other creatures, nor they to him. He possesses the capacity for profound sympathy with everything that lives.[34]

Hence, despite his disregard for Bergson, Breton's poetic-naturalist observations about sympathetic human-nature relations within both urban and rural environments strongly resonate with Bergson's ideas. Whether relating experiences in the city or the natural landscape, Breton's writings brim over with a sense of the constant fluctuations of internal and external necessity, the entanglement of thoughts, perceptions, emotions, sensations, as well as organisms resonating and interacting in such a way that the material and the vital can never be strictly separated.

Although to an extent we might wonder how the organically minded Surrealists missed the implications of *Creative Evolution*, we must remember that these readings of Bergson are relatively recent. Coming only fifty years after Deleuze's reappraisal of

Bergsonism, they have also arisen after developments in eco-philosophy, complexity theory and the shift towards a co-evolutionary paradigm. Bergson's co-evolutionary thinking was not missed, however, by Roger Caillois, and it is precisely this point of shared origins that Caillois exploits in his strategy of "comparative biology", as articulated in his two texts published in *Minotaure*, "The Praying Mantis: From Biology to Psychoanalysis" (1934) and "Mimicry and Legendary Psychasthenia" (1935), wherein he compares insect behaviour to human psychology and myth. That Caillois so perversely exploits Bergson's famous distinction, between the evolutionary tendencies of insects towards instinct and humans towards intelligence, is perhaps not surprising given Bergson's own examples of instinctual sympathy. In the chapter of *Creative Evolution* entitled "Intelligence and Instinct", Bergson describes the ghoulish behaviour of certain insects *as if* they arose from an uncanny knowledge akin to intelligence. He describes a certain wasp that "like a learned entomologist and a skilful surgeon" stings its prey at a certain exact point in the nervous system so as to paralyse but not kill it, in order to retain it alive for nourishment.[35] For Bergson, this kind of instinctual "sympathy" exists because all living things derive from the same evolutionary origin. As he explains: "Thus the instinctive knowledge which one species possesses of another on a certain particular point has its roots in the very unity of life, which is, to use the expression of an ancient philosopher, a 'whole sympathetic to itself.'"[36]

"The gothic beauty of Bergson's nightmarish and florid vision of universal sympathy", as Christian Kerslake calls it,[37] is what I propose influences Caillois to develop his own ghoulish invocation of atavistic human-insect relations. Caillois alludes to Bergson in both the aforementioned articles but without referencing his work, and in the latter compares seemingly self-destructive behaviour in insects and humans with "the inertia of the *élan vital*".[38] In his mantis essay, Caillois bridges mythic and psychological disorders in humans with behaviour in insects, considering the example of castration anxiety as "the vestigial residue in one species of behavioural patterns observed in others".[39] This looks, in fact, very much like the theory of myth as a residual instinct that Bergson proposed in *Two Sources of Religion and Morality*,[40] to which Caillois referred some years later in *Myth and Man*.[41] Thus, Caillois' disconcerting comparative biology in the *Minotaure* essays, what Dawn Ades referred to as "perhaps the most unexpected 'presence' in the whole review",[42] appears rooted in Bergson's ideas.

Given that Breton defined Caillois in 1934 as "the mental compass of Surrealism", it seems conspicuous that he did not follow Caillois' thinking to its source.[43] Arguably then, for Breton to complain that natural philosophy has not advanced in terms of the interest in co-evolutionary behaviours suggests a lamentable lacuna in his awareness of developments in evolutionary philosophy. By shifting from these Bergsonian currents in Breton's and Caillois' thinking, I will next explore how, through evocation, Surrealist artists conveyed the fluidity of time and the Bergsonian idea of experience as duration. Through the blurring of distinctions of form, they also collapsed clear temporal and ontological distinctions.

Salvador Dalí's Clock-Time, Psychological Time, and Memory

If Breton's unfamiliarity with *Creative Evolution* reflects an antipathy to Bergson, his critique of chronological time in *Communicating Vessels* (1932) reveals an even more wilfully ideological position. Breton here addresses the insufficiencies of clock-time for grasping the *vital*: "Time and space are only to be considered here and there, but *equally here and there*, dialectically, which limits the possibilities of measuring in any absolute and vital way by the metre and the clock".[44] However, rather than refer to Bergson, the philosopher so closely associated with the critique of transposing time into the spatial terms of the clock, Breton refers to Lenin: "Time and space in the dream are thus real time and space: 'Is chronology obligatory? No!' (Lenin)."[45] This choice of reference exactly reflects Michael Sheringham's definition of *Communicating Vessels* as "essentially a piece of ideological writing".[46] Turning away from Breton, however, towards a work of visual art that is synonymous with Surrealism, a more straightforwardly Bergsonian reading may be applied to Salvador Dalí's *The Persistence of Memory* (Figure 8.1).

A dark, flat, earthy landscape recedes to a distant horizon in the upper section of the canvas, beyond which a pale sea and sky seamlessly merge, leaving a barren rocky promontory appearing to float indeterminately amidst an ambiguous pale plane of water and air. In the foreground of the canvas is what looks like a section of a wooden box, brown like the earth, its perspectival lines meeting at no specific point on the horizon, just optically signifying depth of space. Over the right edge of the box slips a melting gold-framed pocket watch – its hands resting at what looks like five minutes to seven – upon which sits a fly and its shadow. On the top of the box rests another golden pocket watch, face down, its shiny surface bejewelled with a mass of black ants which seem to be swarming to their own temporal rhythm. To the rear of the box stands a

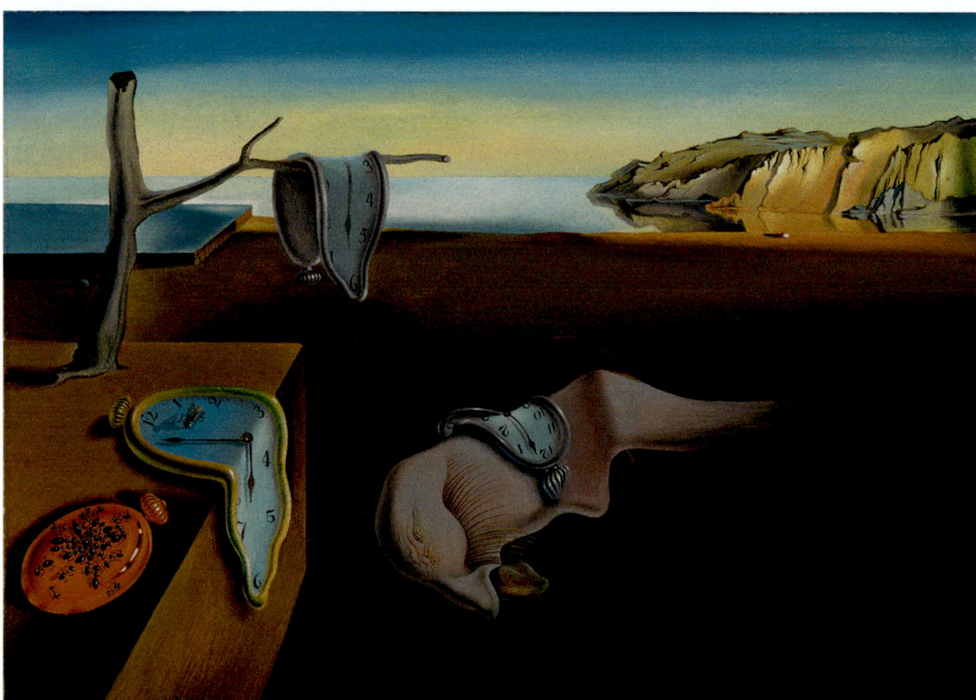

Figure 8.1 Salvador Dalí, *The Persistence of Memory,* 1931, oil on canvas, 24.1 × 33 cm. (Courtesy of The Museum of Modern Art, New York.)

short single-branched dead tree, at once merging into the receding terrain and resting like a fixture on a dressing table used for holding jewellery. Slinking lugubriously over the arm-like branch that points towards the centre of the canvas is another soft pocket watch, silver-framed, its hour hand resting at the numerical six. Floating indeterminately to the right of the box is a pale, amorphous form found in other paintings by Dalí, understood to represent the artist's sleeping head on its side shaped in the form of certain rocks found along the beach in Cadaqués – its long-lashed eye closed, its tongue sticking out. Over the back of the head is draped another melting, silver-framed pocket watch, one finger at twelve, the other out of view.

The sleeping head, the softly melting signifiers of time, the empty, slightly hallucinatory landscape all hint at the dream state with which the Surrealists are historically identified. While the mnemonic nature of the landscape and insects has been identified in some interpretations, others have proposed a connection with Albert Einstein's theories on time and relativity. Dawn Ades first interpreted the painting in relation to the latter: "The soft watches are an unconscious symbol of the relativity of space and time, a Surrealist meditation on the collapse of our notions of a fixed cosmic order."[47] While seemingly a sound interpretation, the *title* of the work is overlooked. The pairing of time and *memory* is identifiable with Bergson's theory of memory as the continuity of the past within the present and anticipations of the future, alongside the connection of fluid time and memory. In fact, in *Time and Free Will*, Bergson writes about an experience of hearing a clock strike and the sensations produced by each chime, "instead of being set side by side, had melted into one another".[48] Below measured time signified by this experiential moment of the clock chiming lies another fluid state, which Bergson defines as "a duration whose heterogeneous moments permeate one another; below the self with well-defined states, a self in which *succeeding each other means melting into* one another".[49] While duration melts together those moments that clock-time distinguishes, the fact that the three watches display different times further implies Bergson's emphasis on the artificially constructed nature of chronometry.

When Bergson first outlined his theory of duration in *Time and Free Will*, he made the distinction between time understood in terms of discontinuous measurement and duration as unbroken continuity, as eucidated by Brauer in Chapter 5 on time and ergonometry).[50] It is the notion of continuous time, that our intellectual habits struggle to grasp, which Bergson reconceives as duration. Measured time or clock time can be understood according to Bergson's idea of the human mind as having evolved to prioritise intellect for the purposes of practical action, for distinguishing between things and ordering them. Although useful, if not essential for organising life, this is an example of the intellect artificially cutting up a reality which exists in no such rationalised measure. For Bergson, this form of clock time is the result of an artificial transposition of time into spatial terms. In his introduction in 1922 to *Duration and Simultaneity, with Reference to Einstein's Theory,* Bergson announced: "Time is the very fluidity of our inner life."[51] Refuting the physicists' imperium of measured time, Bergson argues: "Real duration is *experienced*; we *learn* that time unfolds and, moreover, we are unable to measure it without converting it to space."[52] For Bergson, the basic condition of our intellect mixes recollection and perception just as it mixes time and space. He considers this typical of our tendency to see only differences of degree where in fact there exist only differences in kind. The method by which such differences are distinguished is Bergsonian intuition.

In *Duration and Simultaneity*, Bergson responds to the theory of relativity and describes looking at "the simultaneity between readings given by two separated clocks" in regards to the experiments of relativity on different synchronised clocks.[53] Arguing that "the theory of relativity cannot express all of reality", Bergson refuses to exclude experiential time from this reality.[54] In discussing this issue, Ansell-Pearson and Ó Maoilearca refer to Paul Davies' proposition in his book, *About Time,* that "the greatest outstanding riddle concerns the glaring mismatch between physical time and subjective time or psychological time".[55] This is certainly one of the tensions that can be read in Dalí's painting. The editors of Bergson's *Key Writings* suggest that, contrary to physicists, Bergson insisted that time flows and cannot be reduced to a merely phenomenological drama, as illuminated by his rhetorical question: "If time does not flow, does this mean that its experience is merely the result of a psychological illusion? What conception of time are we left with once we have shown that time does not flow?"[56] Such questions can also be read into Dalí's *The Persistence of Memory.*

The composition of Dalí's painting may be seen to reflect Bergson's distinction between homogenising space signified by the landscape that merges all, and heterogeneous time reflected in the array of watches. This might imply the simultaneity of the different interior durations of psychological time. A clock would seem the perfect representation of the artificial arresting of time which, in an illusionistic painting such as Dalí's, can only be represented in the necessarily static lines of the arrows of the clock face. That the watches themselves are fluid, although *fixed* in their fluidity, points to the illusions of the logico-spatial realm and to the underlying reality of the fluid continuity of life as permeated with memory.

In deciphering *The Persistence of Memory* through Bergson's writings, the tension created by Dalí between static and fluid, extension and interiority, clock time and psychological time, might be regarded as illustrating Bergson's very argument of reality as a temporal fluidity in terms of "duration" that cannot be fixed in static frames: "What is real is the continual *change of* form: *form is only a snapshot view of a transition",* Bergson writes. "Therefore, here again, our perception manages to solidify into discontinuous images the fluid continuity of the real."[57] Given Dalí's interest in illusions, he could have been alluding to the illusions of consciousness and intellect arising from measured time as distinct from psychological time pertaining to, in Bergsonian terms, the interrelationship of perception and memory. Dalí, who placed a lump of sugar in 1931 in his *Object Functioning Symbolically* and who, in *The Persistence of Memory*, configured *melting* clock faces, seems to be explicitly playing with Bergson's illustration of the distinction between measured, mathematical clock time and internal, psychological time through the image of melting clocks and sugar lumps. In an early chapter of *Creative Evolution*, Bergson writes: "If I want to mix a glass of sugar and water, I must, willy nilly, wait until the sugar melts. This little fact is big with meaning. For here the time I have to wait is not that of mathematical time."[58]

The liquefaction of forms, or more precisely, those specific objects used to measure time by spatial analogy, cannot be separated from the implications of the painting's title. Through his juxtaposition of words and images, Dalí deploys a distinctive Surrealist strategy to intensify affect. The juxtaposition of Dali's temporal and spatial implications with his title is highly suggestive of the experiential, durational, mnemonic, and psychological complexity of reality which for Bergson so eludes the grasp of intellect.

Separating memory into two registers, Bergson identifies one as a kind of motor memory utilised for practical action, the other as independent recollection. Clarifying these distinctions in *Matter and Memory*, Bergson acknowledges "there is no perception that is not full of memories".[59] As he explains: "[M]emory, inseparable in practice from perception, imports the past into the present".[60] In *Creative Evolution,* Bergson concludes:

> We trail behind us, unawares, the whole of our past; but our memory pours into the present only the odd recollection or two that in some way completes our present situation.[61]

Bergson's image of experience as an inextricable fusion of involuntary memory, perception and recollection seems encapsulated by Dalí's painting. Long memories of the Catalan landscape and the squirming ants that re-appear in many of his works are intertwined with perceptions and oneiric, imaginary forms. Inner duration symbolised by the melting time pieces reflect the mysterious subjective experience of time that cannot be measured through hours, minutes and seconds. A distorted form of the artist's body may be discerned in the centre of the painting in the manner of an ideogrammatic trope or avatar. Yet while suggesting some stable essence of identity, it seems to drift somnambulistically, as if a mere phantom of transient being.

Tanguy's Evolutionary Time as Memory

Exuding an uncannily animate character, Tanguy's paintings seem to depict invisible organic processes at work that evoke both a Bergsonian sense of *élan vital* and the psycho-biological implications of evolution as a kind of memory. While not depicting any definite life forms, they nonetheless connote life and thus, in Bergsonian terms, temporality, as Bergson observes: "Wherever anything lives, there is, open somewhere, a register in which time is being inscribed."[62] It is in Tanguy's work that Bergson's ideas on the organic nature of different modes of duration appear to come most vividly into play. In their peculiarly psycho-morphological formations, Tanguy's paintings seem to represent a co-mingling of the process of painting as an inner duration precipitating matter and consciousness and the temporal coming-into-being of life forms. Hence, the vital impetus of art as process and product, *natura naturans* and *natura naturata*, can be observed in Tanguy's artwork to reflect the vital impetus of natural processes.

In 1925, Breton welcomed Tanguy into the Surrealist group and championed him as an exceptional automatist painter. Given Tanguy's family origins in Brittany and his time in the merchant navy, his artwork has invariably been correlated with the submarine, but also with the paranormal. In his 1942 text, "Genesis and Perspective of Surrealism in the Plastic Arts", Breton intertwined these two genealogies, writing that Tanguy represents "the Neptunian light of clairvoyance".[63] Tanguy's painting *Et Voilà! (La veille au soir)/And there it is! (The Evening Before)* (Figure 8.2) can also be read in terms of a deep underwater impression, totally absent of any Apollonian daylight, signifying the absence of perception that enables the discerning of distinct objects logically appointed in space. The strong downward pull of figures and lines, rocks, vegetation, billowing gases, and star-like phosphorescence seem to stir in a strange confusion of scale and the profusion of microcosm and macrocosm. Certain white

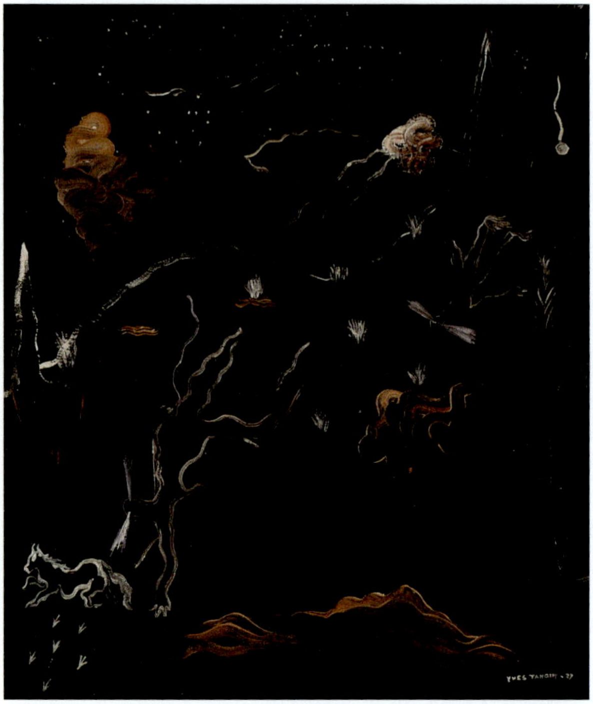

Figure 8.2 Yves Tanguy, *Et Voilà (La veille au soir)* (And there it is! [The Evening Before]), 1927, oil on canvas, 65.4 × 54.3 cm. (Courtesy of The Menil Collection, Houston.)

contours outline hybrid humanoid forms, one in the bottom left of the painting seemingly in the act of chasing a horse, a classic signifier of the dream signalled in the title. Even in such relatively early works as *The Storm (Black Landscape)* (1926), Tanguy represented this mix of imaginary, mnemonic, and oneiric material. Seemingly far removed from, in Bergsonian terms, the logical world of distinct solids, its sinuous lines, diagonals and horizontals, sprouting flora, spermatazoic organisms and cloudy gaseous emissions, *And there it is!* seems to ripple with vital energy and a poetic pleasure in ambiguity.

In *Creative Evolution,* Bergson created a picture of life as duration and a continuity in which everything partakes. He also extended his theory of duration into organic nature, which he depicted as a kind of universal memory, as Deleuze observes: "Everything happens as if the universe were a tremendous Memory".[64] Evolution, Bergson proposes, began in a state of primordial unity and, via the different tendencies of the *élan vital*, directed itself through matter to develop into different species and natural kingdoms, each characterised by different degrees of the two tendencies of instinct and intelligence. It is this original state of evolutionary nature that Bergson suggests accounts for the instinctual sympathies between species, morphological recurrences, and behavioural patterns. It is owing to shared origins that he deduces distinct and individuated identity as "not fully realised anywhere".[65] While the ontological implications of this evocatively entangled image of nature follows a broad Darwinian view on the difference between species as one of degree and not of kind, it also presents a vivid image of involution, harking back to deep ancestral states of primordial indistinction, as Bergson indicates:

> Where, then, does the vital principle of the individual begin or end? Gradually we shall be carried further, further back, up to the individual's remotest ancestors: We shall find him solidary [sic] with each of them, solidary with all that little mass of protoplasmic jelly which is probably at the root of the genealogical tree of life. Being, to a certain extent, one with this primitive ancestor, he is also solidary with all that descends from the ancestor in divergent directions. In this sense each individual is said to remain united with the totality of living beings by invisible bonds.[66]

It is this strange sense of evolutionary anteriority which I think can be read into the artwork of Tanguy. Rather than identify Tanguy's artworks with Freudian concepts of the unconscious, they seem more accessible through Bergsonian concepts of memory, which accounts for their implication of *déjà vu*. In Bergsonian terms (but not exclusively), the correlation between depth and interiority in Tanguy's *And there it is!* (Figure 8.2) may be perceived alongside the curious life forms and points of energy signifying a unification of matter with an implicit, animating, and pulsating consciousness. The solitary spermatozoon would seem to imply a picture of life as pre- or non-human anteriority in the process of becoming, with the darkness making of space an infinite temporality.

The theme of temporality runs through Tanguy's œuvre recurring, for example, in *The Furniture of Time* (1939) and *Time and Again* (1942). This sense of temporality in Tanguy's work is always uncanny and is, implicitly, connected to the artist's cultivation of a mediumistic aura to his work, most pronounced in *The Invisibles* (1951) while also conveyed in *The Certitude of the Never Seen* (1933). Tanguy's phantasmagorical morphologies thus hint at uncanny dimensions of both space and time. They are at once strangely familiar concrescences of worlds beyond perception and precipitations of temporalities beyond our experiential horizon.[67] In converging these dimensions, the peculiar quality of anteriority in Tanguy's paintings would make him a medium of deep time, of duration beyond the subjective into the realm of Bergsonian memory, reaching back to the pre-natal and primordial. Bergson's image of involution leading back to a primordial whole implies that organic unity belongs to a deep past, a regression to which, Kerslake notes, "involves a divinatory intuition of a recollected, common past".[68] In viewing Tanguy's paintings as implying ambiguous and amorphous phantasms of the living and organic becomings of matter and consciousness, they may be related to what Kerslake has called Bergson's "gothic philosophy of memory".[69] In his chapter in *Creative Evolution* on "Duration", Bergson presents a strikingly evocative image of the hauntological persistence of memory from prenatal experiences onwards, if not from the genesis of homo-sapiens that correlates evolution with uncanny narratives:

> These memories, messengers from the unconscious, remind us of what we are dragging behind us unawares. But even though we may have no distinct idea of it, we feel vaguely that our past remains present to us. What are we, in fact, what is our *character*, if not the condensation of the history that we have lived from our birth – nay, even before our birth, since we bring with us prenatal dispositions?[70]

Tanguy's paintings seem to represent flashes of this kind of memory. Bergson's uncanny image of evolution as connecting each individual to a pre-natal existence, one which even reaches beyond the presence of the human species, is echoed in Tanguy's *Hérédité des caractères acquis (Heredity of Acquired Characteristics)* (Figure 8.3), which connotes a direct relation to evolutionary theory.

The title of the work directly alludes to Neo-Lamarckian evolutionary theory, which was one of the linchpins of Freud's metapsychological theory of the structure of neuroses as originating in the experiences of early humans.[71] Transposing homogeneous space into deep time, Tanguy's biomorphic populace implies a diverse collection of organic characteristics. Although clearly delineated, the identity of these forms is indistinct and unrecognisable. For Caillois, Tanguy's work represents an "unfinished fauna" which seems "neither terrestrial nor human". As he explains: "It does not have any familiar or intelligible form [...] It functions by threatening us with distant biologies."[72] Two further works by Tanguy from the 1930s might corroborate this observation, *The Extinction of Species I* (1936) and *The Extinction of Species II* (1938).

In his Surrealist writings of the mid-1930s Roger Caillois followed Georges Bataille in highlighting the human anxiety of the *informe* as a kind of ontological horror alerted by perception. In his 1935 essay on mimicry and psychasthenia, Caillois would advance this theme by underlining the question of *distinction* in broader terms. "Ultimately," he asserts, "from whatever angle one may approach things, the fundamental question proves to be that of *distinction* [...] no distinction is more pronounced than the one demarcating an organism from its environment".[73]

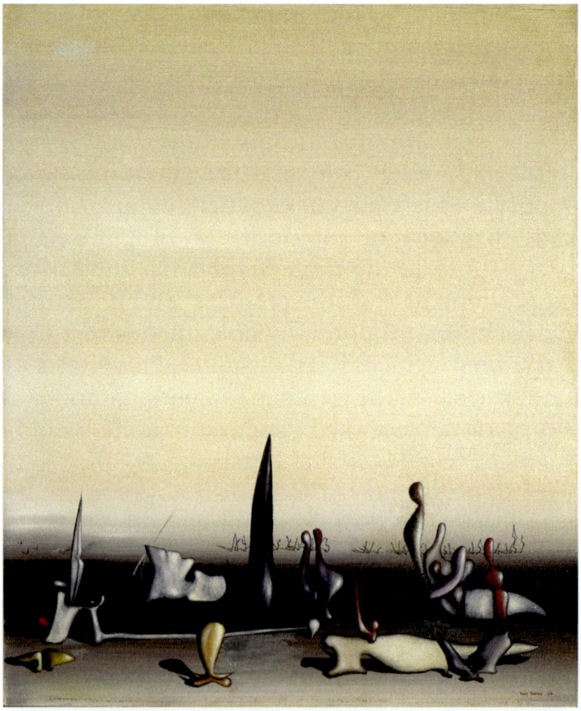

Figure 8.3 Yves Tanguy, *Hérédité des caractères acquis* (Heredity of Acquired Characteristics), 1936, oil on canvas, 41 × 33 cm. (Courtesy of The Menil Collection, Houston.)

The above statement clearly relates to the unease which Caillois later expresses about the indistinction he finds in Tanguy's paintings. Although pejorative, Caillois' response to Tanguy's paintings is nonetheless insightful. What seems to unnerve Caillois in Tanguy's evocation of "distant biologies" is not only their indistinct morphology but also what they appear to suggest about a form of life (vaguely human) inherent in the implicitly organic shapes strewn across primal landscapes. What is *informe* here, then, arises from a deeply mnemonic sense of the biological uncanny, one which returns us to some primordial evolutionary precedent. Moreover, it also relates to the way in which Tanguy dissolves a clear phenomenological demarcation between interiority and exteriority, between the organic and the psychological, or between matter and *élan vital*. However Tanguy manages to convey this, his work thereby seems to present the life of mind and matter in an undivided state of becoming and emergence, as in *Heredity of Acquired Characteristics*, or in a state of dwindling, as in *The Extinction of Species*. This character of Tanguy's paintings calls to mind Bergson's comments on early animations of life, such as the primitive micro-organism *Infusoria*, in which a dynamic of consciousness is understood to be at work: "Even in these humblest manifestations of life they discover traces of an effective psychological activity."[74] Tanguy's paintings may then be seen as echoing such primal concrescences of the *élan vital* at work in forming matter. Whether most directly as an act of painting and the creative mind itself or as a reflection of universal organic processes, Tanguy's work may then be seen in Bergson's terms as "life evolv[ing] before our eyes as a continuous creation of unforeseeable form".[75]

If Tanguy's work is to be discussed in terms of mediumism, it should include this Bergsonian capacity for entering into the depths of life and its process. If this process needs to be fixed, as painting requires, then it appears to be done so in the most indeterminate way. Following Bergson's particular sensibility with mediumistic consciousness, as illuminated in Chapters 4 and 5, and his perspective on artistic insight into the depths of life, Tanguy's paintings may then be viewed as the channelling of not only personal and other worldly simulacra but also as an imaginary morphology of ontologically undifferentiated life forms.

Conclusion

These examples of Surrealist theories and art practices in relation to Bergsonism as a vitalist evolutionary philosophy and a philosophy of experience, show that what connects them is a view of life as a whole within which mind and nature, the subjective and the material, cannot be fundamentally separated. For Bergson, life's vital impetus, in all its creative possibility, diversity and intensity, tends to become subsumed by the urgency or habitual grind of practical necessity, oriented by the very nature of human intelligence in its basic evolutionary tendency for survival. This was why philosophy for Bergson writing in his final work, *La Pensée et le mouvement* (*The Creative Mind*), was required to develop thought beyond the basic demands of the human condition.[76] Arguably then, the ethically aspirational conception of Surrealism, as offering a creative and open-ended mode of life beyond the values of mere utility, closely follows that of Bergson's philosophy.

Perhaps because of Breton's antipathy to Bergson, he missed the bio-philosophical implications of *Creative Evolution*. Yet whether direct influence exists or not, it is possible to offer Bergsonian readings of Surrealism as illuminated by the explorations of Breton and Caillois and the artwork of Dalí and Tanguy in this chapter. Both Bergson and the Surrealists insisted upon the open-endedness of vital activity. Both sought a more nuanced grasp of the complex dynamics between freedom and determination and both expressed conviction in a creative and ethically formative impulse, rooted within natural life itself. Both partake in that radically reformulating and vitalist spirit of the Modernist era – the *Vitalist Modernism* signified by the title of this book – which sought to overcome restrictive, logocentric, and deterministic accounts of human potentiality and to redefine the human subject in terms of its sympathetic place within the natural world.

Notes

1 For the influence of Bergson on modern art and Modernism. see Mark Antliff, *Inventing Bergson: Cultural Politics and the Parisian Avant-Garde* (Princeton NJ: Princeton University Press, 1993); Paul Ardoin, S.E. Gontarski and Laci Mattison, eds., *Understanding Bergson, Understanding Modernism* (London and New York: Bloomsbury, 2013); Paul Atkinson, *Henri Bergson and Visual Culture: A Philosophy for a New Aesthetic* (London and New York: Bloomsbury, 2020).

2 I refer to the tract that famously implores the reader "Read" and "Don't Read" published by José Corti in a catalogue of Surrealist publications, 1931; see Krzysztof Fijalkowski & Michael Richardson, eds., *Surrealism: Key Concepts* (London & New York: Routledge, 2016) 3. For Breton's comments on Bergson, see Breton's 1934 lecture, "What is Surrealism?" in *What is Surrealism? Selected Writings*, ed. Franklin Rosemont (New York: Pathfinder, 1978) 113, hereafter referred to as *WIS*; and the 1952 radio interviews with André Parinaud in André Breton, *Conversations: The Autobiography of Surrealism*, trans. Mark Polizzotti (New York: Marlowe & Co., 1993) 15. As mentioned in the *Introduction*, in 1917 Bergson had been enlisted by the French Government to negotiate with President Woodrow Wilson a deal whereby if the United States entered the First World War on the side of the Allies, after the War France and Britain would support the creation of a League of Nations dedicated to maintaining world peace. Successfully negotiated, Bergson was appointed President of the League's International Committee for Intellectual Cooperation.

3 Henri Bergson, *Creative Evolution* (1907); trans. Arthur Mitchell (Mineola, NY: Dover Publications, 1998) 87; hereafter referred to as *CE*.

4 Maurice Merleau-Ponty, *Nature: Course Notes from the Collège de France*, trans. Robert Vallier (Evanston, Illinois: Northwestern University Press, 2003) xviii.

5 Cohen, "Philo, Spinoza, Bergson: the rise of an ecological age", in *The New Bergson*, ed. John Mullarkey (Manchester and New York: The University of Manchester Press, 1999: 18–31) 25; hereafter referred to as *TNB*.

6 Cohen, *TNB*, 25.

7 Bergson *CE*, 163.

8 Cohen, *TNB*, 28.

9 Elizabeth Grosz, *Becoming Undone: Darwinian Reflections on Life, Politics, and Art* (Durham and London: Duke University Press, 2011) 19.

10 Bergson, *CE*, 167.

11 Gilles Deleuze, *Bergsonism*, trans. Hugh Tomlinson and Barbara Habberjam (New York: Zone Books, 1991) 77.

12 See Donna Roberts "Surrealism: The Ecological Imperative", *Surrealism: Key Concepts*, eds. Krzysztof Fijalkowski and Michael Richardson (London and New York: Routledge, 2016) 216–227.

13 Elizabeth Grosz, *Becoming Undone*, 33. See also Keith Ansell Pearson, *Bergson: Thinking Beyond the Human Condition* (London and New York: Bloomsbury, 2018) 100–101.

14 Monica Greco, "Vitalism Now – A Problematic" in *Theory, Culture, and Society*, Vol. 38:2 (2021) 47–69.

15 Greco, "Vitalism Now – A Problematic", 47–69.
16 William Bechtel and Robert C. Richardson, "Vitalism" in Edward Craig, ed., *Routledge Encyclopaedia of Philosophy* (London: Routledge, 1999) as quoted in Greco, Vitalism Now – A Problematic", 47.
17 Greco, "Vitalism Now – A Problematic", 47–69.
18 Greco, "Vitalism Now – A Problematic", 56.
19 André Breton, *Mad Love*, trans. Mary Ann Caws (Lincoln and London: University of Nebraska Press, 1987) 11.
20 Jane Bennett, *Vibrant Matter: A Political Ecology of Things* (Durham and London: Duke University Press, 2010) vii.
21 Greco, "Vitalism Now – A Problematic", 48–49.
22 Robert Mitchell, *Experimental Life: Vitalism in Romantic Science and Literature* (Baltimore: John Hopkins University Press, 2013) 2.
23 Mitchell, *Experimental Life*, 7–8.
24 Raihan Kadri has placed the question of life at the centre of Surrealist critique and practice. With a philosophical focus on correlating Surrealism with Friedrich Nietzsche's reconception of life beyond the bounds of utility, Kadri defines the basic body-centred approach of early Surrealism as a form of vitalist subjectivity which then develops into a more complex form of ever-shifting intersubjectivity which he terms *epivitalism*. See Kadri *Reimagining Life: Philosophical Pessimism and the Revolution of Surrealism* (Plymouth: Farleigh Dickinson University Press, 2011).
25 Julien Gracq, *André Breton, quelque aspects de l'écrivain* (Paris: José Corti, 1948) 178; my thanks to Timo Kaitaro for this reference. Georges Sebbag has also drawn comparisons between Bergson's and the Surrealists' thinking about the dream and creative imagination. See Sebbag, "Dream: A Manifesto of the Manifest Dream", in *Surrealism: Key Concepts*, 163–170.
26 For the influence of Bergson on Reverdy see Jean-François Puff, "Le travail du sujet dans la poésie de Pierre Reverdy", *Littérature* (No. 183) 2016: 40–50.
27 Henri Bergson, "The Perception of Change", *Henry Bergson: Key Writings*, ed. Keith Ansell Pearson and John Ó Maoilearca (London: Bloomsbury, 2002) 307; hereafter referred to as *KW*.
28 Henri Bergson, *Laughter: An Essay on the Meaning of the Comic* (New York: The Macmillan Company) 156.
29 Bergson, *Laughter*, 156–157.
30 Keith Ansell-Pearson, "Bergson and Creative Evolution/Involution: Exposing the Transcendental Illusion of Organic Life," *TNB*, 147.
31 Breton, *WIS*, 201.
32 Ansell-Pearson, *TNB*, 148.
33 Bergson, *CE*, 178.
34 Peter A.Y. Gunter, "Bergson and the War Against Nature," *TNB* (168–182) 174.
35 Bergson, *CE*, 146.
36 Bergson, *CE*, 167.
37 Christian Kerslake, *Deleuze and the Unconscious* (London and New York: Continuum, 2007) 62.
38 Roger Caillois, *The Edge of Surrealism: A Roger Caillois Reader*, ed. Claudine Frank, trans. Claudine Frank and Camille Naish (Durham and London: Duke University Press, 2003) 102.
39 Caillois, *The Edge of Surrealism,* 81.
40 Henri Bergson, *Les Deux Sources de la morale et de la religion* (Paris: Librairie Félix Alcan, 1932).
41 Roger Caillois, *Le mythe et l'homme* (Paris: Boivin, 1935).
42 Dawn Ades, *Dada and Surrealism Reviewed* (London: Arts Council of Great Britain 1978) 280.
43 See Donna Roberts, "An Introduction to Caillois' Stones and Other Texts", *Flint Magazine* (Iss.1+2) 2018.
44 André Breton, *Communicating Vessels*, trans. Mary Ann Caws and Geoffrey T. Harris (Lincoln and London: University of Nebraska Press, 1990) 48.

45 Breton, *Communicating Vessels*, 48.
46 Michael Sheringham, "Vitalism and the theory and practice of subjectivity in Breton's writings", *André Breton: The Power of Language* (Exeter: Elm Bank Publications, 2000) 13.
47 Dawn Ades, *Dalí* (London: Thames and Hudson, 1982) 145.
48 Bergson, *KW*, 87.
49 Bergson, *KW*, 88.
50 Bergson, *KW*, 3.
51 Bergson, *Durée et Simulanéité: à propos de la théorie d'Einstein*, 1922; *Duration and Simultaneity: With reference to Einstein's theory*, English trans. Leon Jacobson (Bobbs-Merril Company Inc. USA, 1965) 251.
52 Bergson, *KW*, 265.
53 Bergson, *KW*, 259.
54 Bergson, *KW*, 267.
55 Bergson, *KW*, 32; Paul Davies, *About Time: Einstein's Unfinished Revolution* (Penguin Books, 1995).
56 Bergson, *KW*, 26.
57 Bergson, *CE*, 302.
58 Bergson, *CE*, 9.
59 Bergson, *Matter and Memory*, trans. Nancy Margaret Paul & W. Scott Palmer (London: George Allen & Unwin, [1911] 1970, 24.
60 Bergson, *Matter and Memory,* 80.
61 Bergson, *CE*, 167.
62 Bergson, *CE*, 16.
63 Breton, *WIS*, 225.
64 Deleuze, *Bergsonism*, trans. Hugh Tomlinson and Barbara Habberjam (New York: Zone Books, 1991) 77.
65 Bergson, *CE*, 12.
66 Bergson, *CE*, 43.
67 The meaning of "concrescence" here is inflected with its use by Alfred North Whitehead as a process of "growing together". Concrescence specifically refers to the process within space and time of components in various stages of development and potentiality (subjective or material) coming together to form a novel entity: "the "production of novel togetherness" is the ultimate notion embodied in the term "concrescence""; Whitehead *Process and Reality: An Essay in Cosmology*, eds. David Ray Griffin and Donald W. Sherburne (New York: The Free Press, 1978) 21.
68 Kerslake, *Deleuze and the Unconscious*, 62.
69 Kerslake, *Deleuze and the Unconscious*, 16.
70 Bergson, *CE*, 5.
71 For more details on the prevalence of Neo-Lamarckian thought in France in the early twentieth century, see Fae Brauer, "Eroticizing Lamarckian Eugenics: The Body Stripped Bare during French Sexual Neoregulation", Chapter Three, *Art, Sex and Eugenics: Corpus Delecti*, eds. Fae Brauer and Anthea Callen (Hampshire, UK and Burlington, USA: Ashgate Publishing, 2008) 97–138; Fae Brauer, "Becoming Simian: Devolution as Evolution in Transformist Modernism", Chapter 7, *Picturing Evolution and Extinction: Degeneration and Regeneration in Modern Visual Culture* (Newcastle upon Tyne: Cambridge Scholars Publishing, 2015) 127-156.
72 Caillois, "Surrealism as a World of Signs," *The Edge of Surrealism*, 330.
73 Caillois 'Mimicry and Legendary Psychasthenia', *The Edge of Surrealism*, 91.
74 Bergson, *CE*, 35.
75 Bergson, *CE*, 30.
76 Henri Bergson, *Le Pensée et le mouvement* (Paris: Félix Alcan, 1934).

9 Bergson, Creativity and the Vitalist Left

Egoism, Syndicalism, Communism

Mark Antliff

The study of Henri Bergson's influence on the radical left has focused largely on the anarchist-syndicalist writings of Georges Sorel and his close ally Edouard Berth, but we have yet to consider that impact in tandem with Bergson's surprising popularity among intellectuals affiliated with the anarchist individualist and communist movements. In so doing, I will consider three figures central to the development of an alternative radicalism to that espoused by Sorel: The self-styled philosopher André Colomer (1886–1931), who merged the Egoist philosophy of Max Stirner with that of Bergson before and during World War One, and the polemicists Eden and Cedar Paul, two prominent Marxists who aligned Bergson's ideas with those of the Communist Party of Great Britain in the aftermath of that war.

In the Paris-based journal *Action d'Art* (1913–1919), Colomer promoted his notion of radical individualism, and his theories subsequently influenced important cultural figures among the Futurist, Dadaist and Surrealist movements. *Action d'art* had distributers throughout Europe, North Africa and the Americas assuring that Colomer's radical views reached an international audience.[1] Eden and Cedar Paul had a comparable impact through their book-length *apologia* for Bergsonian communism, *Creative Revolution: A Study of Communist Ergotocracy* (1920), which appeared in numerous subsequent editions.[2] This book served as a textbook for the militant workers' education organization The Plebs League (founded 1908–1909), which became allied to the communist movement in 1920.[3] The organization's 1924 constitution proclaimed its mandate to develop "the class consciousness of workers, by propaganda and education, in order to destroy wage-slavery and to win power."[4] The League's program resulted in the pre-war establishment, with union support, of Labour Colleges throughout Britain,[5] and their first pedagogical initiative was to commission the Pauls to write an accessible handbook applying their "new psychology" to Marxist ideology.[6] *Creative Revolution* was the result, and in their foreword to the Plebs edition, the authors unapologetically described the text as "a book for revolutionists," composed of a heady synthesis of the thought of Bergson, Freud, Marx and Sorel with the militant tenets of Marxist Leninism.[7] The British Communist Party's weekly newspaper, *The Communist,* praised *Creative Revolution* as a volume "vibrating with revolutionary enthusiasm," and it was quickly introduced into labour colleges throughout Britain at a time when enrolment ballooned to the tens of thousands.[8]

In considering *Creative Revolution,* Eden and Cedar Paul's deployment of Bergson's concept of *élan vital* to justify their theory of revolution shall be analyzed, followed by their Bergsonian critique of rationalism to attack democratic institutions; their related appropriation of a Bergson-inspired theory of the image as a catalyst to revolutionary

DOI: 10.4324/9781003045595-13

consciousness; their definition of the Soviet as synonymous with a Bergsonian concept of a vital order, and their assimilation of Bergson's theory of artistic perception and of the personality with that of the communist revolutionary, most prominently Lenin himself. Simultaneously, I examine Colomer's usage of these same Bergsonian paradigms to critique both syndicalism and communism and to promote his alternative theory of radical nominalism and creativity, based on a merger of Bergson's thought with that of Max Stirner, author of the individualist manifesto, *Der Einzige und sein Eigentum (The Ego and Its Own)* (1844). By reconsidering Sorel's Bergsonian vitalism in light of these alternative political and aestheticized visions, I will point to the instability of Bergson's metaphysics when applied to ideology, as well as his transformative impact on competing leftist movements during the early twentieth century.

Bergson's *L'Ėlan Vital* and Historical Materialism

Throughout *Creative Revolution,* references to the *élan vital* serve to define voluntarism as integral to the Marxist doctrine of historical materialism. The authors inform their working-class readers that "though Man *wills* certain things, *what* he wills is decided by the influence upon him of his environment, which in turn is modified by his actions," thus implying that while social and material relations may shape humans, they possess the potential to shape events and transform these conditions.[9] In their chapter on "Creative Revolution," they make plain the Bergsonian import of their thesis:

> No left-wing socialist who grasps the meaning of the vital impulse will contest the value of the conception (when applied to social evolution) as the obverse of the materialist conception of history. The idea of human conation, of human impulse and desire, acting and reacting on material conditions of production, operating throughout history as the instrument of creative evolution and manifesting itself in times of crisis as the quasi- omnipotent force of creative revolution ... have we not here a truth which is the supplement and logical development of Marxism?[10]

In *Creative Evolution,* Bergson described the vital impulse as a durational force synonymous with invention, the creation of forms, and the continual elaboration of the absolutely new, noting that it is humanity's special role, by virtue of its intuitive ability to empathize with duration, to intervene in order to shape the very process of creation itself.[11] In a section of this book cited by the Pauls, Bergson cast this argument in cosmic terms. Speaking of the vital impulse, Bergson wrote that "life as a whole, from the initial impulsion that thrusts it into the world, will appear as a wave which rises, and which is opposed by the descending movement of matter." The resulting "current is converted by matter into a vortex" but "at one point alone it passes freely ... At this point is humanity; it is our privileged situation."[12] The Pauls evidently appropriated this parable to tell another epic story: That of historical materialism. Human activity in their narrative is likewise "an instrument of creative evolution." However, such change is now cast in terms of the creation of new social systems and concomitant material relations from Feudalism to Industrial Capitalism, culminating in a final evolutionary push from Democracy ("people's rule") to Ergatocracy, or "worker's rule."[13] Further, they note that in moments of historical crisis — such as World War One — humanity has the potential to instigate radical rather than incremental change, to precipitate a *qualitative* and therefore *highly creative* break from the past.[14] Such revolutionary transformation is premised on the heightened class-consciousness of its instigators who see themselves as transforming history, and throughout *Creative Revolution* those instigating actors are identified as the organized proletariat. The Pauls describe the pre-war "syndicalist" and "shop-steward" movements in France and Britain as precursors to the post-war European "workers committees" and Russian "soviets": All are manifestations of this "vital impulse" and as such are part of an evolutionary process now given a revolutionary impetus. Drawing on Sigmund Freud, they also distinguish an impulse to revolt as an expression of "human conation ... of the *libido*" from the purposive acts of a self-conscious, organized working class, whose behavior is premised on "Bergsonian indeterminism."[15] Thus "the urge of the unconscious will" can become an agent for creative revolution only if *free will* is brought into play in transforming the existing social order. The Pauls further inform us that "the quickened pulse of war has but accelerated a transformation which was already in progress before the war, and of which the syndicalist movement was an early inchoate expression."[16]

Prominent leaders in the Communist Party of Great Britain thought a revolution was indeed impending due to the explosion in union membership following the war (an increase of 3 million by 1920), widespread rioting within the military and the rapid proliferation of strike activity throughout England.[17] These workers were instructed by the Pauls to look for inspiration in the "splendid vital impetus of fighting revolutionism" that emerged in Russia in 1917.[18] While this war may have signalled "the nadir of capitalism," it was "human conation" marshalled in the service of free will that made revolution possible. "An advance peculiar to our age," they claim, "is that social evolution is becoming self-conscious."[19] As a result humankind was no longer subject to "the long agony of unconscious drift at the mercy of blind economic and social forces" but was instead undergoing "the opening act in the great drama of purposive, creative revolution," exemplified by the "ergatocratic" trend initiated by the Russian Revolution.[20]

Reading Stirner through Bergson: The Durational Ego

Whereas the Pauls reconfigured the *élan vital* in terms of Marxist historical materialism, Colomer adapted Stirner's radical nominalism to his Bergsonian conception of the free-standing ego, divorced from collective notions of class or the materialist forces of history. To understand Colomer's anarchist gambit we must first begin by examining Stirner's *The Ego and its Own*.[21] The first part of Stirner's book focuses on the ideological means through which individuals are pressured by social forces to deny their own self-interest; the second part seeks to define "ownness," the condition of freedom from coercive influences. Stirner repeatedly defines the self as embodied, as motivated by irrational sensations of physical desire, as parcel and parcel of the material world, and as a temporal being undergoing constant change, both physical and psychological. This "egoist" self was described as "the unique one," a particular being unlike any other.[22] Thus our ego was a "corporeal ego" with self-realization only achievable when the individual "has fallen in love with his corporeal self and takes a pleasure in himself as a living flesh-and-blood person."[23] This radical nominalism leads Stirner to describe the "transitory ego" as undergoing constant change characterized as an unending process of willed, creative-destruction.[24] The enemy of this heterogeneous self is any institution or belief system, whether religious or secular, that directs individuals away from their own embodied interests and desires in favour of subservience to spectral abstractions.[25] Among emerging civilizations, such abstractions first took the form of imaginary Gods, but following the rise of Cartesianism in the seventeenth century, Christian faith in an absolute, eternal "pure spirit" metamorphosed into a fetishized veneration of rationalism.[26]

This notion of "pure spirit" subsequently took on other myriad forms. In addition to Cartesianism and concepts of the "Sacred," other such transcendental "essences" include "Morality," the "State," the rule of "Law," "Justice," "Humanity," the "Citizen" and the "Fatherland."[27] All are characterized as "fixed ideas" of the "mere word,"[28] divorced from the temporal flow of embodied individual existence, and Stirner routinely capitalizes them to mock their supposed status as concrete entities. He argues that such ideas are invariably deployed by vested interests to subordinate flesh-and-blood in the service of an abstract ideal. An individual's self-interest cedes to a collective "common interest" and individual welfare is sacrificed in the name of the "general good."[29] "What is it, then, that is called a 'fixed idea'? An idea that has subjected the man to itself."[30] "God, immortality ... humanity are drilled into us from childhood as thoughts and feelings which move our inner being more or less strongly ... ruling us without our knowing it" – thus they "are always not aroused, but imparted, feelings."[31] Morality for instance is "nothing else than loyalty [to] fulfillment of the law" — "one must carry in himself the law, the statute, and he who is most legally disposed is the most moral." Under the auspices of the State judicial system, Christian precepts are transferred to the secular realm as ethical absolutes, not to be transgressed. Thus "monogamy" and other restrictions on sexual freedom become "*a dogma of faith*" and "every Prussian carries his *gendarme* in his breast."[32]

Morality even quells creative self-expression for "in its first and most unintelligible form morality shows itself as *habit*. To act according to the habit and usage (mores) of one's country — is to be moral there." "*Innovation*" by contrast "is the deadly enemy of *habit*, of the old, of *permanence*," thus behaviour departing from social norms is to be censored.[33] This self-subjugation reaches an extreme in the case of the citizen soldier, called to offer his very life in the service of another collective abstraction, that of the "Nation." "A good patriot brings his sacrifice to the altar of the fatherland; yet it cannot be disputed that the fatherland is an idea, since for beasts incapable of mind, or children as yet without mind, there is no fatherland and no patriotism." The "egoist" by contrast is one who "instead of living to an idea, that is, a spiritual thing, and sacrificing to it his personal advantage, serves the latter."[34] On an even more subversive note, Stirner calls on us to ignore concepts of good and evil and related notions of "illegality" as moral categories inhibiting "self-ownership" and the freedom to respond "to the full energy of the will."[35] On this basis, individuals would be free not only to engage in "unwedded cohabitation" but also in a full-blown "insurrection" against the State.[36]

Clearly Stirner's pronouncements on the "unique," "disembodied abstraction," "fixed ideas," and "habit" as opposed to "innovation" and creativity had a strong resonance with Bergson's metaphysics. In the broadest sense Stirner's method is not unlike that of Bergson inasmuch as he engages in an introspective meditation on the nature of individual being, grounding the self in the temporal flux of sensate experience, and thereby privileging heterogeneity over the homogenous, embodied consciousness over disembodied abstractions, irrationalism over rationality, and temporal change over concepts devoid of temporality. Bergson's notion of duration — the bedrock of his metaphysics — has all of these characteristics. Stirner then identifies modes of thinking that alienate us from ourselves. He also anticipates Bergson by critiquing Cartesianism for misrepresenting individual being and the world, robbing both of their temporality.

For Colomer, a creative reading of Stirner's egoism through the lens of Bergson's metaphysics proved especially compelling as a justification for anarchist-individualism. Following Stirner, Colomer endorsed the idea of a self-devouring subjectivity which he related to Bergson's critique of the notion of an underlying subject as opposed to an enduring subject. Bergson succinctly summarized his thesis as follows: "There are changes, but there are underneath the change no things which change: change has no need of a support. There are movements, but there is no inert or invariable object which moves: movement does not imply a mobile."[37] Colomer added intuition to the equation as the means by which an egoist could tap into his own inner duration. Intuition enabled the individual to resist the pernicious effects of analysis and its offspring of fixed abstractions, both condemned by Bergson and Stirner for denying temporality and particularity. Colomer followed Bergson in claiming that these abstractions, like science, had a "practical role" with no other purpose "than to serve our action" with regard to our "use of matter."[38] Science therefore was "a marvelous instrument" allowing us to grasp the material world to meet our "external necessities" but it was ill suited to grasp individual duration.[39]

The Ideology of Rationalism: Democracy vs Class-consciousness

Colomer deployed Stirner and Bergson's critique of rationalism and abstractions to justify his notion of the liberated ego. By contrast, Eden and Cedar Paul endorsed Georges Sorel's Bergsonian critique of rationalism in its ideological and rhetorical manifestations as an impediment to the revolutionary *élan vital* native to class consciousness. In the realm of politics, this obstruction took two forms: Those parliamentary socialists who relied on rationalist argumentation to support calls for social reform and those plutocrats who used the Enlightenment precepts undergirding democratic ideology as a pernicious weapon to undermine class-consciousness and prop up the bourgeois status-quo.[40] The latter strategy was premised on the institution of parliamentary democracy, a normative system of representation designed to suppress class identity in favour of a chimeric vision of universal suffrage.

Sorel first developed this critique in a series of articles published in 1906 in the radical socialist journal, *Le Mouvement socialiste*, which later made up his *Reflections on Violence* (1908).[41] In the book, Sorel deployed Bergson to criticize the Republican principle of "one man, one vote" for falsely positing political equality among all citizens on the basis of Enlightenment ideals. Sorel drew a contrast between degenerative and vital social forces, premised on a Bergsonian division between social structures emanating from intellectual modes of thought and those tied to intuition, expressive of a creative force opposed to intellectualism. In Sorel's view, Republican ideology subsumed all classes into its atomized concept of citizenship; he countered this homogenization by asserting the heterogeneity of class difference and identifying vital qualities unique to each class.[42] In *Le Mouvement socialiste* and his two books *Reflections on Violence* and *Illusions of Progress* (1908), Sorel critiqued the deterministic, mechanistic and materialist aspects of both capitalism and parliamentary democracy, which in turn inspired him to posit a spiritualist road to revolution in order to galvanize both the bourgeoisie and the proletariat. For Sorel, if the proletariat actively resisted the bourgeois politics of appeasement exemplified by parliamentary reform, the bourgeoisie would have no choice but to engage in the cataclysmic class war hoped for by Marx. Sorel designated that war a myth since it could not be predicted along Marxist lines, but had to be artificially induced.[43] This theory was also part of Sorel's broader critique of orthodox Marxists whose overly deterministic interpretation of historical materialism had undermined the role of human agency in Marx's political theory. These Marxists, many of whom condoned parliamentary politics and its system of representation, argued that the inequality fostered under capitalism made increased class conflict and a future revolution inevitable. Sorel's crucial insight was to analyze the role of rationalist ideology and democratic institutions in quelling that antagonism and forestalling revolutionary change. Sorel and his partisan disciple, Edouard Berth, put a premium on Marx's emotive appeal to class war. For instance, Berth's essay "Marx inédit!" (*Le Mouvement socialiste,* 1904) identified Marx as "a man of action, a revolutionary militant" who championed "the battle of classes" as the principle means of achieving social change. In *Reflections on Violence,* Sorel lauded Marx's vision of class conflict culminating in a "catastrophic revolution" as the mythic essence of his anti-rationalist agenda.[44] The Pauls endorsed this view in *Creative Revolution*, citing Berth's "Marx inédit!" with reference to the "creative will" of a "class conscious minority" of revolutionary proletarians, while lauding Sorel's thesis on "violence" and intuition as affirming Marx's call for "the purposive, conscious use of class struggle as a means of bringing about revolution."[45]

Intercine Warfare: Colomer on Communism

In his writings for *Action d'Art,* Colomer did not address Berth and Sorel's diatribe against the rationalist underpinnings of parliamentary democracy, but he did utilize Bergson to critique another adversary, the theory of anarchist communism championed by the prominent Russian, Peter Kropotkin (1842–1921) and his ally, Jean Grave, editor of *Les Temps Nouveaux* (1895–1921). Colomer's critical appraisal coincided with Grave's and Kropotkin's orchestration of the anarchist individualists' ouster from the official congress of the *Fédération Communiste-anarchiste* held in Paris in August 1913. Kropotkin followed that dismissal with a critique of Bergson's newest book, *L'Évolution créatrice,* in *Les Temps Nouveaux.*[46] Kropotkin, basing his anarchist theory on the scientific study of cooperative behaviour among nonhuman species, allied Bergson's philosophy to a more generalized revolt on the part of the bourgeoisie and their clerical allies against the rise of scientific materialism and its logical outcome: Worker emancipation. Kropotkin argued that recent discoveries in the sciences overturning previous assumptions did not warrant a wholesale refutation of the scientific method of inductive thinking, as Bergson asserted.[47] To Kropotkin, Bergson's alternative method of intuition was little more than a weak form of inductive thinking lacking in intellectual rigor and based on mere "analogies" and "metaphors" with no real basis in scientific facts. Having proclaimed intuition superior to reason, Bergson was then accused of catering to a gullible public by making "elegantly fantastical assertions" in support of *a priori* assumptions derived from "the fiat of Genesis."[48] Thus Kropotkin made plain his unapologetic allegiance to current scientific methods, a position he encapsulated in his *Modern Science and Anarchism* (1912). In this book, Kropotkin argued that the only viable form of anarchism embraced "the scientific inductive-deductive method" and parted "forever with metaphysics," including the "Hegelian metaphysical" line of Stirner and his followers.[49] Kropotkin dismissed Stirner as an unreconstructed metaphysician, and his attack on Bergson in *Les Temps Nouveaux* was a carefully calculated campaign against the Bergsonian anarchism of Colomer and his allies.

In response, Colomer publicly mocked Grave and Kropotkin in an essay entitled "Art, Anarchy and the Christian Soul," by casting these anarchists as the staunch defenders of Christianity, one of Stirner's spectral abstractions.[50] "Christianism taught human fraternity, altruism," wrote Colomer; "anarchist-communists today" reportedly preached a similar line by following "this humanitarian and altruistic ideal, this belief in a universal concord and in an egalitarian fraternity to which the individual must devote himself."[51] Colomer dismisses such universals as "idols more tyrannical than those of Divinity and Royalty," for "Anarchy must be individual or it does not exist."[52] This stinging rebuke was later expanded by Colomer into a Bergsonian critique of communist notions of collectivity and the communitarian groups Grave and his colleagues wished to generate. Colomer claimed that these "constructors of a social future" called on individuals to subordinate their egos to an abstract conception, whether in the guise of a utopian vision or the *a priori* set of moral principles outlined above.[53] Their concept of a union premised on communist ideals constituted a set of rigid precepts to which individuals must adapt if they are to gain membership in the group. "The communist idea," therefore, "ends by taking the place of patriotism," for the individual "is condemned to suffer in the company of human beings whom s/he does not like for the well-being of the Cause—for the prosperity of the Colony."[54] Such "Causes" or "theories" are "only empty frames," and those who would subordinate themselves to such abstractions are not true anarchists, for they have "never fathomed the reason for their anarchism," namely the cultivation of their "personality."[55] "Communist colonies" therefore constitute "hybrid ensembles of inharmonious elements," a random gathering of individuals lacking in empathy who are predisposed for conflict.[56] Form and content — the communist ideal and the individuals who bring about its realization — are therefore bifurcated in the anarchist-communist imagination.

Colomer's polemic brings to mind Bergson's own critique in *Creative Evolution* of the intellect's propensity to fabricate such empty frames in order to foster pragmatic activity. "The intellectual faculty," states Bergson, focuses solely on "relations" rather than on "things" and as such "possesses naturally only an external and empty knowledge, but it has thereby the advantage of supplying a frame in which an infinity of objects may find room in turn."[57] Unaware of the utilitarian origins of such thinking, some mistakenly identify the intellect as a faculty designed for pure speculation and take its general frames for reality itself. This radical bifurcation of abstract form from material content is something Bergson advocated overcoming by means of intuition, a faculty of willed empathy that enables us not only to grasp inner duration, but to develop pliable forms of representation moulded to this durational content. Thus, intuitive knowledge can produce representational forms that are integral to the content they wish to represent, and as such they are akin to the vital order of living organisms. To Colomer's mind, they were analogous to a Stirner-inspired rejection of the anarchist-communist ideal envisioned by Kropotkin and Grave.

Plebeian Sorelians

Colomer's pre-war Bergsonian critique of rival leftists found no echo in the writing of the Pauls, who instead deployed Sorelian ideology to set their sights on a new nemesis: Marxists allied to Britain's resurgent Labour Party. Speaking generically of the continuing appeal of rationalist discourse among reform-minded Marxists in Britain, the Pauls noted that while Freud and Bergson may have overthrown "the superstition that man is essentially a rational animal," leftists had been slow to grasp the ideological import of their critique due to a blind faith in the supposed inevitability of revolution, premised on the fatal contradictions inherent in the system of capitalist production analyzed by Marx.[58] These socialists effectively claimed that Marxism "reduces the entire content of history to an automatic process wherein the consciousness of human units plays no part."[59] As we have seen, the Pauls approved the Sorelians' claim that Marx was an avid anti-determinist and that his theory of historical materialism was premised on the critical role of class-consciousness as a prerequisite for revolutionary change. Thus, the authors of *Creative Revolution* endorsed the assertion in *Le Mouvement socialiste* that "Marx and Bergson" shared an unequivocal hostility "to every intellectualist doctrine."[60] They then followed their French colleagues in developing a theory of class-consciousness premised on a radical reading of Bergson's theory of creativity.

Like Sorel, the Plebs League was adamantly opposed to democratic institutions and those socialists who sought change through an incremental process of reform using the instrument of parliament. The League endorsed the so-called "Impossibilist" agenda promoted by the Socialist Labour Party (founded in 1903) and the British Socialist Party (founded in 1911), both of whom rejected the parliamentary strategies of the Labour Party, constituted in 1906. Following the Russian Revolution in 1917, these Impossibilists rallied to the Bolshevik cause and formed the core of the newly minted Communist Party of Great Britain in July 1920.[61] The prominent Labour Party politician, Ramsay MacDonald, was especially reviled in these circles for his open hostility to Bolshevism, which MacDonald condemned as yet another manifestation of the extremism of the "Impossibilists" and their anarchist-syndicalist counterparts in France.[62]

Creative Revolution took up this anti-parliamentary agenda but gave it a Bergsonian and Sorelian twist through the critique of rationalism. The Pauls identified Ramsay MacDonald as the English counterpart of those orthodox Marxists and parliamentary socialists condemned by Berth and Sorel. Left-wing advocates of the "chimera" of social solidarity, including MacDonald, held that "the political institutions of bourgeoisdom are, or can be made, effective expression of the needs of all classes conceived as forming an integral society."[63] Moreover the faith MacDonald put in democratic institutions as a vehicle for change was deemed the product of "an obsolete psychology, the characteristic psychology of the democratic age" which is "based upon the belief that reason is the main motive force of human action."[64] While MacDonald argued that progressive politics could only be achieved through "the acceptance of programs by reason of their rationality," the Pauls mocked his acceptance of this "liberal and democratic tradition" by questioning how he would go about persuading an omniscient bourgeoisie "that socialism is sweetly reasonable."[65] Like Sorel, they dismissed parliament as a tool of the plutocratic class and 'citizenship' as a rationalist abstraction antithetical to class-consciousness.[66] In contrast to MacDonald, the Pauls exalted class identity as found in the "syndicat," "workers' committee" or "soviet," as based on intersubjective relations of "men and women working together in a particular place at a particular time."[67]

Feeling Class Struggle; Intuiting Socialism: Activist Images

Having identified rationalism as an obstruction to revolutionary consciousness, the Pauls next turned to the function of the image as the emotive and volitional means of instilling a revolutionary spirit among the proletariat. Due to the "new psychology," revolutionaries now recognized that "while reason may serve at times to light the path" to revolution, "the drive along that path comes preponderantly from the urge of the unconscious will."[68] They therefore call on revolutionaries to marshal support among the working class by evoking the "tactic of class struggle" instead of making "rhetorical appeals to reason."[69] The Pauls identify this vision of class struggle as a "mental entity," a form of aspiration that exists "in the realm of art and feeling, in the realm of impulse and desire," which on a conscious level gains further affirmation "as a result of our intellectual criticism of capitalism."[70] Thus images of class struggle operate first and foremost on an emotional level; they are activist images — "emotionally and in the realm of art it is a feeling that we can replace [the existing order] by a better order that shapes itself in the imagination ... volitionally, or in the realm of will it is an endeavour to create in the world of objective fact what we have already conceived in the intellectual and artistic imagination."[71]

Such statements bring to mind Sorel's famous appropriation of Bergson's theory of intuition and images, in his description of the myth of the general strike as able to arouse class consciousness and, with it, insurrection. In *Reflections on Violence,* Sorel called on revolutionary leaders to abandon rational argumentation when addressing the proletariat and instead conjure up "a body of images" highlighting the "drama of the general strike" that, "by intuition alone, and before any considered analyses are made, is capable of evoking ... the mass of sentiments which corresponds to the different manifestations of the war undertaken by Socialism."[72] For workers such images stir up memories of past strike actions that "have engendered in the proletariat the noblest, deepest, and most moving sentiments that they possess"; moreover, "the general strike groups them all into a coordinated picture [and] gives to each one of them its maximum of intensity; appealing to their painful memories of particular conflicts, it colours with intense life all the details of the composition presented in consciousness." "We thus obtain," concluded Sorel, "that intuition of Socialism" and "we obtain it as a whole, perceived instantaneously."[73]

In his *Introduction to Metaphysics,* Bergson celebrated intuition as a form of "intellectual sympathy"[74] able to overcome the limits imposed by the intellect in grasping duration and engaging in creative acts fully expressive of our unfolding personalities. Perceptual images, which Bergson defined as comingling memory and matter, were considered as forms of representation that are integral to human actions and by definition are closer to immediate or "concrete" experience, unlike the abstractions generated by analytical reasoning.[75] The emotive combination of many different kinds of images, he explained, "will be able, through the convergence of their action, to direct consciousness to the precise point where there is a certain intuition to be seized." He added, if "they all require from the mind the same kind of attention, and in some sort the same degree of tension, we shall gradually accustom consciousness to a particular and clearly defined disposition."[76] For Bergson, intuition inspired artistic creation, philosophic reflection and scientific discovery, while for Sorel and the authors of *Creative Revolution*, its chief manifestation was the overthrow of democracy and the nation-state through class struggle.

By contrast, for Colomer, intuition and Bergson's theory of activist images served as justification for an individualist revolt, untethered to the Sorelian or Marxist ideology of class conflict. Drawing on Bergson's metaphorical method, Colomer compared the life of a human being to "a circle whose destiny is represented by the circumference," and whose creative actions take the form of "rays." Colomer argued that this "free and creative center" was only discernible through the inward turn of intuition, which allowed individuals to grasp "being itself, its consciousness, its life."[77] The egoist, in discovering his or her own uniqueness, simultaneously revolutionized consciousness by overthrowing the yoke of habit and adherence to abstractions in favour of actions expressive of an individual's creative capacities. This is what Colomer meant by action d'art: the cultivation of the self was to his mind a form of artistic creation.

Creative Revolutionists: Artists, Soviets and The Vital Order

Where the Pauls differed from both Colomer and Sorel was in arguing that class consciousness needed to be shaped by an elite group of "artist-thinkers" so that this revolutionary impulse could be moulded into a vital order, exemplified by the Soviet and the oligarchic Dictatorship of the Proletariat. In *Reflections on Violence,* Sorel called on union orators to marshal the myth of the general strike, but crucially he did not give them a directive role as leaders, preferring instead to fold them into his Bergsonian concept of the syndicat as a vitalist manifestation of societal order. Indeed, Sorel openly rejected any notion of a "dictatorship of the proletariat," declaring it a "system of slavery."[78]

Bergson declared all forms of vital order integral to duration, which he frequently compared to a melody.[79] In *Matter and Memory,* he described the material universe as composed of multiple durations whose rhythmic pulse measured the degree of freedom native to different kinds of consciousness.[80] For example, the living organism as the product of creative evolution had been separated by nature itself, given a particular rhythm and durational arc, and composed of parts that form an undivided whole. However, despite their qualitative differentiation, durational organisms are not wholly separate; rather they are joined by their sympathetic relation to each other. Identifying sympathy in a mother's relation to her child as evidence, Bergson concluded that "the living being is above all a thoroughfare and that the essence of life is in the movement by which life is transmitted."[81] For Bergson, this *élan vital* found its fullest expression in homo-sapiens, the only species capable of intuitive consciousness, and with it, the ability consciously to shape this vital order. To Bergson, the will to create and the internal impulse driving personal development is itself an expression of a larger evolutionary process. "Each personality," states Bergson, "is a creative force; and there is every appearance that the role of each person is to create, just as if a great Artist had produced as his work other artists."[82]

For Sorel, this creative élan took the form of the intuitive sympathy integrating class-conscious workers into the vital order of the syndicat, whose protean energy produces revolution. This voluntary association and coordinated discipline meant that these individuals acted in consort — whether while on strike or in the workplace.[83] These anarchist syndicalists also shared an "artist's taste" born of technical innovation. "The idea of the general strike," for Sorel, therefore bends "all the energies of the mind to that condition necessary to the realisation of a workshop carried on by free men, eagerly seeking the betterment of industry."[84]

While Sorel defined this vital order in terms of the intuitive sympathy uniting militant syndicalists, Colomer drew on these same philosophical premises to define the basis on which anarchist individualists formed a vitalist Union of Egoists. Colomer argued that such a "band," in contrast to the "empty frame" governing the anarchist-communist collective, "is not a fixed form," "it is not an Entity" nor "a Cause" for it cannot exist apart from the individuals who constitute it. "The band has nothing *a priori* about it. It is formed by the force itself of the individualities who compose it. It can transform itself or dissolve following this same force." Colomer defines his collective as a heterogenous grouping that nurtures qualitative difference, rather than homogeneity. "In order to form a proper band, the individualities who compose it do not search for a common ideal, a common theory, common conceptions." Instead this "community of temperaments" arises spontaneously by virtue of an "intuitive sympathy which draws these individuals towards each other ... without even having planned it." What serves to unite them is a sense of "harmony" manifest as friendship and shared enthusiasm, which will nurture "the greatest realization of each individual" who composes the band.[85] Thus discerning one's own inner harmony can lead to a sense of harmony with others, while creativity, rather than morality or an abstract cause, provides the basis for this contingent union. Further aestheticizing his conception, Colomer described this union of egoists as "a community blazing with colours of various hues," echoing Bergson's own use of colour as a metaphor for duration in such texts as *Matter and Memory* and *The Introduction to Metaphysics*.[86]

The emotive register able to generate such cohesion was not the militant *esprit de corps* uniting Sorel's syndicat, but instead a shared feeling of joy that permeated the band. As seminal to this theory,[87] Colomer cited Bergson's *Time and Free Will* (1889), the text in which Bergson aligned the human experience of joy to the sensations of grace, rhythm and harmony integral to works of art.[88] Foremost for Colomer, "*joie-de-vivre*" arose when the beholder experienced the creative, internally-harmonious actions of an anarchist animated by intuition. Individuals inspired by this life-affirmative joy would, argued Colomer, spontaneously attract like-minded individualists to join them in the formation of a contingent union of egoists. What united these radical individualists was an "intuitive sympathy" and mutual enthusiasm that enabled them to delight in each other's radical difference. Hence Colomer prioritized "*joie-de-vivre*" in his theory of art as part of a larger celebration of sensory embodiment. Not only did it entail an emotive register of an intuitive state of mind but also an exaltation of all cultural manifestations of novelty, intensity, rhythm and harmony as fundamentally anarchist by virtue of their durational and life-affirming properties.

The Pauls by contrast rejected Sorel's and Colomer's anarchist assumptions, substituting the 'vital order' of the Soviet for that of the syndicat or union of egoists, and declaring artistic creativity the sole province of a revolutionary elite, rather than an attribute of the proletarian on the factory floor or those anarchist individualists whose creativity found expression in 'art actions.' Having chastised the syndicalist movement as "lacking in the constructive side"; they promoted the Soviet as the "new instrument" of such construction, since it combined "industrial" and "political" functions.[89] They viewed soviet dictatorship as constituting a form of direct democracy, since it was made up of a structural hierarchy of elected workers, starting with those delegated from local workers' committees and culminating in an "all British Soviet." [90] The Pauls then used vitalist metaphors to describe the Soviet as "a new creation, flexible, mobile, and yet persistent as are the thought and will it expresses."[91]

Moreover, the Soviet structure reportedly nurtured individual freedom while simultaneously fostering "a voluntary self-discipline" and group cohesion.[92] While Sorel regarded this intuitive "self-discipline" as sufficient to assure the success of the syndicalist revolution, by contrast the Pauls argued that such self-coordination needed to be further orchestrated from above by an elite group of "artist-thinkers," most prominently Lenin himself.[93] They made the Bergsonian import of this formulation clear by prefacing their chapter on artist-thinkers with an epigraph on Bergson's theory of artistic perception, culled from Swedish philosopher, Algot Ruhe's 1914 monograph on the philosopher. In his text, Ruhe skilfully outlined the full import of Bergson's theory of art, noting the role of intuition in enabling artists to plumb duration to develop their personalities to the fullest, and to create aesthetic forms integral to duration.[94]

The Pauls drew on Ruhe's text to politicize Bergson's notion of artistic creativity. For the Pauls, the Soviet, born of the revolution, may have constituted a vital order, "flexible" and "mobile" like duration itself, but that organism's development still needed to be shaped following the artistic dictates of benevolent leaders like Lenin who are themselves "self-conscious workers" free "from all taint of exploitation."[95] As they surmised in *Creative Revolution:* "For Lenin was reserved the privilege of exercising his genius upon the plastic material of humanity."[96] Lenin, to paraphrase Bergson, was the "Great artist" producing by means of creative revolution not exactly "other artists," but instead pliant revolutionaries. In this regard, the Pauls note that "the efficient self-discipline of enlightened workers" within the Soviet "will become more and more to resemble the harmonious self-discipline of an orchestra or a choir." However, this Bergsonian metaphor for self-coordinating rhythms and melodic structure is then tempered by the presence of the dictatorial elite, which "will become more like the guiding will and inspiration of a competent orchestral conductor or choral master."[97] "Fundamentally poetic, yet gifted with a due share of efficiency," members of this elite "are, before all, creative revolutionists." Lenin, as "the archetype of the creative revolutionist in action" is able "to find an artist's self-expression and to delight the artistic sensibilities of all true revolutionists" by virtue of his ability to sculpt "the plastic material of humanity." Trotsky, in his turn, "feels an abiding delight in the making and the management of the Red Army," and taken together the leaders of the Soviet Union are "all persons who secure a supremely congenial mode of self-expression in their task of large-scale manipulation of man as the social being."[98] In the conclusion of their book, Eden and Cedar Paul claimed that the privileged role bestowed on humanity in Bergson's *Creative Evolution* found its supreme expression in the unfolding "movement towards freedom" initiated by the advent of the Soviet Union.[99] As subsequent history made clear, however, their Panglossian vision of an elite cadre of "artist-thinkers" had little to do with the repressive reality of this newly-minted totalitarian dictatorship.[100]

Notes

1 On *Action d'art* and on Colomer's impact, see Theresa Papinikolas, *Anarchism and the Advent of Paris Dada: Art and Criticism, 1914-1924* (London: Ashgate, 2010), 105–112; and Richard Sonn, *Sex, Violence and the Avant-Garde: Anarchism in Interwar France* (University Park: The Pennsylvania State University Press), 2010, 198–209.

2 Eden and Cedar Paul, *Creative Revolution: A Study of Communist Ergatocracy* (London: Plebs League, 1920). The book was republished in Britain by Allen and Unwin and in the United States by Thomas Seltzer Publishers.

3 On Eden Paul (1865–1944) and Cedar Paul (1880–1972), the Plebs League and the Communist Party, see entries in *The Labour Whos Who: a Biographical Directory* (London: Labour Publishing Company, 1927); Jonathan Rée, *Proletarian Philosophers: Problems in Socialist Culture in Britain, 1900-1940* (Oxford: Clarendon Press, 1984) 40–41; and Andy Miles, "Workers' Education: The Communist Party and the Plebs League in the 1920's" *History Workshop* (Autumn, 1984), 102–114.

4 Lawrence Golman, "Education as Politics: university Adult Education in England since 1870" *Oxford Review of Education* (March-June 1999), 89-101; and Rée, *Proletarian Philosophers*, 20–21.

5 Miles, "Workers' Education," 103; Rée, *Proletarian Philosophers*, 33–34.

6 Rée, *Proletarian Philosophers*, 23-45; Eden and Cedar Paul, *Creative Revolution*, 5–6.

7 Paul, *Creative Revolution*, 5–6.

8 *The Communist* No. 20 (16 December 1920) 5; cited in Rée, *Proletarian Philosophers*, 48. On the Labour colleges' ballooning enrolment see Rée, *Proletarian Philosophers*, n. 22, 141.

9 Paul, *Creative Revolution*, 5.

10 Paul, *Creative Revolution*, 197.

11 Henri Bergson, *Creative Evolution*. Trans Arthur Mitchell (New York: Henry Holt and Company, 1911, rpt. 1931), 87–97.

12 Paul, *Creative Revolution*, 217; Bergson, *Creative Evolution*, 269.

13 Paul, *Creative Revolution*, 13.

14 Paul, *Creative Revolution*, 22.

15 Paul, *Creative Revolution*, 19, 37.

16 Paul, *Creative Revolution*, 65.

17 Paul Ward, *Red Flag and Union Jack: Englishness, Patriotism and the British Left, 1881-1924* (Bury St. Edmunds: The Boydell Press, 1998), 148–166.

18 Paul, *Creative Revolution*, 98.

19 Paul, *Creative Revolution*, 193.

20 Paul, *Creative Revolution*, 85.

21 Max Stirner, *The Ego and its Own* (translation by Steven Byington, 1907, rpt. London: Rebel Press, 1993) For an analysis of Stirner's philosophy see John Clark, *Max Stirner's Egoism* (London: Freedom Press, 1976).

22 Stirner, *The Ego and its Own*, 362–366.

23 Stirner, *The Ego and its Own*, 13, 363

24 Stirner, *The Ego and its Own*, 182, 366.

25 Stirner, *The Ego and its Own*, 182.

26 Stirner, *The Ego and its Own*, 82.

27 Stirner, *The Ego and its Own*, 26, 29, 32, 44, 46–48.

28 Stirner, *The Ego and its Own*, 31

29 Stirner, *The Ego and its Own*, 85.

30 Stirner, *The Ego and its Own*, 43

31 Stirner, *The Ego and its Own*, 65.

32 Stirner, *The Ego and its Own*, 46, 52–53.

33 Stirner, *The Ego and its Own*, 66.

34 Stirner, *The Ego and its Own*, 29–30.

35 Stirner, *The Ego and its Own*, 51–55

36 Stirner, *The Ego and its Own*, 54–55; 316.

37 Henri Bergson, "The Perception of Change" (1912), in *The Creative Mind*, translated by Mabelle L. Andison (New York: Philosophical Library, 1946), 173. The original French is as follows: "Il y a des changements, mais il n'y a pas de choses qui change: le changement n'a pas besoin d'un support. Il y a des mouvements, mais il n'y a pas nécessairement des objets invariable qui se meuvent: le movement n'implique pas un mobile." Henri Bergson, "La Perception du changement," in *Mélanges* (Presses Universitaires de France, 1972), 904–905. On the centrality of this concept to Bergson's philosophical method, see Garrett Barden, "Method in Philosophy," in Ed. John Mullarkey, *The New Bergson* (New York: St. Martin's Press, 1999), 32–40.

38 André Colomer, ""La Science et l'Intuition: Leurs rôles dans l'individualisme," *L'Action d'art* (10 May 1913), 3.

39 Colomer, ""La Science et l'Intuition: Leurs rôles dans l'individualisme,"3.
40 Paul, *Creative Revolution*, 27–42.
41 Georges Sorel, *Reflections on Violence* (1908). Trans, T.E. Hulme (New York: Peter Smith, 1915, rpt. 1941). For Sorel's involvement with *Le Mouvement socialiste* before his break with the journal in 1909, see Marion de Flers, "*Le Mouvement socialiste* (1899-1914)," *Cahiers du Georges Sorel* No. 5 (1987), 49–76.
42 Mark Antliff, *Avant-Garde Fascism: the Mobilization of Myth, Art and Culture in* France (Durham: Duke University Press, 2007), 73–81.
43 Vernon, *Commitment and Change: Georges Sorel and the Idea of Revolution* (Toronto: University of Toronto Press, 1978), 51–52.
44 Edouard Berth, "Marx Inédit!," *Le Mouvement socialiste* (1 November, 1904), 98–99; Sorel, *Reflections on Violence*, 22, 34–35.
45 Paul, *Creative Revolution*, 5, 136, 151.
46 Mark Antliff, *Inventing Bergson: Cultural Politics and the Parisian Avant-Garde* (Princeton: Princeton University Press, 1993), 142–143; Peter Kropotkin, "La Croisade contre la Science de M. Bergson," *Les Temps Nouveaux* (15 October, 1913), 2–4.
47 Henri Bergson, *Creative Evolution*, 211–216.
48 Kropotkin, "La Croisade contre la Science de M. Bergson," *Les Temps Nouveaux* (15 October, 1913), 3.
49 Peter Kropotkin, *Anarchism and Modern Science* (London: Freedom Press, 1912), 38–39; 50; 69–70; 92–93.
50 Stirner, *The Ego and its Own*, 182; André Colomer, "L'Art, l'anarchie, et l'ame Chrétienne," *L'Action d'art* (15 April 1913), 1–2.
51 Colomer, "L'Art, l'anarchie, et l'ame Chrétienne,"1.
52 André Colomer, "L'Art, l'anarchie, et l'ame Chrétienne," *L'Action d'art* (15 April 1913), 1–2.
53 André Colomer, "Illusions sociales et delusions scientist," *L'Action d'art* (25 August 1913), 2.
54 André Colomer, "La Bande," *L'Action d'art* (10 November 1913), 2.
55 André Colomer, "La Bande," 2.
56 André Colomer, "La Bande," 2.
57 Bergson, *Creative Evolution*, 149–150.
58 Paul, *Creative Revolution*, 198.
59 Paul, *Creative Revolution*, 191–192, 198.
60 Paul, *Creative Revolution*, 45. The authors took this quotation from Algot Ruhe, *Henri Bergson: An Account of his LIfe and Philosophy* (London: MacMillan, 1914), 50. It is taken from a survey on Bergson's influence published by *Le Mouvement socialiste* over the course of 1911–1912.
61 Rée, *Proletarian Philosophers*, 12–14, 20–22, 46–47.
62 Ward, *Red Flag and the Union Jack*, 49–51, 88–89, 155–156.
63 Paul, *Creative Revolution*, 28, 31.
64 Paul, *Creative Revolution*, 36.
65 Paul, *Creative Revolution*, 36–37.
66 Paul, *Creative Revolution*, 47.
67 Paul, *Creative Revolution*, 172–173.
68 Paul, *Creative Revolution*, 37.
69 Paul, *Creative Revolution*, 41–42.
70 Paul, *Creative Revolution*, 19–20.
71 Paul, *Creative Revolution*, 19–20.
72 Sorel, *Reflections on Violence*, 137.
73 Sorel, *Reflections on Violence*, 137.
74 Henri Bergson, *An Introduction to Metaphysics* (1903) Trans. T.E. Hulme, Introduction by John Mullarkey (London: Palgrave MacMillan, 2007), 5.
75 On Bergson's theory of images see Frédéric Worms, "*Matter and Memory* on Mind and Body," in Mullarkey, *The New Bergson*, 88–98; and John Mullarkey, *Bergson and Philosophy* (Notre Dame: University of Notre Dame Press, 2000), 31–37.
76 Bergson, *An Introduction to Metaphysics*, 10–11.
77 Colomer, ""La Science et l'Intuition: Leurs rôles dans l'individualisme," 3.

78 Sorel, *Reflections on Violence*, 191–195.
79 Henri Bergson, "The Perception of Change," 174–176
80 Henri Bergson, *Matter and Memory* (1896) Trans. N.M. Paul and W.S. Palmer (New York: Humanities Press, 1978), 295–296.
81 Bergson, *Creative Evolution*, 128.
82 Henri Bergson, *Mélanges*, 1071. The statement comes from Bergson's lectures on the personality delivered at the University of Edinburgh in April-May 1914.
83 Sorel, *Reflections on Violence*, 280, 284–285.
84 Sorel, *Reflections on Violence*, 287, 294.
85 Colomer, "La Bande," 2.
86 Bergson, *Matter and Memory*, 268-9; Bergson, *Introduction to Metaphysics*, 221.
87 André Colomer, "M Bergson et les Jeunes Gens d'Aujourd'hui," *Action d'art* ((1 March, 1913), 1.
88 André Colomer, "M Bergson et les Jeunes Gens d'Aujourd'hui,"1; Henri Bergson, *Time and Free Will* (1889), trans. F.L. Pogson, 1910 (rpt. New York: Harper & Row, 1960), 10–15.
89 Paul, *Creative Revolution*, 53, 65, 67. Eden and Cedar Paul's critique of syndicalism followed that of Lenin in his proclamation to the Second Congress of the Communist International published in Spring 1920. See V.I. Lenin, "Left Wing Communism, An Infantile Disorder," in Karl Marx, Friedrich Engels, V.I. Lenin, *Anarchism and Anarcho-Syndicalism* (New York: International Publishers, 1972), 304–315.
90 Paul, *Creative Revolution*, 171–174.
91 Paul, *Creative Revolution*, 158.
92 Paul, *Creative Revolution*, 158.
93 Paul, *Creative Revolution*, 155, 196.
94 Algot Ruhe, *Henri Bergson*, 70–87, 91–95.
95 Paul, *Creative Revolution*, 133.
96 Paul, *Creative Revolution*, 200.
97 Paul, *Creative Revolution*, 135.
98 Paul, *Creative Revolution*, 155–156.
99 Paul, *Creative Revolution*, 218.
100 For an overview of the repression of political dissent under Lenin, with special attention to the mass imprisonment and killing of leftists opposed to Bolshevism, see G.P. Maximoff, *The Guillotine at Work: Twenty Years of Terror in Russia,* (Sanday: Cienfuegos Press, 1979).

10 Revitalizing Traumatised Soviet Soldiers

Art, Psychology and "Creative Darwinism"

Pat Simpson

Introduction

This chapter investigates the "modern", vitalist implications of an art exhibition and multiple illustrated lecture presentations, at the prestigious M. V. Frunze Military Academy hospital in Moscow during the Second World War.[1] The primary audience comprised wounded, traumatised Soviet soldiers – possibly both male and female.[2] These events were mainly provided by the Directors of the State Darwin Museum in Moscow, the ornithologist Dr Aleksandr Kots and his wife, the zoo-psychologist Nadezhda Ladygina-Kots. They continued to do hospital presentations, and then also presentations to the wounded at the Darwin Museum until 1945. However, this chapter focuses on their work from 1941 to 1942.

The "modern" element of these provisions did not reside in the visual or technical radicalism of the artworks shown. All obeyed the current Soviet rulings on Socialist Realism laid down in 1934, in terms of legibility of subject matter, presentation and themes.[3] Vitalist theory, in the senses most familiar to the West, had been excoriated and shut down by the Soviet government under Stalin by that year. As I will argue, however, elements of vitalism continued to exist covertly in Soviet wartime culture, partly in the attempt to follow through some key ideas extant within "modern" and past Soviet psychology that were potentially relevant to the situation. These vitalist elements also partly resided in the Darwin Museum's focal topic: Charles Darwin's theories of evolution – albeit ideologically re-moulded to fit with the recent Soviet discursive context on Darwinism. These theories were central to contemporary Soviet constructs of modernism and modernity, both in science and in politics. Thus, they were equally central to the policies and activities of the Darwin Museum itself.

In the late nineteenth and early twentieth centuries, vitalist and neo-vitalist ideas circulated within the Russian and early Soviet intelligentsia much as they did elsewhere, as exemplified by other contributions to this volume.[4] Russian cultural interest in aspects of vitalistic thought was fuelled by acquaintance with avant-garde Modernist art and particularly facilitated by the influence of the international Symbolist artistic and literary movements of the 1880s to 1910s.[5] In this context, all of the writings of Henri Bergson had been translated into Russian by 1914 and had provided a popular basis for intellectual debate.[6] Indeed, the ideas and art practices of avant-garde artists of the 1910s until the early 1920s, including Kazimir Malevich and Vasilii Kandinsky, can be seen to relate to Bergson's vitalistic theories of "creative evolution", as has been indicated in Chapter 7 of this volume.[7] Moreover, Hans Dreisch's *Vitalism: Its History and System* was published in Russian in 1915, making his ideas freely available.[8]

DOI: 10.4324/9781003045595-14

However, as the political historian of Russian/Soviet science, Aleksandr Vucinich, has indicated, Russian and early Soviet experimental bio-scientists and natural historians were largely much less interested in the metaphysically orientated forms of vitalism and neo-vitalism than their western counterparts.[9] Additionally, following the Stalinist shut-down of avant-garde cultures and ideas, from the early 1930s onwards the words "vitalist" and "vitalism" became terms of abuse, indicating that those thus accused, whether artists or scientists, were enemies of the state and their publications should be (and were) banned, as will be demonstrated in this chapter. Nevertheless, as will be argued, this did not preclude the development of several covert elements of "vitalist" ideas in Soviet culture, largely in the name of "materialist" science.

By the Great Patriotic War against Nazi Germany, what was being offered by the Darwin Museum team to the wounded soldiers at the Frunze hospital, I will suggest, was a form of "evolutionary psychotherapy". This approach primarily appeared to use aspects of current Soviet ideas on military psychology, particularly in relation to the field of physiological and psychological trauma. The team's apparent aim was to help "revitalize" the soldiers and restore them to a condition in which they could resume battle against the Nazi invaders currently looming close to Moscow. The specific vitalist elements drawn upon for this purpose were not, at least overtly, derived from Bergsonian or other earlier European ideas of vitalism, but arguably from two major and potentially conflicting sources peculiar to the specific cultural and historical context within the USSR in the early 1940s.

The first of these sources was the vitalistic and metaphysical aspects of Russian Orthodox concepts of collective engagement and empathy, which may have impacted on public understandings of the Soviet requirements for collectivism at the time. The second was the equally unacknowledged, neo- or crypto-Lamarckian elements of vitalism connected to the exertion of human will, both individual and collective, as an evolutionary socio-political force. This was a notion that had become embedded in Soviet Darwinism by the 1940s, particularly through the activities of Trofim Lysenko and his supporters. Both elements, as I will argue, were fundamental to the propagation of the enduring aspirational ideal of the "New Man". This was a heroic type of collectivized humankind unknown in the capitalist West, but which had been prophesied to emerge triumphantly in the USSR.[10]

The chapter starts by considering some significant aspects of the complex political and scientific context for the Darwin Museum's activities, focusing particularly on the established Soviet psychiatric and psychological approaches to battle trauma. Here I suggest that elements of these approaches, especially as relating to covertly vitalistic aspects of psychology linked to the contemporary Soviet emphasis on collectivism helped to give rise to the Darwin Museum's involvement with the Frunze and other military hospitals in this period.

The second section investigates the nature of the exhibition and other Darwin Museum presentations at the Frunze military hospital and how they might be perceived to offer a range of "revitalising" experiences for the wounded, in relation to contemporary Soviet military psychology. The final section examines significant aspects of the Darwin Museum's research and hospital presentations that may also be seen to be grounded in two areas of professional concern for the museum staff. These were "comparative psychology" – banned in 1930 for its "vitalistic" notions of "mind" and "instinct" – and the covert elements of vitalism within Lysenko's notion of "Creative Darwinism", which followed Friedrich Engels' emphasis on individual and collective will as means for humans to conquer nature, including their own.

The conclusion will first summarize my main points about the covertly existing and arguably interrelated "threads" of vitalistic thought in Soviet culture 1941–1942. It will then draw attention to the ironic relationship of Lysenko's construct of "Creative Darwinism" to the other observable elements of Soviet vitalism that may have fed into the Darwin Museum's activities. It will end with a postscript citing potential evidence to suggest that, with the support of the Frunze hospital, the Soviet government took a positive view of the impact of the Darwin Museum's contributions to the war effort.

There is a caveat, however, that the following arguments are partially speculative. On my original research visit, although I found archival sources listing the topics of the talks, I could not find the lecture drafts/notes. This is unusual for the Darwin Museum archive, as it normally preserved such data meticulously. However, these documents may well have been stored elsewhere, and I have had no opportunity to return to Moscow to double-check due to the recent/ongoing Covid-19 pandemic. But there is also a possibility that these documents may have been "purged" by the museum for their vitalistic elements, alongside other apparent "purges" of other archival materials that did not accord with the dominant Lysenkoist view of Darwinism in the late 1940s.[11]

The "Great Patriotic War": Soviet Psychiatry, Psychology and Vitalism

For the Soviet Union, the "Great Patriotic War" was marked by the German invasion of Soviet territory that began on 22 June 1941. Having left a trail of carnage, German troops arrived near Moscow in November 1941.[12] The Frunze hospital, like all Soviet military hospitals, was at the forefront of developments in the Soviet disciplines of psychiatry and psychology dealing with battle trauma.[13] However, as the conflict came closer to Moscow, the Frunze obviously could not cope alone with the number of casualties being brought back from the front. Supplementary makeshift military hospitals were set up at secret locations all over Moscow, known only by code numbers. All the casualties brought back to the hospital appear to have been traumatised by their experiences of battle to some extent, whether both physically and mentally or purely mentally. Indeed, Paul Wanke in 2005 convincingly suggested that this level of battle trauma constituted "psychoneurosis", previously well-known in Russia and elsewhere as "shell-shock", now labelled Post Traumatic Stress Disorder (PTSD).[14] This may well have resulted from the increase in the more impersonal, "industrialized" nature of warfare during the Second World War, which was underscored by the apparently barbaric treatment meted out to Soviet soldiers and civilians by the invaders.[15]

A key indicator of battle trauma to both Soviet psychiatrists and psychologists was the high level of introversion exhibited by the sufferers, which these scientists called "egocentrism".[16] Later western theorists of PTSD highlighted it as an inability to empathise.[17] This condition entailed a lack of ability to engage or empathise with a wider collective effort. In the USSR, engagement with collective endeavour had been systematically laid down in the post-Revolutionary period as the Soviet citizen's primary duty. It was further emphasized in the Stalin era, particularly during the Second World War. However, as Catherine Merridale argued in 2000, this emphasis on collectivism, although coming from a Marxist-Leninist political stance after October 1917, had powerful historical resonances that reached back to centuries of Russian Orthodox religion.[18]

Russian Orthodoxy had stressed the importance of *sobornost'*, the spiritual, psychic and empathetic bond of the congregation of believers, that should also extend into any secular collective activity or organisation with which they were involved.[19] The religious basis for this idea was that human minds were collectively connected to the divine and therefore had access to the "truest knowledge",[20] whereas the individual human body did not. This divide of mind and body was largely regarded, even in the 1920s, as unscientific, occultist, mystical and thus "vitalist", inimical to the self-professed "materialist" Soviet regime. It was arguably, however, a powerfully enduring, covert vitalistic presence in the Russian psyche regarding the notion of the Soviet collective, particularly under Stalin in the wartime period.

In relation to this, it is significant that the Russian Orthodox church wholeheartedly supported the Soviet war effort from 1941 onwards. In 1943, it was rewarded by being given Stalin's permission to re-establish the Orthodox Patriarchate.[21] This curious scenario in an atheist state can be understood as a concessionary move by the Soviet Communist Party. It implicitly acknowledged the deep Russian Orthodox roots embracing the sense of collectivity that the Party had propagandized as central to Soviet existence since 1917, albeit from a different ideological direction. At the time, it also acknowledged the shared goal of revitalising the combatants to focus on victory at any cost. As will be seen, the exhibition and lecture/presentations by the Darwin Museum team appear to have offered reinforcements of this goal. By 1941, psychiatry and psychology in the USSR seem to have had three basic concepts in common that are important to this chapter. The first of these was the absolute necessity for promoting collectivism of thought and behaviour, as already indicated. The second was Lenin's "Theory of Reflection", which held that the mind was a "reflection" of the external, social world and in the mind's natural/normal condition these "reflections" would be "true" in relation to the Communist Party's propagandized visions of this world. Thus, psychological or emotional problems that did not mesh with this vision, were to be understood as resulting from faulty reflections of, or from, experienced material reality.[22]

Ivan Pavlov, whose work was highly regarded in the USSR at the time, supported this "materialist" view of the psyche. In his laboratory experiments, Pavlov had demonstrated that damage to parts of the brain could cause symptoms of "neurosis", as exemplified in some of his experiments with dogs.[23] His theories, along with those of Lenin, also generated the idea of education and/or training in alternative work as a means to rehabilitate traumatised individuals and reintegrate them into the collective of Soviet society.[24] Indeed, such ideas had been used, for example, in Alexei Gastev's experimental notions of "psychotechnics" in the 1920s–1930s.[25]

Soviet psychiatry and psychology, however, had different approaches to the problems around shell-shock and therefore had different solutions to offer, especially in the context of 1941–1942. This is illustrated by the ways in which contemporary Soviet military psychiatry, following Pavlov, regarded battle trauma as resulting from physiological causes, particularly "micro damage to the brain" and nervous system which then "caused other physiological changes to the whole organism".[26] This form of diagnosis was aligned with the "materialist" thrust of Imperial Russian and Soviet psychiatry. Although Freudian psychoanalysis had enjoyed some popularity in the 1900s and 1920s, by the early 1930s it had been banned along with other modes of studying the human or animal mind that were deemed to be too "metaphysical" or "vitalist" for Soviet science.[27]

Military psychiatry could offer "conditioning" of the body and mind, mainly by a number of physically invasive means including insulin shock, electrotherapy, prolonged sleep therapy (a potentially long-term mixture of hypnosis and sedatives) or tranquilizers.[28] Because of the numbers of wounded in Moscow by 1941–1942, it is likely that the in-house psychiatrists could only offer potentially recoverable soldiers shock treatments or tranquilizers, alongside medical and surgical treatment for their physical wounds, in order to get them back to the front within the shortest time possible. Sleep therapy could take months.[29]

Military psychology, however, was regarded as a "softer" discipline, concerned with "psychological aspects of training and education of personnel, the patterns of mental activity of men and the spiritual life of man and the collective".[30] The apparently related notions of "spiritual life" and "the collective" here, are particularly interesting in this definition, because it harks back to the mystical Russian Orthodox tradition of *sobornost'* while being a statement that supports Soviet collectivism as paramount, particularly in the military.

While the discipline of Soviet psychology paid heed to Pavlov's work, it also had roots in the work of the Russian neurologist, Vladimir Mikhailovich Bekhterev (1857–1927). Bekhterev had maintained that in order to recover, traumatised people needed to be placed in pleasant, relaxing environments that were hygienic, but not as regimented or clinical as hospitals and which offered aspects of a more normal social life. Indeed, he set up a successful private institution of this sort after the Russo-Japanese War in 1904–1905. In full knowledge of the rehabilitation centres set up after the First World War in Britain, France and Italy, Behkterev's ideas were revived after the Russian Civil War from 1917 to 1922.[31] In 1923, for example, a resort named "Red Star" was set up for traumatised Red Army combatants in Yalta, on the Black Sea. Unfortunately, this facility ran out of funding and was closed in 1926, but its ideas endured into the 1940s.[32]

The Frunze hospital had two non-clinical social spaces, a "club-room" and a large foyer in which the mobile inmates could meet, socialize and, in so doing, have a chance to become "revitalized" by recovering their sense of being part of the military community and of the broader Soviet collective. As in the case of military psychiatry, however, given the numbers of wounded pouring into Moscow in 1941–1942, the psychologists attached to the Frunze Military hospital in Moscow probably could not have dealt alone with the numbers of traumatised soldiers. The Darwin Museum, however, had staff versed in aspects of contemporary Soviet psychology, particularly the zoo-psychologist, Nadezhda Ladygina-Kots, and were able to step into this breach.

Art, Empathy and "Evolutionary Psychology": The Darwin Museum at the Frunze Military Hospital

The museum's work in military hospitals in Moscow began in November 1941 with an exhibition of paintings entitled *Animals in War in a Historical Overview*, shown in the Frunze hospital's foyer (Figure 10.1).[33] The exhibition was displayed

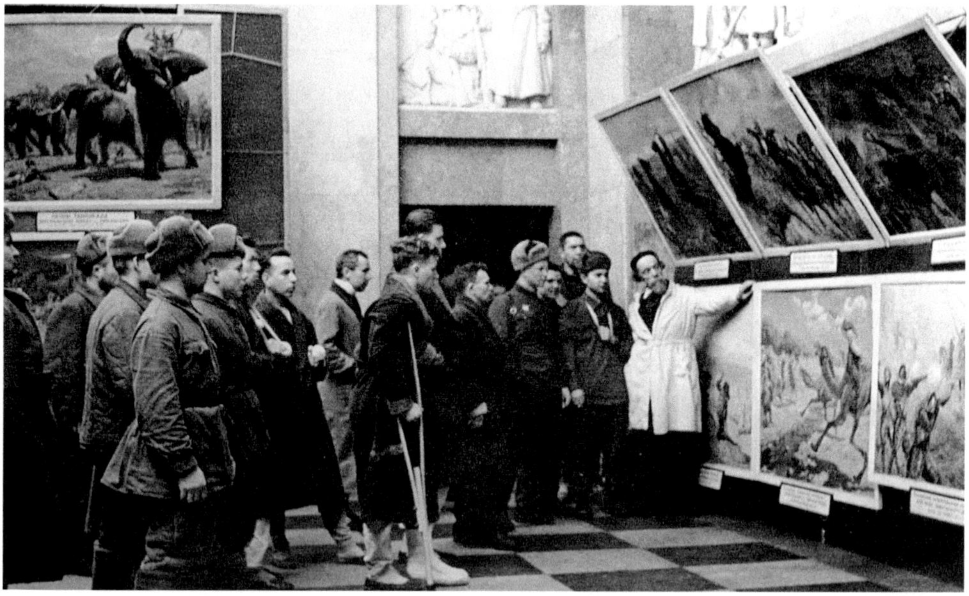

Figure 10.1 *Aleksandr Kots lecturing at the exhibition, Animals in War an Historical Overview, in the Frunze Military Academy Hospital foyer, Moscow,* c.1941. (Kalacheva, "Gosudarstvennyi darvinovskii muzei", p. 39. Photo Mikhail Alekseevich Sirotkin. ©State Darwin Museum, Moscow, 2021.)

from November 1941 to March 1942 as part of the defiant commemoration of the 24th anniversary of the October Revolution that coincided with the Nazi's attempt to besiege the city. The exhibition was accompanied by explanatory talks given by Aleksandr Kots in the foyer. Following the exhibition's opening, other less formal illustrated talks were also given by Kots and his staff, both in the hospital's "clubroom" for the mobile soldiers (Figure 10.2) and in the wards occupied by the more seriously damaged, bedridden inmates (Figure 10.7).

Before the exhibition and other activities started, however, Kots gave an advisory talk to his staff about the different priorities that should be observed when dealing with the mobile and the severely wounded patients:

> If for the mobile wounded, the success of the lectures is desirable, for the bed-bound seriously wounded, success is OBLIGATORY [Kots' capitalisation]. Hence the basic requirement for the talks is to encourage maximum engagement, captivation, absorption [and] fascination, through sincerity.[34]

Thus, it seems that one of the main "revitalising" functions of the illustrated lectures was to take the soldiers out of themselves, to enable them to begin to empathise not only with the subject matter or themes of the imagery presented but, ultimately, also with the broader context of the beleaguered "Motherland" for which they had recently fought.

Significant clues to the vitalist elements in this line of thinking exist in a later short film by Rudi Kots entitled *Farewell Greeting, From Aleksandr Kots* (c.1957).

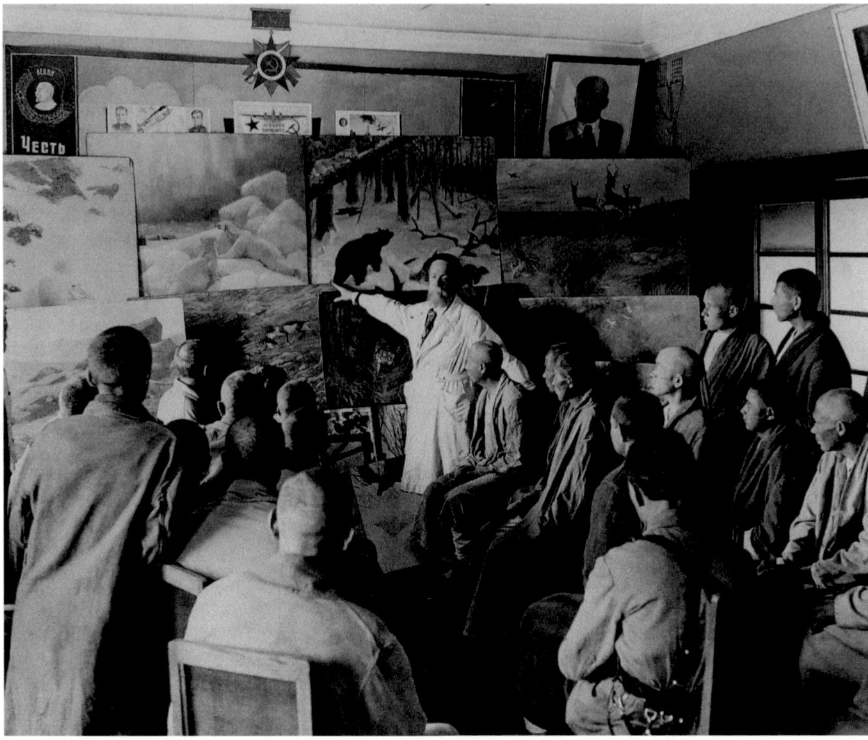

Figure 10.2 Aleksandr Kots lecturing in the Frunze Military Hospital "clubroom", 1941–1942. (Kalacheva, "Gosudarstvennyi darvinovskii muzei", p. 40. Photo Mikhail Alekseevich Sirotkin. ©State Darwin Museum, Moscow.)

Kots' inspiring speech to a group of schoolchildren visiting the Darwin Museum offered them a "spark" of his vital driving force and source of his "enthusiasm" for nature and evolutionary theory, in a "covenant". He stated his belief that "without a complete and profound knowledge of nature, it is not possible to fashion a modern and free scientific worldview". However, this "scientific worldview" was not to be understood as solely applicable to academia, museums or laboratory science but as also pertinent to "all matters of daily life". In this regard, he asserted that the "spark" may lie dormant in beholders for a long time until "fatigue or reflection overcomes you, and you may ask yourself what could the theory of evolution offer to [your] life?" Then he stirringly asserted: "And then remember these walls and these words. And let my little spark begin to smoulder in you and take hold of you, and burn and give warmth, and give you strength to continue to serve our great motherland".[35] His statement relates to the covertly vitalistic idea of "transference" that became embedded in Soviet educational theory in the 1930s.

This is an idea that also appears – albeit in a very different context – to lurk in Brandon Taylor's account of Hans Arp's potentially vitalistic notion of art practice in Chapter Seven and its impact on the viewer.[36] In terms of Soviet "materialistic" science, the idea of "transference" was, however, essentially a belief that education was a "dynamic force" which, once internalized by the person being educated, is capable of creating physio-chemical changes in the brain.[37] This being so, it might even lead to new "centres" in the cerebral cortex being established for certain cognitive and physical behaviours to replace those "centres" that had been damaged by, for example, shell-shock in relation to Pavlov's ideas of the brain's potential capacity to create such new "centres". This is potentially what the exhibition at the Frunze Hospital set out to do.

All the works shown in the Frunze exhibition were newly commissioned by Kots from the sole remaining museum artist at that time: the zoologist and "animalist" sculptor and painter, Konstantin Flerov. At the outbreak of war, Kots gave Flerov a number of war-related themes to work on, such as "animals as a means and weapon of war" and "animal camouflage and its meaning for wartime camouflage".[38] Despite the difficult times, Flerov produced over a hundred works related to Kots' themes over the wartime period.[39]

The formal exhibition in the Frunze foyer focused on the theme "animals as a means and weapon of war" (Figure 10.1). However, only two works relating to the theme can be clearly identified in the photograph of the exhibition (Figures 10.5 and 10.6). By contrast, the photograph in Figure 10.2 exemplifies the more informal presentations given in the hospital "clubroom", which still used brightly coloured paintings but unlabelled, unframed and propped against each other, almost as they would be in an artist's studio, to be pulled out of a stack to make a point. This image refers to talks given by Kots on the theme of "animal camouflage and its meaning for wartime camouflage", as may be seen from his dramatic gesture towards Flerov's painted image of polar bears collaboratively hunting on an ice floe. This presentation addressed a very topical issue at the time. Because it was winter, Soviet troops, particularly snipers, were using white camouflage to blend in with the snowy environment, just like the polar bears, in a collaborative effort to pick off their prey.

In the dangerous context of 1941–1942, it might seem unusual for a natural history museum to stage an art exhibition or indeed to commission art works at all. Yet, as the Darwin museum was – and still is – a unique museum, it may be perceived as a natural response to the situation. From its inception in 1907, its adoption by the Moscow University Higher Women's Courses in 1914 and then by the Soviet Moscow State University in 1917, the Darwin Museum had laid great stress on the use of art works. Paintings, sculptures and taxidermy (itself a form of sculpture) provided exciting visual means to capture their audiences' attention and engagement. Basically, these artworks were meant to function as illustrations to the pre-booked group lecture tours, which were the only means of public access to the collection.[40] However, the images were also meant to act as stimuli to the audiences' imagination and understanding. This was arguably an important factor in the hospital presentations, as it would have potentially helped the patients to re-orientate their health concerns back to the collective and away from themselves.

Regarding the dramatic narrative paintings that can be identified in the photograph of the exhibition in the Frunze foyer (Figure 10.1), these were full of dynamically executed but legibly depicted, imagined military scenes from the histories of ancient western and eastern civilisations that the soldiers may have learned about in high school.[41] The exotic subject matter – far removed from the alienating, industrialized warfare that the Soviet soldiers had experienced – was delivered in attractive bright colours to capture their audience's attention, as demonstrated by Figures 10.3 to 10.6. Two paintings related to the exhibition theme are *Fighting Dog of the Ancient Greeks* and *Fighting Dog of the Assyrians* (Figures 10.3 and 10.4).

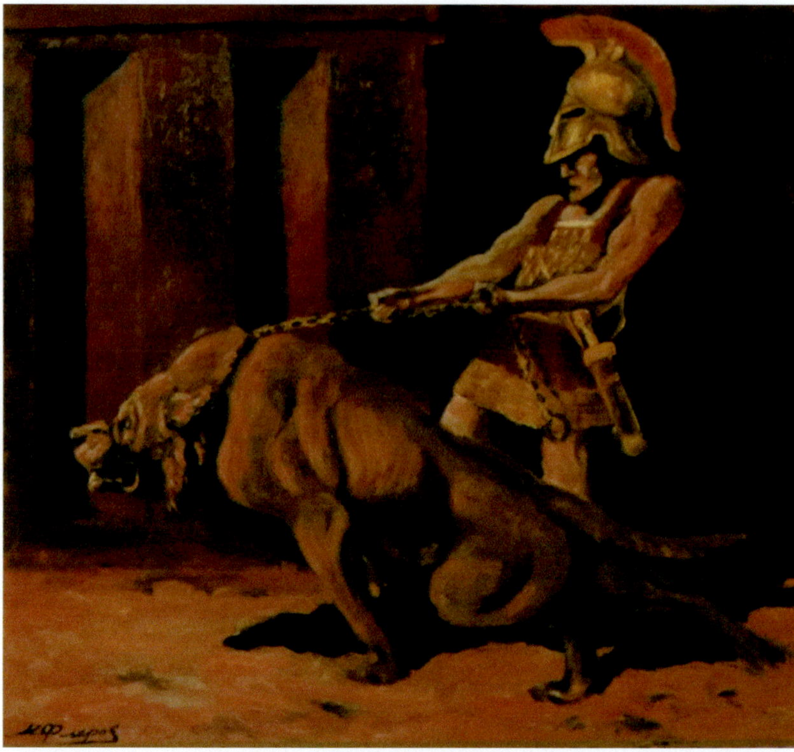

Figure 10.3 Konstantin Flerov, *Fighting Dog of the Ancient Greeks*, 1941–1942, oil on canvas, 131 × 123 cm. (Udal'tsova, p. 51, plate 129. © State Darwin Museum, Moscow, 2021.)

These creatures are represented as weapons of war, hugely muscled and obediently poised to spring at the enemy on their handlers' command. The dogs look like large types of mastiff (Molossos, the type of dog used by the Assyrians for lion hunting, as exemplified by friezes on the North Palace wall in Nineveh, its image also appearing as a protective object under or around doorways in Nineveh).[42] The same type of dog was also used by the ancient Greeks for war, as supported by the sculpture held at the British Museum named "Jenning's Dog".[43] It is not clear what visual sources Flerov used, but as a zoologist and a keen hunter, he may well have modelled the images on the large Russian mastiffs (Caucasian Shepherd Dogs, also known as Russian Bear Dogs) that were used for bear hunting and security watch-dogs in the USSR, so may have been familiar to some of the Russian soldiers.[44]

The image of the Greek dog-handler (Figure 10.3) appears to offer a military role model of stoic, patriotic and heroically unswerving determination, as shown in the treatment of the facial expression and depicted stance. Both paintings also offer models of superhuman physiological strength. These were all attributes of the Soviet ideal of the "New Man",[45] which was propagated in the USSR from the late 1920s, as a heroic aspirational model for Soviet citizen's approach to work. In 1938, for example, the President of the Russian Soviet Republic, Mikhail Kalinin, had declared that: "In our

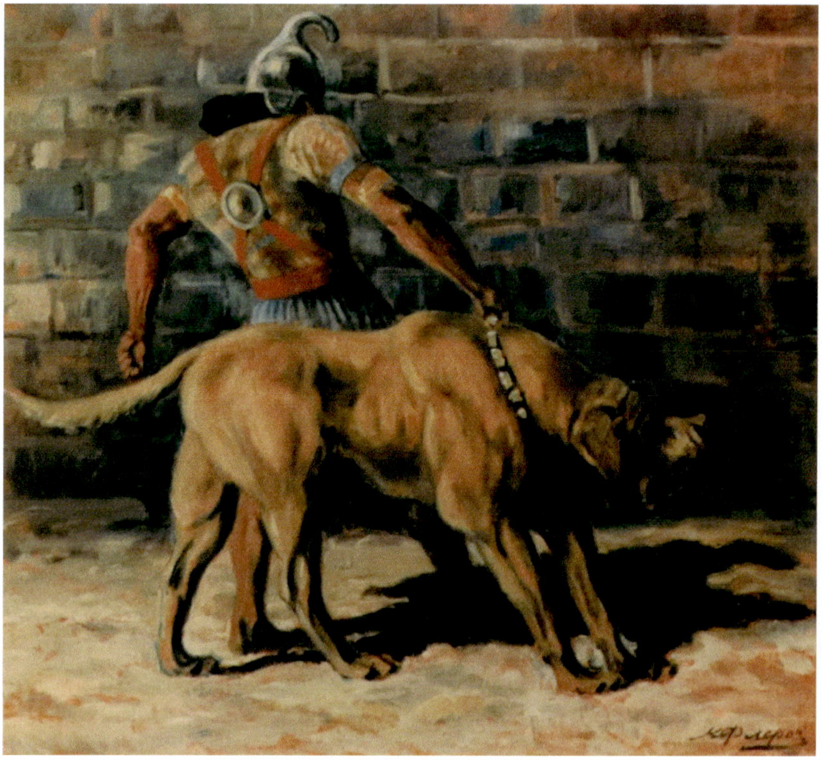

Figure 10.4 Konstantin Flerov, *Fighting Dog of the Assyrians*, 1941–1942, oil on canvas, 131 × 123 cm. (Udal'tsova, p. 52, plate 130. ©State Darwin Museum, Moscow, 2021.)

country, work is a matter of honour, glory, valour and heroism".[46] Hence, the possession of a "free personality",[47] in his terms, was regarded as being guaranteed by fulfilling patriotic and partisan duties to the Soviet motherland, the Communist Party and the broader collective – these being all connected by the vitalistic element of *sobornost'*.

The work of a soldier was, clearly, to defend the Soviet "motherland". That of the wounded soldier was to reconnect with his/her duties, either by returning to the front or, if that were impossible, to embark on vocational retraining for a different kind of useful work in the sense defined by Kalinin. This was regarded as part of the job of the Soviet psychologist to provide or organize. In a sense, these paintings then implicitly presented a clarion call to the wounded soldiers to "revitalize" themselves in order to carry on the patriotic fight. This was a process that would require the assertion of individual will, something that particularly related to Soviet "Creative Darwinism" and its covert vitalistic elements. This will be discussed further in the next section.

A slightly different stimulus to "revitalisation" was offered by two other paintings by Flerov displayed in the Frunze exhibition that may also be partially seen in the photograph, Figure 10.1. These are *Camels Outside Sardis* (Figure 10.5 – fragment in 10.1 bottom right) – and *Hannibal's Battle Outside Trebbia* (Figure 10.6 – fragment

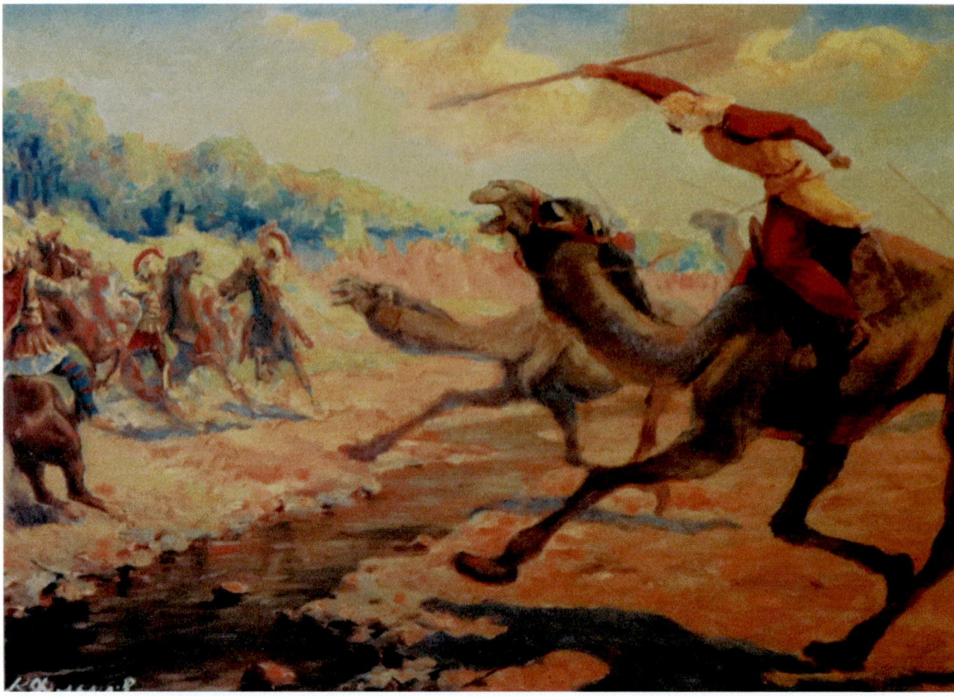

Figure 10.5 Konstantin Flerov, *Camels outside Sardis,* 1941, oil on canvas, 153 × 111 cm. Udal'tsova, p. 52, plate 131. (©State Darwin Museum, Moscow, 2021.)

in Figure 10.1 top right). Unlike Flerov's images of the *Fighting Dogs* (Figures 10.3 and 10.4), they depict scenes of ancient historical battles. The mode of depiction is romanticized so that despite the paintings' subject matter, it does not show the blood and agony that the Soviet soldiers would have experienced but rather illustrates educational, historical examples of successful, heroic and determined collective opposition to a common foe.

The image in Figure 10.5 represents a scene from the battle of Thymbra in 546 BC that resulted in the victory of the Persians led by King Cyrus the Great over the Lydians and Egyptians who were led by the Lydian King Croesus outside the town of Sardis.[48] The dramatic scene shows camel riders of the Persian army attacking and routing a group of Lydian cavalry, whose horses are depicted as terrified by the onslaught of the camel riders. Another equally victorious scene is represented in Figure 10.6. This, however, refers to a different and historically later war. Set in 218 BC, it depicts a decisive moment in the defeat of a Roman army, commanded by Tiberius Sempronius Longus, by the Carthaginian army commanded by Hannibal.[49] According to Vera Uldal'tsova, the Darwin Museum's archivist of art works, this captures the moment following the terrorisation of the Romans' horses by the Carthaginian elephants, causing the horses to dispose of their riders under the elephants' feet, with fatal consequences. This victory led to the Carthaginians' (temporary) defeat of the mighty forces of Imperial Rome.[50] In both paintings, the focal animals signify both the means to facilitate warfare as transport for the warriors and as obedient auxiliary battle weapons for intimidating, trampling on and barging into the enemy.

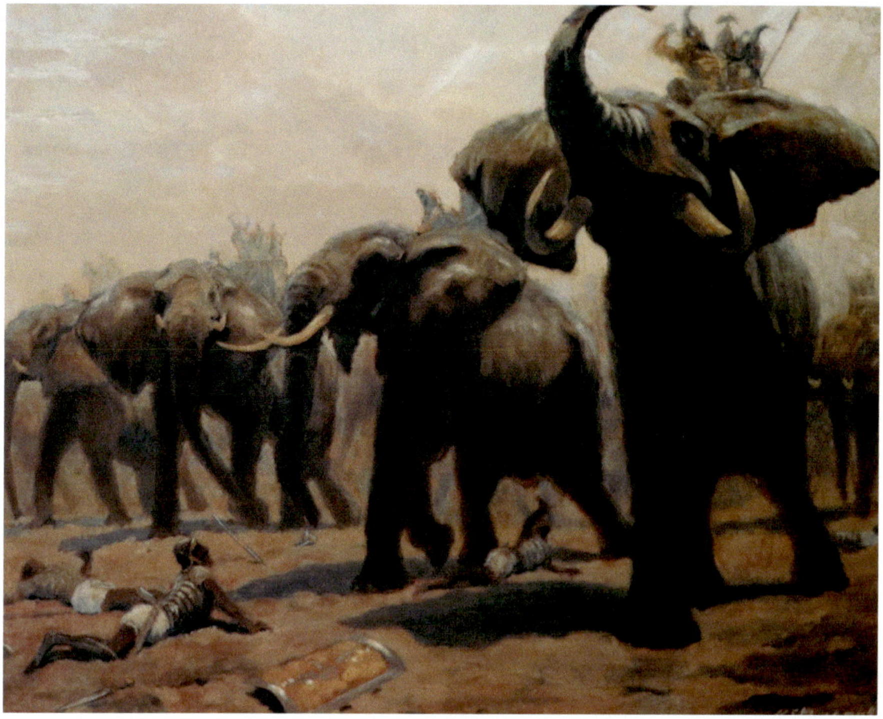

Figure 10.6 Konstantin Flerov, *Hannibal's Battle Outside Trebbia*, 1941, oil on canvas, 150 × 123 cm. (Udal'tsova, p. 52, plate 132. ©State Darwin Museum, Moscow, 2021.)

These paintings would have been used to illustrate the stories that Kots told the Soviet soldiers about the historical battles represented in the paintings and the roles played by animals in them. Kots' stories, however, would also have included intimations of who were to be seen either as heroes or villains of the encounters and thus indicated with whom the soldiers should feel empathy or *sobornost'*. This is important because according to McGleish (1975), the Soviet concept of *sobornost'* was not "a sentimental love for all people without exception. Soviet man cannot love the enemies of the working class nor anyone who stands across the path to the happy future. It is a virtue to hate the oppressor or the aggressor".[51] Given that the underlying, "revitalising" purpose of the exhibition was to help motivate the audience to re-engage in battle with the German invaders, it is extremely likely that they were invited by Kots to empathise with the victorious historical exploits of the Persians (Figure 10.5) and the Carthaginians (Figure 10.6), rather than with the actions or motivations of their slaughtered foes.

Such robust materials with their overtly military themes or references may have been considered by the Darwin Museum team as suitable for the walking-wounded who were soon to be sent back to the front. However, there seems to have been a lower key and less militarized approach adopted in both the "comradely" behaviour and the subject matter of illustrated talks provided to the more seriously wounded, bed-ridden soldiers by Nadezhda Ladygina-Kots, as shown in Figure 10.7.

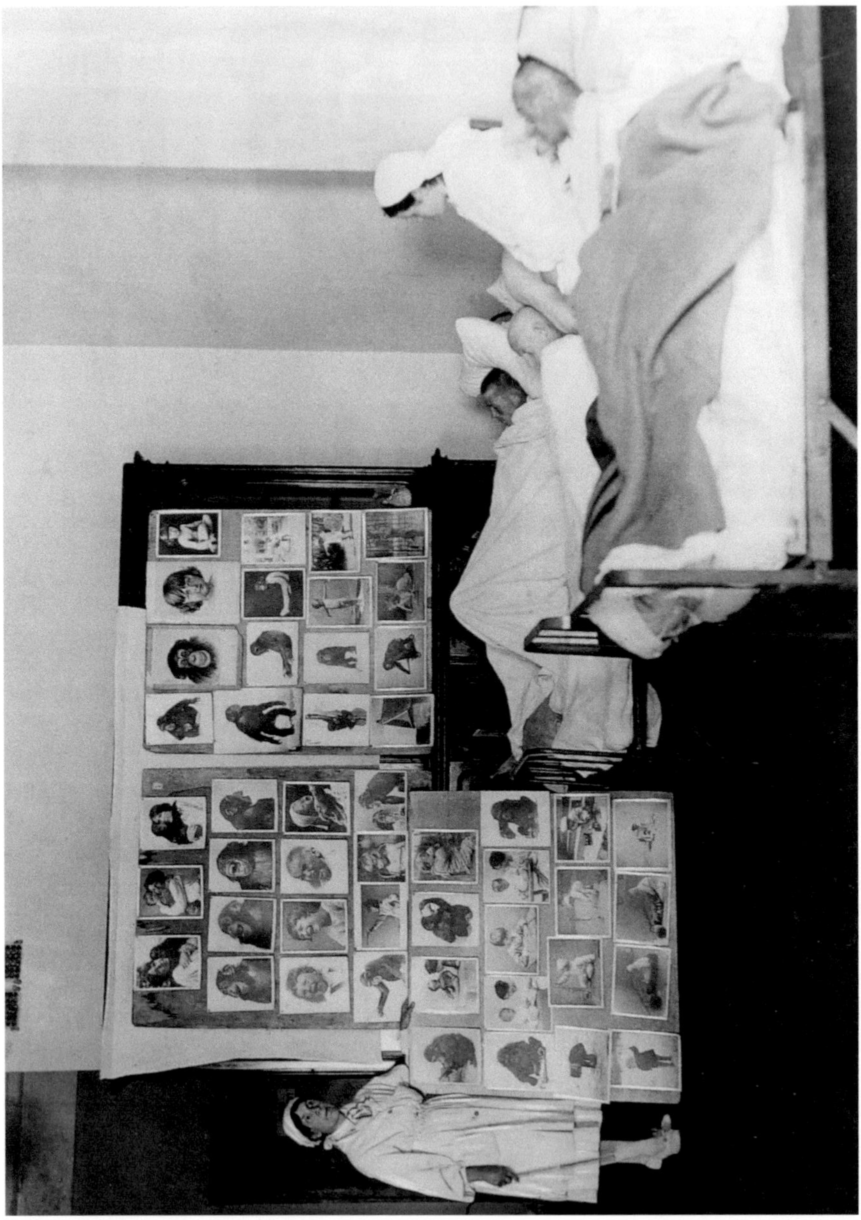

Figure 10.7 *Nadezhda Ladygina-Kots lecturing to seriously wounded soldiers on "The Mental/Emotional Worlds of Animals and Humans", at the Frunze Military Academy Hospital Moscow*, 1941–1942. (Kalacheva, "Gosudarstvennyi darvinovskii muzei", p. 41. Photograph Mikhail Alekseevich Sirotkin. ©State Darwin Museum, Moscow, 2021.)

It has been suggested by Olga Krylova (2001) that psychological "healing" for severe emotional trauma was largely perceived in the USSR as the responsibility of the home or, failing that, institutional environments and provided by women as "social therapists of traumatised male souls".[52] It would seem that Nadezhda Ladygina-Kots partly took on this feminized role at the Frunze hospital. As well as giving illustrated talks relating to her current and past professional research,[53] she also read versified messages of congratulation to decorated, wounded soldiers. She would also answer any questions, privately or publicly, whether or not they were related to her lectures, while helping soldiers to write to their friends and families.[54] This basic, psychologically boosting enterprise was also supported by the official museum photographer, Mikhail Alekseevich Sirotkin, who was available to take and process photographs of the patients to send home, if they so wished.[55] According to Irina Kalecheva, the Darwin Museum's current archivist, Ladygina-Kots' hospital lecture on "The Mental/Emotional Worlds of Animals and Humans" represented in Figure 10.7 was "a form of psychological therapy by a professional psychologist".[56] The lecture had been designed particularly, but not exclusively, for severely wounded women, "aimed at relieving their sufferings".[57] The image in Figure 10.7, however, shows her giving the talk to injured men, which raises the (currently unanswerable) question as to whether the women did not want to be photographed by Sirotkin in their current state.

The lecture used images from Ladygina-Kots' comparative psychological research study (1913–1935) of a young chimpanzee, "Joni", and her infant son, "Rudi", published in 1935 as *Infant Chimpanzee and Human Child*.[58] Superficially, the lecture reinforced the current Soviet role model for women, particularly as it was delivered by a well-dressed woman holding a highly professional job, who was also a mother and therefore contributing productively both to the Soviet economy and to expansion of the collective workforce.[59] It also potentially provided a fascinating and heart-warming educational experience about the comparative mental and emotional, developmental capacities of young chimpanzees and human children. This might have enabled her audiences to focus interest away from their current sufferings in an empathetic sense, in parallel to Kots' lectures and presentations, by delivering the vitalistic "spark" necessary to enable their re-entry to the Soviet collective, but in a different way.

The conclusions of her study, inevitably iterated in her talk, would also have given a particular, propagandistic message. At the end of *Infant Chimpanzee and Human Child* (1935), Ladygina-Kots had emphatically stated that her study showed that the modern chimpanzee was effectively a degenerate species from the anthropoid ancestors of humans. Thus, she argued, contemporary human capacities for emotional and mental development were infinitely superior to those of modern chimpanzees. The reason for this was the chimpanzee's lack of orientation towards working at learning by means of the power of individual will, in order to assimilate with the will of the collective.[60] This element of the lecture would have reinforced the vitalistic elements of the tenets of contemporary Soviet psychology as well as the overarching metaphysical and vitalistic Soviet concept of "*sobornost*", which were all crucial to the process of "revitalising" the soldiers. Additionally, it can be argued that non-metaphysical Soviet vitalism was associated not only with the discipline of zoo-psychology, but also with the revised construct of Darwinism propagated by Trofim Lysenko and his supporters from the 1930s onwards.[61]

Vitalism, Zoo-Psychology and Soviet Darwinism in the 1930s–1940s

The research behind Ladygina-Kots' "revitalising" lecture to the bed-ridden soldiers had been motivated initially by the work of the Russian/Soviet zoo-psychologist and field researcher, Vladimir Vagner (1849–1934). Indeed, in the early 1920s, Vagner became Ladygina-Kots' mentor and published her comparative behavioural psychological study of wolves and dogs in his journal *Studies in Comparative Psychology*.[62] There are many books and papers by him in her personal library held at the State Darwin Museum archives but, unsurprisingly, there are no specific references to him or to the term "comparative psychology" in her book. By the time the book went to press in 1935, not only was Vagner dead but his scientific endeavours had already been discredited, his journal shut down and his other publications banned.

What is important about Vagner's influence on Ladygina-Kots is that in the late nineteenth-century, he became the co-founder, alongside the eminent nineteenth-century plant-physiologist, Aleksandr Famintsyn, of a new bio-scientific discipline initially called "behavioural evolutionism".[63] This was an organicist form of comparative psychology that recognized the unity of body and psyche. In this sense, it had some parallels with Pavlov's theories. Unlike Pavlov, however, Vagner and Famintsyn sought for scientific means to demonstrate that despite this unity, "mental life is not reducible to the laws of mechanics".[64] Additionally, this irreducibility to physiology and chemistry alone was posited as key to the process of evolution even in the lowest of life-forms. They also suggested that "mind" -which they also called "instinct" - was capable of being studied from a separate perspective from bio-physio-chemical studies of brain responses.[65]

Basically, they were suggesting that the mind/psyche was an intangible thing that included will and could be informed by instinct. Moreover, they believed that "mind" could sometimes command the actions/reactions of the body,[66] rather than the body always commanding the activities of the mind as in Pavlov's theories. To their peril, their theory seemed compatible with Bergsonism. Even through Bergson's philosophy had proven popular from 1900 to the 1920s with numerous Russian philosophers and Modernists, as indicated earlier, his philosophy of intuition became systematically delegitimized after the October Revolution until 1930, the year that Hilary L. Fink (1999) has called "the death of intuition" in the USSR.[67]

Prior to 1930, however, the ideas of Vagner and Famintsyn gained some scientific popularity under the new title of "comparative psychology". It attracted both a small number of Russian/early Soviet scientists orientated towards neo-vitalism,[68] as well as an equally small number of Russian/early Soviet psycho-Lamarckist scientists, who believed that "behaviour makes animals and plants participants in their own evolution".[69] As a result of this level of support, after the 1917 Revolution some experimental laboratories were established, including one at Moscow zoo overseen by Ladygina-Kots. In addition, courses were set up at various universities to study comparative psychology with publications briefly abounding in this new field.[70] However, in the late 1920s early 1930s, the Soviet government began to exert more strict controls over Soviet science and, indeed, over all aspects of culture in the USSR. These controls resulted in comparative psychology increasingly becoming the butt of accusations of "vitalism" and "metaphysics" because of its implication that "mind" was able to operate influentially, whether autonomously or semi-independently, from the physio-chemical processes and activities of the body. The criticisms eventually prompted Vagner to publish a scornfully defensive riposte in 1927:

> The tendency to call everyone a vitalist if they refuse to follow blindly the doctrines of materialistic schools, à la Molleshot and Buchner, and he who speaks of psychology as if it is more than the data of chemistry, anatomy, and physics, is so zealously propagandized that the word "vitalist" sounds to a real scientist the way the word "heretic" sounded to a real religious person in the Middle Ages.[71]

This defensive statement was to no avail. By the end of the 1930s, the discipline of comparative psychology had been abolished and obliterated along with its leading theorists, associated publications, university courses and laboratories, including that of Ladygina-Kots at Moscow Zoo. This action followed on from a broad-spectrum Communist Party Central Committee resolution "On Pedagogical Perversions in the Narkompros System",[71] which attacked several scientific and educational disciplines for being insufficiently "materialist" to continue to exist in the USSR.[72]

Ladgygina-Kots' research presentations to severely wounded soldiers arguably did not refer explicitly to the ideologically suspect "vitalistic" work of Vagner and Famintsyn – as indeed her book had not. The reason for this is, I suggest, that the Stalinist purges of Soviet bio-scientists only ended late in 1940 and by the start of the war in 1941, there was still no news of what had happened to those scientific colleagues who had been accused of ideological misdemeanours and subsequently "disappeared" into prisons and gulags.[73] Thus she and her Darwin Museum colleagues would have had to continue to tread very carefully, in an ideological sense, to retain their status and positions. The new language of Soviet "Creative Darwinism" propagated by Trofim Lysenko and his supporters from 1934 to the 1950s, however, arguably offered a convenient and politically correct means for Ladygina-Kots and her colleagues to deliver their covertly "vitalist" messages to the war-wounded.

Aleksandr Kots' lectures certainly deliberately referred to Lysenko's idea of Soviet "Creative Darwinism". The evidence for this lies in a letter of profound thanks written to the Darwin Museum in November 1943 by a member of a group excursion to the Museum from a military sanatorium 40 kilometres outside Moscow. The letter's punchline, obviously gathered from the lecture tour, was: "We will remember the teaching of Michurin: We must not ask favours of nature but wrest them from her – this is our task".[74] Lysenko's published account of Michurin's enduring motto was worded slightly differently: "We cannot wait for favours from Nature, we must wrest them from her".[75] Nevertheless, it is clear from the similarity between the soldier's recollection from Kots' lecture and Lysenko's published statement that Kots had referred to it in his lecture tour. Arguably, another significant clue to this Lysenkoist connection is to be found in the title of one of Kots' hospital lecture topics: "The power of humankind over nature". This was an important contemporary idea, rooted in the Soviet acceptance of Friedrich Engels' interpretations of Darwin, that were also forcefully illustrated in Flerov's paintings (Figures 10.3 to 10.6).[76] It was intimately connected with the focus on individual and collective "will" in contemporary Soviet psychology that potentially supplied a politically sound model of language to the Darwin Museum team in which to frame a collection of related vitalistic ideas from various sources.

Conclusion

This chapter has argued that, while western modernism and overtly vitalistic theory were anathematized in the USSR, there remained some covert threads of vitalism running through Soviet culture that can be detected in the exhibition and presentations delivered to traumatised soldiers at the Frunze Military hospital by the Darwin Museum staff, 1941 to 1942. The aim of these events was to "revitalise" the soldiers, to enable them to return to the front or to enter re=training for some other occupation useful to the state.

Amongst the vitalistic threads alluded to, the Russian Orthodox notion of *sobornost'* (spiritual collectivism) was the most metaphysical in relation to its long-standing religious roots in Russia. It was tolerated, however, for the extra boost it could give to the Communist Party's contemporary emphasis on its own atheistic construct of collective solidarity at a time when the Soviet capital, Moscow, lay under threat of siege by Nazi troops. The other threads were ostensibly less metaphysical. These were largely entangled with scientific disciplines viewed in Soviet terms as the apex of modernity: Military psychology, educational theory and Lysenko's construct of "Creative Darwinism". These disciplines were also permeated by the enduring Russian/Soviet aspirational ideal of the "New Man", a new type of humankind that would be generated by the collectivized environment of the USSR.

There is, however, a delicious irony regarding "Creative Darwinism" in that "will", whether individual or collective, was a concept that was just as intangible and non-physiological as the Russian Orthodox concept of *sobornost'*, the educational notion of "transference" and Vagner and Famyntsin's construct of "mind". In parallel with the latter's idea of "mind", it could even be positioned as having the potential to require the physiological aspects of the brain and body to act for the common good, despite whatever damage the body had sustained. Thus, paradoxically, it seems that both Ladygina-Kots and Aleksandr Kots could easily have been still working to the vitalistic script partly supplied by Vagner, while using the currently more acceptable descriptive language offered by Lysenko.

The impacts of the Darwin Museum's "revitalising" endeavours regarding specific patients at the Frunze 1941–1942 are unknown. It is significant, however, that while all the team members were awarded various medals for their wartime work, only Aleksander Kots was awarded the state medal: "Excellent Worker of Public Health".[77] The accolade implies that with the support of the Frunze Hospital, the Soviet government recognized the Darwin Museum's "evolutionary psychotherapy" activities as having been both medically and ideologically significant to the recovery and "revitalisation" of traumatised soldiers.[78]

Notes

1 Irina Kalacheva, "60 let velikoi pobedy: darvinovskii muzei v gody velikoe otechestvennoe voini", *Trudy gosudarstvennogo darvinovskogo muzeia*, issue VIII (2007): 112.

2 Svetlana Alexievich, Keith Hammond and Ludmilla Lezhneva, "'I Am Loth to Recall': Russian Women Soldiers in World War II", *Women's Studies Quarterly*, vol. 23, nos 3/4, "Rethinking Women's Peace Studies" (Fall/Winter, 1995): 78–84, https://www.jstor.org/stable/40003502.

3 Andrei Zhdanov, "Soviet Literature – The Richest in Ideas, The Most Advanced Literature", in H.G. Scott, ed., *Soviet Writers' Congress 1934: The Debate on Socialist Realism and Modernism in the Soviet Union. Gorky, Radek, Bukharin, Zhdanov and Others* (London: Lawrence and Wishart, 1977), 15–24.

4 Alexander Vuchinich, *Darwinism in Russian Thought* (Berkeley, Los Angeles, Oxford: University of California Press, 1989), 178–185: Isobel Wünsche, *The Organic School of the Russian Avant Garde: Nature's Creative Principles* (Farnham Surrey, and Burlington, VT: Ashgate, 2016). See also Pat Simpson, "Imag[in]ing Post-Revolutionary Evolution: The Taylorised Proletarian, 'Conditioning' and Soviet Darwinism in the 1920s", in Barbara Larson and Fae Brauer, *The Art of Evolution: Darwin, Darwinisms and Visual Culture* (Lebanon NH: Dartmouth College Press/University of New England Press, 2009), 226–261.

5 See James West, *Russian Symbolism: A Study of Vyacheslav Ivanov and the Russian Symbolist Aesthetic* (London: Methuen, 1970), 11, 12, 17, 19, 50–51, 92, 100, 118, 121–122, 126–128, 132ff., 146, 151, 157, 159–160, 165, 176; Wassily Kandinsky, *Concerning the Spiritual in Art* (1912), trans. Michael Sadlier 1914, with re-translation by Francis Golfing, Michael Harrison and Ferdinand Ostertag (New York: George Wittenborn Inc., 1970), 59.

6 Ludmilla Rudova, "Bergonism in Russia: The Case of Bakhtin", *Neophilogus*, vol. 80, no. 2 (April 1996): 175–188.

7 See for example: Ryan Drake, "Aristotelian Aisthesis and the Violence of Suprematism", *Epoché: A Journal for the History of Philosophy*, 18, no. 1 (2013): 50; Chris Short, *The Art Theory of Wassily Kandinsky 1909-1928: The Quest for Synthesis* (New York: Peter Lang, 2010), 88.

8 Nikolai Kol'tsov, "Review of Professor Hans Driesch, *Vitalism: Its History and System*, authorised translation by Professor A.G. Gurevich (Moscow: Nauka Publishing House, 1915)", in *Priroda*, vol. 12 (1914): 1528–1530; Igor Popov, *Orthogenesis Versus Darwinism*, tr. Natalia Lentsman (Cham, Switzerland: Springer Nature Switzerland AG, 2018), 189.

9 Vucinich, *Darwinism in Russian Thought*, 178–185.

10 John McGleish, *Soviet Psychology: History, Theory, Content* (London: Methuen, 1975), 179–180.

11 See: P. Simpson, "Lysenko's Michurinism and Art at the Moscow Darwin Museum 1935–1964", in W. DeJong-Lambert and N. Krementsov, *The Lysenko Controversy as a Global Phenomenon*, vol 1 (Cham, Switzerland: Palgrave Macmillan/Springer Nature, 2017), 130, 160.

12 Geoffrey Hosking, *A History of the Soviet Union 1917-1991: Final Edition* (London: Fontana, 1990), 162.

13 Paul Wanke, *Russian/Soviet Military Psychiatry 1905-1945* (Abingdon, Oxon., and New York: Routledge, 2005), 1–3.

14 Ibid., 1.

15 Ibid.

16 Catherine Merridale, "The Collective Mind: Trauma and Shell-Shock in Twentieth Century Russia", *Journal of Contemporary History*, vol. 35, no. 1 (2000): 40.

17 Gabriella Nietlisbach, Andres Maerker, Wulf Rossler and Helena Haker, "Are Empathic Abilities Impaired in Post-traumatic Stress Disorder?", *Psychological Reports*, vol. 106, no. 3 (2010): 832–844.

18 Merridale, "The Collective Mind", 40.

19 Merridale, "The Collective Mind", 45; see also McGleish, *Soviet Psychology*, 18–19, 129.

20 McGleish, *Soviet Psychology*: 18.

21 Hosking, *A History of the Soviet Union*, 236.

22 McLeish, *Soviet Psychology*, 178–179.

23 Ibid. 117.

24 Ibid. 174–177; Sarah D. Phillips, "'There Are No Invalids in the USSR!' A Missing Soviet Chapter in the New Disability History", *Disability Quarterly*, vol. 29 (3) 2009 (unpaginated) https://dsq-sds.org/article/view/936/0.

25 Simpson, "Imag[in]ing Post-Revolutionary Evolution", 226–261; Wünsche, *The Organic School*.

26 Wanke, *Russian/Soviet Military Psychology*, 1–3.

27 Elena Kazennaya and Ekaterina Divid, "Chronic Trauma and Dissociation in Russia: A Brief Overview of the History of Psychological Services and Psychotherapy in Russia", European Association for Trauma and Dissociation (ESDT) (unpaginated and no date), https://www.estd.org/countries/russia.

28 McGleish, *Soviet Psychology*, 176.

29 Ibid., 177.
30 Soviet Ministry of Defence, *Soviet Military Psychology*, 1972, cited in Wanke, *Russian/ Soviet Military Psychology*, 2.
31 Michael Jabara Carley, "Review: Allied Intervention and the Russian Civil War, 1917-1922", *The International History Review*, vol. 11, no. 4, November 1989): 689–700.
32 Merridale, *The Collective Mind*, 42–43.
33 Kalacheva, "60 let velikoi pobedy", 91.
34 Irina Kalacheva, "Gosudarstvennyi darvinovskii muzei v gody otechestvennoi voiny 1941-1945", in Andei Trofimov, ed., *Darvinovskii muzei: 100 let so dnia osnovanie, 1907-2007* (Moscow: Izdatel'skaia programma "Interrossa", 2007), 40.
35 Rudolf Kohts, *Proshchal'noe privetstviia Aleksandra Feodorovicha Kots*, short film, c.1957, with English subtitles translated from the Russian by Sunny M. Boscoc: https://www.youtube.com/watch?v=jyBuSOn7h_w .
36 Brandon Taylor, "Was Dadaism Vitalistic?", Chapter Seven, of the present volume.
37 McGleish, *Soviet Psychology*, 171.
38 Kalacheva, "Gosudarstvennyi darvinovskii muzei", 40.
39 Kalacheva, "60 let velikoi pobedy": 112.
40 Aleksandr Kots, *Muzei evoliutsionnoi istorii moskovskikh vysshikh kursov za 1913-1914*, Moscow (printed pamphlet, no publisher named, 1914). Arkhiv gosudarstvennyi darvinovskii muzei (AGDM) f.12430, o.32, ed.khr.165, 16pp.
41 Hugh Graham, "The Significant Role of the Study of Ancient History in the Soviet Union", *The Classical World*, vol. 61, no. 3 (1967): 85–97.
42 "Canines", *The Oxford Handbook of Animals in Classical Thought and Life,* ed. Gordon Lindsey-Campbell (Oxford and New York: Oxford University Press), 286–287. See also "Nimrud dogs", Cambridge University Museum of Archaeology and Anthropology: http://oracc.museum.upenn.edu/nimrud/livesofobjects/catordog/index.html.
43 "The Jennings Dog/the Duncombe Dog/ The Dog of Alcibaiades", The British Museum, https://www.britishmuseum.org/research/collection_online/collection_object_details.aspx?objectId=467443&partId=1.
44 Samsonova, A. "Kavkazskaia Ovtcharka" (Caucasian Shepherd Dog), revised by Renee Sporre-Wiles, 2010, Federation Cynologique Internationale (AISRL), 02/03/20111/En, FC Standard no.328, https://www.fci.be/Nomenclature/Standards/328g02-en.pdf, 1-9. For some extra history and images, see also: American Kennel Club, "Caucasian Shepherd Dog", 2022, https://www.akc.org/dog-breeds/caucasian-shepherd-dog/.ab.1.0.0j0i22i30l9.5201.13684.
45 See Simpson: "Imag[in]ing Post-Revolutionary Evolution", 226–261.
46 Mikhail Kalinin, "On Communist Education" (1938), cited in McGleish, *Soviet Pyschology*, 166.
47 Ibid.
48 Paul Davis,"Thymbra 546 BC.", *100 Decisive Battles from Ancient Times to the Present* (London and New York: Oxford University Press, 2001), 6–9.
49 Paul Hunt, "Battle of the Trebbia River: Roman-Carthaginian History", *Encyclopedia Britannica* (2014), https://www.britannica.com/event/Battle-of-the-Trebbia-River.
50 Vera Udal'tsova "Konstantin Konstantinovich Flerov", *Gosudarstvennyi darvinovskii muzei. Sokhrovishcha russkogo iskusstva* (Moscow: Belyi gorod, 2007), 52.
51 Mcleish, *Soviet Psychology*, 165.
52 Olga Krylova'Healers of Wounded Souls: The Crisis of Private Life in Soviet Literature, 1944-1946", *The Journal of Modern History*, vol. 73 (2001): 307–331, http://dsq-sds.org/article/view/936/1111#endnoteref20.
53 Kalacheva, "60 let velikoi pobedy": 113, 118.
54 Kalacheva, "Gosudarstvenyi darvinovskii muzei", 41.
55 Ibid.; Kalacheva, "60 let velikoi pobedy": 118.
56 Ibid.
57 Ibid.
58 Ladygina-Kohts, *Infant Chimpanzee and Human Child*, 956. See also Rudolf Kohts, *Opyt c shimpanze, raboty N. Ladygina-Kots*, short film from the collection of the State Darwin Museum, Moscow, no date: https://www.youtube.com/watch?v=8imskoHgAb0.

59 Ashwin, S. "Gender, State and Society in Soviet and Post-Soviet Russia", *Gender, State and Society in Soviet and Post-Soviet Russia*, ed. S. Ashwin (London: Routledge, 2000) 11–12.
60 Ladygina-Kots, *Infant Chimpanzee and Human Child*, 393–398.
61 Nikolai Krementsov, "Darwinism, Marxism and Genetics in the Soviet Union", *Biology and Ideology from Descartes to Darwin,* eds. Denis Alexander and Ronald Numbers (Chicago and London: University of Chicago Press, 2010), 215–246.
62 Nadezhda Ladygina-Kots, "Avtobiografiia", in *Gosudarstvennyi darvinovskii muzei* (Moscow: Mozhaiskii poligraficheski kombinat, 1993), 79.
63 Vucinich, *Darwinism in Russian Thought*, 156–176.
64 See Valerian Lunkevich,"Nereshennye problem biologii", St Petersburg (1904) cited in Vucinich, *Darwinism in Russian Thought*, 181.
65 Ibid.
66 Nikolai Krementsov. "V. A. Wagner and the origin of Russian Ethology", *International Journal of Comparative Psychology*, vol. 6, no. 1 (1992): 61–70.
67 Hilary Fink,"The Death of Intuition", *Bergson and Russian Modernism 1900-1930* (Evanston, Illinois: Northwestern University Press, 1999), 101–111.
68 Vucinich, *Darwinism in Russian Thought*, 156, 162, 170–177.
69 Ibid. 160.
70 Krementsov, "V.A. Wagner", 64–65.
71 Vladimir Vagner, *Vozniknovenie i razvitie psikicheskikh sposobnost'ei*, 9 vols, 1925–1929, 1927 volume: 18, cited in Krementsov, "V.A. Wagner", 67. On Jacob Moleschott (1822–1893) and Ludwig Buchner (1824–1899), see Daniel P. Todes, *Ivan Pavlov: A Life in Science* (New York: Oxford University Press, 2014), 29, 36.
72 Krementsov, "V.A. Wagner", 67.
73 Ibid.
74 See, for example, Peter Pringle, *The murder of Nicolai Vavilov: The story of Stalin's persecution of one of the great scientists of the twentieth century* (London: Simon and Shuster, 2008).
75 Unsigned letter from a member of the Kratovo Republican Internat of Invalids of the Patriotic War to the State Darwin Museum, Moscow, November 8, 1943, cited in Kalacheva, "60 let velikoi pobedy": 38.
76 Trofim Lysenko, "The Situation in Biological Science. Address given to the Vaskhnil Session, 31 July 1948", *Agrobiology. Essays on problems of genetics, plant breeding and seed growing* (Moscow: Foreign Languages Publishing House, 1954), 552.
77 For a more detailed explanation of this see: Pat Simpson, "Revolutionary Evolution in Apes and Humans in the 1920s: Sculpture and Constructs of the New Man at the Darwin Museum Moscow", *The art and science of making the new man in early twentieth century Russia*, eds. Nikolai Krementsov and Yvonne Howell (New York, Oxford, New Delhi, Sydney: Bloomsbury Academic, 2021), 155–174.
78 Kalacheva, "Gosudarstvennyi darvinovskii muzei", 40–41.

Bibliography

Introduction

Antliff, Mark. "Cubism, Futurism, Anarchism: The 'Aestheticism' of the *Action d'Art* Group, 1906–1920", *Oxford Art Journal*, 21/2 (1998): 99–120.

———. "Revolutionary Immanence: Bergson among the Anarchists", *Bergson and The art of Immanence: Painting, Photography, Film*, eds. John Ó Maoilearca and Charlotte de Mille (Edinburgh: Edinburgh University Press, 2013) 94–111.

Atkinson, Paul. *Henri Bergson and Visual Culture: A Philosophy for a New Aesthetic* (London, etc.: Bloomsbury Academic, 2021).

Bergson, Henri. "De la simulation inconsciente dans l'état d'hypnotisme", *Revue philosophique*, tome XXII (November 1886): 525–531.

———. *L'énergie spirituelle: essais et conférences* (Paris: Librairie Félix Alcan, 1919).

———. *L'Évolution créatrice* (Paris: Félix Alcan, Éditeur, 1907); *Creative Evolution*, trans. Arthur Mitchell (Mineola, New York: Dover Publications, Inc., 1998).

———. *Matière et Mémoire: Essai sur la relation du corps à l'esprit* (Paris: Félix Alcan, Éditeur, 1896); *Matter and Memory*, trans. Nancy Margaret Paul and W. Scott Palmer (New York: Zone Books, 1988).

Bianco, Giuseppe. "Henri Bergson's Failed Career as a Psychologist: Dealing "Seriously" with Philosophy?", *Academia.edu* (2018).

Boccioni, Umberto et al. "Futurist Painting: Technical Manifesto", April 1910; *Futurist Manifestos*, ed. Umbro Apollonio, trans. Robert Brain et al. (New York: Viking Press, 1970).

Brauer, Fae. "Vitalist Cubisms: The Biocultures of Virility, Militarism and *La Vie Sportive*", *Sport and the European Avant-Garde (1900–1945)*, eds. Andreas Kramer and Przemyslaw Strozek (Leiden/Boston: Brill Avant-Garde Critical Studies, 2022) 19–56.

———. "Capturing Unconsciousness: The New Psychology, Hypnosis and the Culture of Hysteria", *A Companion to Nineteenth Century Art*, ed. Michelle Facos (London et al.: John Wiley & Sons, Inc., 2019) 243–262.

———. "Becoming Simian: Devolution as Evolution in Transformist Modernism", *Picturing Evolution and Extinction: Degeneration and Regeneration in Modern Visual Culture*, eds. Fae Brauer and Serena Keshavjee (Lady Stephenson Library, Newcastle upon Tyne: Cambridge Scholars Publishing, 2015) 127–156.

Cage, John. *Lecture on Nothing* (1949; Incontri Musicali, August 1959).

Chastenet, Jacques. *La France de M. Fallières* (Paris: Fayad, 1949).

Cohn, Jesse S. *Anarchism and the Crisis of Representation: Hermeneutics, Aesthetics, Politics* (Selisgrove: Susquehanna University Press, 2006).

Cohen, Richard A. "Philo, Spinoza, Bergson: The Rise of an Ecological Age", *The New Bergson*, ed. John Mullarkey (Manchester: Manchester University Press, 1999) 18–31.

Coubertin, Pierre de. *L'Éducation en Angleterre: Collèges et Universités* (Paris: Librairie Hachette et Cie, 1888).

Darwin, Charles. *The Descent of Man, and Selection in Relation to Sex* (London: John Murray, 1871).

Dalrymple Henderson, Linda. "Rethinking Modern Art, Science and Occultism in Light of the Ether of Space: Wassily Kandinsky, Umberto Boccioni, and Kazimir Malevich", *The History of Art and 'Rejected Knowledge': From the Hermetic Tradition to the 21st Century*, ed. Anna Korndorf (Moscow: The State Institute of Art Studies, 2018) 218–237.

Deleuze, Gilles and Félix Guattari. *Anti-Oedipus*, trans. Robert Hurley, Mark Seem and Helen R. Lane (London and New York: Continuum, 2004).

Driesch, Hans. *The History and Theory of Vitalism*, trans. C. K. Ogden (London: Macmillan and Co., Limited, 1914).

———. "Psychical Research and Philosophy", *The Case for and against Psychical Belief*, ed. Carl Murchison (Worcester: Clark University, 1927).

———. *The Science of the Super-normal* (London: Bell, 1933).

Fink, Hilary L. *Bergson and Russian Modernism 1900-1930* (Evanston: Northwestern University Press, 2012).

Freyhofer, Horst Heinz. *The Vitalism of Hans Driesch: The Success and Decline of a Scientific Theory* (European University Studies, 1982).

Grogin, Robert C. "Henri Bergson and the University Community, 1900–1914", *Historical Reflections*, 2/2 (1976): 220–222.

Hvidberg-Hansen, Gertrud and Gertrud Oelsner, eds. *The Spirit of Vitalism: Health, Beauty and Strength in Danish Art, 1890–1940* (Chicago: The University of Chicago Press, 2011).

James, William. *The Principles of Psychology* (New York: Henry Holt and Company, 1890).

Janet, Pierre *L'automatisme psychologique, essai de psychologie expérimentale sur les forms inférieures de l'activité humaine*, Thèse d'État, Faculté des lettres (Paris: Félix Alcan, 1889).

Krage, Helge. *An Introduction to the Historiography of Science* (New York: Cambridge University Press, 1989).

Larson, Barbara. "Gauguin: Vitalist, Hypnotist", *Gauguin's Challenge: New Perspectives after Postmodernism*, ed. Norma Broude (London and New York: Bloomsbury Visual Arts, 2018).

Lehan, Stephen. "Bergson and the Discourse of the Moderns", *The Crisis in Modernity: Vitalist Controversy*, eds. Frederick Burwick and Paul Douglass (Cambridge: The Cambridge University Press, 2010).

Lundy, Craig. *Deleuze's Bergsonism* (Edinburgh: Edinburgh University Press, 2018).

Maoilearca, John Ó and Charlotte de Mille. eds. *Bergson and the Art of Immanence: Painting, Photography, Film* (Edinburgh: Edinburgh University Press, 2013).

Marcel, Gabriel. *The Bergsonian Heritage*, ed. Thomas Hanna (New York: Columbia University Press, 1962).

Michael, Michael T. "Unconscious Emotion and Free-Energy: A Philosophical and Neuroscientific Exploration", *Frontiers in Psychology* (21 May 2020) https://doi.org/10.3389/fpsyg.2020.00984.

Munch, Edvard. *Konkurransen om den Kunstneriske Utsmykning av Universitetets Nye Festsal (Competition for the Artistic Decoration of the University's New Festival Hall)*, pamphlet, 1911.

Nietzsche, Frederick. *The Birth of Tragedy, Or: Hellenism and Pessimism*, trans. Douglas Smith (1872; Oxford: Oxford University Press, 2008).

———. *Twilight of the Idols, or, How to punish with a Hammer*, trans. Duncan Large (Liepzig: 1889; Oxford: Oxford University Press, 2008).

Ostwald, Wilhelm. "The Modern Theory of Energetics", *The Monist*, 17/4 (1907): 481–515.

———. *Der energetische Imperativ* (Leipzig: Akademische Verlagsgesellschaft m.b.H., 1912).

Pareti, Germana. "Hans Driesch's Interest in the Psychical Research. A Historical Study", *Medical Historica*, 1/3 (2017): 152–162.

Richet, Charles. "Faut-il laisser la France périr?", *Revue bleue* (1896): 620–622.

Seillière, Ernest. *Un Artisan d'Énergie Française: Pierre de Coubertin* (Paris: Henri Didier, Librairie-Éditeur, 1917).

Simpson, Patricia. "Imag[in]ing Post-Revolutionary Evolution: The Taylorized Proletarian, "Conditioning", and Soviet Darwinism in the 1920", *The Art of Evolution: Darwin, Darwinisms and Visual Culture*, eds. Barbara Larson and Fae Brauer (Hanover, NJ: University Press of New England, 2009) 226–261.

Stahl, Georg Ernst. *Theora medica vera* (1708).

Taine, Hippolyte. *Les Origines de la France contemporaine: L'Ancien Régime*, vol. 1 (Paris: Hachette, 1876; 1896).

Vitalism and the Scientific Image in Post-Enlightenment Life Science, 1800–2010, eds. Sebastien Normandin and Charles T. Wolfe (New York: Springer, 2013).

Walden, Paul. *Wilhelm Ostwald* (Kessinger Publishing: 1904).

Walusinski, Olivier and Julien Bogousslavsky. "Charcot, Janet and the Models of Psychopathology", *European Neurology* (17 June 2020): 333–340.

Wilhelm Wundt, *Grundzüge der Physiologischen Psychologie* (Leipzig: W. Engelman, 1874).

Williams, Elizabeth A. *A Cultural History of Medical Vitalism in Enlightenment Montpellier* (London and New York: Routledge, 2017).

Wünsche, Isabel. "Creative Intuition: The Russian Interpretation of Henri Bergson's Metaphysics", *Experiment*, 23/1 (October 2017): 215–228.

Chapter One

Adler, Kathleen and Marcia Pointon (eds), *The Body Imaged: The Human Form and Visual Culture Since the Renaissance* (Cambridge and New York: Cambridge University Press, 1993).

Amanda O'Neill, *The Life and Works of Munch* (Bristol: Parragon Book Service, 1996).

Berger, Martin A., "Painting Victorian Manhood", *Thomas Eakins: The Rowing Pictures, Exhibition Catalogue*, ed. Helen A. Cooper (New Haven: Yale University Art Gallery, 1996) 102–123.

Bergson, Henri. *Creative Evolution* (Paris, 1907), tr. Arthur Mitchell (New York: Henry Holt and Company, 1911).

Brauer, Fae, "Flaunting Manliness: Republican Masculinity, Virilized Homosexuality and the Desirable Male Body", 'Masculinities', *Australian and New Zealand Journal of Art*, 6/1 (2005), 23–42.

———. "Virilizing and Valorizing Homoeroticism: Eugen Sandow's Queering of Body Cultures Before and After the Wilde Trials", *Visual Culture in Britain*, 18/1 (2017): 35–67.

Brauer, Fae and Anthea Callen (eds). *Art, Sex and Eugenics: Corpus Delecti* (London: Ashgate, 2008; Routledge, 2015).

Brauer, Fae and Barbara Larson. *The Art of Evolution: Darwin, Darwinisms and Visual Culture*, Lebanon: The University Press of New England, 2009.

Brauer, Fae and Serena Keshavjee. *Picturing Evolution and Extinction: Regeneration and Degeneration in Modern Visual Culture* (Newcastle upon Tyne: Cambridge Scholars Publishing, 2015).

Braun, Marta, *Picturing Time: The Work of Etienne-Jules Marey (1830–1904)* (Chicago: University of Chicago Press, 1994).

Briffault, Eugène, illus. Bertall, *Paris dans l'eau* (Paris: J. Hetzell, 1844).

Burwick F., and P. Douglass (eds.), *The Crisis in Modernism: Bergson and the Vitalist Controversy* (New York: Cambridge University Press, 1992).

Callen, Anthea. "Muscularity and Masculinity: Dr Paul Richer and Modern Manhood", *Paragraph*, 26 (March & July 2003): 1–2, special issue: 'Men's Bodies', ed. Judith Still, 17–41.

———. *The Spectacular Body: Science, Method and Meaning in the Work of Degas* (London and New Haven: Yale University Press, 1995).

———. *Looking at Men: Art, Anatomy and the Modern Male Body* (London and New Haven: Yale University Press, 2018).

Chapman, David L. *Sandow the Magnificent: Eugen Sandow and the Beginnings of Bodybuilding* (Urbana, IL: University of Illinois Press, 2006).

Cooper, Emmanuel. *The Life and Work of Henry Scott Tuke* (Heretic Books, 2003).

Cooper, Helen A. "Rowing in Art and Life", *Thomas Eakins: The Rowing Pictures*, exhibition catalogue, ed. Helen A. Cooper (New Haven: Yale University Art Gallery, 1996) 24–78.

Coubertin, Pierre de. *Essais de psychologie sportive* (Grenoble: Jérôme Millon, 1992).

———. *Essais de psychologie sportive* (Lausanne and Paris: Payot et Cie 1913).

———. *L'Éducation en Angleterre: Collèges et Universités* (Paris: Librairie Hachette et Cie, 1888).

Desbonnet, Edmond. *Comment on devient athlète*, with a preface by Pierre Loti (Paris: Librairie Athlétique de "la Culture Physique", 1909.).

———. *La Force physique: culture rationnelle* (Paris: Berger-Levrault, 1906).

———. *La Méthode Desbonnet, culture physique rationnelle* (Paris: Chaponet, 1906).

Fonsmark, Anne-Birgitte et al. *Gustave Caillebotte*, exhibition catalogue (Bremen Kunsthalle, 2008).

Forth, Christopher E. "Gender, Empire, and the Politics of Regeneration", *French Politics, Culture & Society* 36/2 (2018): 149–156.

Forth, Christopher E. and Bertrand Taithe. *French Masculinities: History, Culture and Politics* (Basingstoke and New York: Palgrave MacMillan, 2007).

Forth, Christopher E. and Peter Cryle (eds). *Sexuality at the Fin de Siècle: The Making of a 'Central Problem'* (Cranbury, NJ: Associated University Presses, 2008).

Foucault, Michel. *History of Sexuality*, volume I, *An Introduction* (Paris, 1976; London: Allen Lane, 1979).

Garb, Tamar. "Muscularity, Masculinity and Modernity in Caillebotte's Male Figures", *Visible Touch: Modernism and Masculinity*, ed. Terry Smith (Chicago: The University of Chicago Press, 1997) 53–74.

Grüring, Jaimee. "Dirty Laundry...", PhD dissertation, Arizona State University, 2011.

Hilaire, Michel and Paul Perrin (Dir.). *Frédéric Bazille: La Jeunesse de l'impressonnisme*, exhibition catalogue, (Montpellier: Musée Fabre, 2016).

Katz, Jonathan and David C. Ward. *Hide/Seek: Difference and Desire in American Portraiture*, exhibition catalogue (Washington DC: National Portrait Gallery and Smithsonian Institution, 2010).

Kershner, R. Brandon. "The World's Strongest Man: Joyce or Sandow", *James Joyce Quarterly*, 30–31: 30/4–31/1, Joyce and Advertising (Summer–Fall, 1993): 667–693.

Sandow's Magazine of Physical Culture (London: Eugen Sandow, 1899–1907).

Michael Hatt. "Muscles, Morals, Mind: The Male Body in Thomas Eakins' *Salutat*", *The Body Imaged: The Human Form and Visual Culture Since the Renaissance*, eds. Kathleen Adler and Marcia Pointon (Cambridge and New York: Cambridge University Press, 1993) 57–69.

———. "The Male Body in Another Frame: Thomas Eakins' *The Swimming Hole* as a Homoerotic Image", *The Body*, eds. Andrew Benjamin, special edition, *Journal of Philosophy and the Visual Arts* (London: Academy Group, 1993) 8–21.

Morton, Mary and George T. Shackelford. *Gustave Caillebotte: The Painter's Eye*, exhibition catalogue, National Gallery of Art, Washington and Kimbell Art Museum, Fort Worth, (Washington DC: National Gallery of Art, 2015).

Nye, Robert A. *Masculinity and Male Codes of Honour in Modern France* (Berkeley and Los Angeles: UCLA Press, 1992).

Park, Sun-Young. *Ideals of the Body: Architecture, Urbanism, and Hygiene in Postrevolutionary Paris* (Pittsburgh, PA: University of Pittsburgh Press, 2018).

Poivert, Michel. "Variété et vérité du corps humain: l'esthétique de Paul Richer", *L'Art du nu au XIXe siècle: le photographe et son modèle*, ed. Sylvie Aubenas, exhibition catalogue, (Paris: Bibliothèque nationale de France, 1997).

Prideaux, Sue. *Edvard Munch: Behind The Scream* (New Haven, CT: Yale University Press, 2005).

Revenin, Régis. 'Homosexualité et virilité' in Alain Corbin (dir.), *Histoire de la virilité. 2. Le tromphe de la virilité. Le XIXe siècle* (Livres de référence - L'Univers historique; Seuil, 2011) 375–407.

Robinson, Cicely. *Henry Scott Tuke*, exhibition catalogue, Watts Gallery (Compton, Surrey) 2021.

Ruiz-Gomez, Natasha. 'The "Scientific Artworks" of Doctor Paul Richer', *Medical Humanities*, 39/1 (June 2013): 4–10; available online, DOI: http://dx.doi.org/10.1136/medhum-2012-010279.

Sandow, Eugen. *The Construction and Reconstruction of the Human Body*, Foreword by Sir Arthur Conan Doyle,(London: John Bale Sons, and Daniellson Ltd., 1907).

Simpson, Marc. *Thomas Eakins* (Philadelphia: Museum of Fine Art, 2001).

Sohn, Anne-Marie. *"Sois un Homme!": La construction de la masculinité au XIXe siècle* (Paris: Seuil, 2009).

Tate, Don. *Strong As Sandow: How Eugen Sandow Became The Strongest Man On Earth* (London: Charlesbridge Children, 2017).

Vigarello, Georges. 'Hygiène du corps et travail des apparences' in Alain Corbin (dir.), *Histoire du corps. 2. De la Révolution à la Grand Guerre* (Paris: Seuil, 2005), part 2, 307–320; Corbin, Alain, *The Foul and the Fragrant: Odor and the French Social Imagination* (Leamington Spa, Hamburg and New York: Berg Press, 1986).

———. *Histoire des pratiques de santé: La sain et le malsain depuis le Moyen-Age* (Paris: Seuil, 1987, 2015).

———. *Le proper et le sale: L'hygiène du corps depuis le Moyen Age* (Paris: Seuil, 1985).

Wallace, Catherine. *Catching the Light: the art and life of Henry Scott Tuke 1858–1929* (Edinburgh: Atelier Books, 2008).

Waller, David. *The Perfect Man: The Muscular Life and Times of Eugen Sandow, Victorian Strongman* (Brighton: Victorian Secrets, 2011).

Woloshyn, Tania. 'Regenerative Tanning: pigmentation, Neo-Lamarckian eugenics and the visual culture of the cure de soleil', *Picturing Evolution and Extinction: Regeneration and Degeneration in Modern Visual Culture*, eds. Fae Brauer and Serena Keshavjee (Newcastle-on-Tyne: Cambridge Scholars Publishing, 2016).

———. *Soaking Up the Rays. Light Therapy and Visual Culture in Britain, c. 1890–1940* (Manchester: Manchester University Press, 2017).

Chapter Two

Berman, Patricia G. "Monumentality and Historicism in Edvard Munch's University of Oslo Festival Hall Paintings", unpublished Ph.D. dissertation, New York University, 1989.

———. "Body and Body Politic in Edvard Munch's Bathing Men", *The Body Imaged. The Human Form and Visual Culture since the Renaissance*, ed. Kathleen Adler and Marcia Pointon (N.Y. and Cambridge: Cambridge University Press, 1993) 71–83, 195.

———. "Dionysus with Tan Lines: Edvard Munch's Discursive Skin". *A Fine Regard: Essays in Honor of Kirk Varnedoe*, ed. Patricia G. Berman and Gertje Utley (London: Ashgate Press, 2008) 68–85.

———. "Edvard Munch's 'Modern Life of the Soul'", *Edvard Munch and the Modern Life of the Soul*, (exh. cat.), ed. Kynaston McShine (New York: Museum of Modern Art, 2006) 34–51.

———. and Peder Anker. *Edvard Munchs Aulamalerier: Fra Kontroversielt Prosjekt til Nasjonalskatt* (Oslo: Messels Forlag, 2011).

———. "From Folk to a Folk Race: Carl Arbo and National Romantic Anthropology in Norway", *Constructing Race on the Borders of Europe: Ethnography, Anthropology, and Visual Culture, c. 1850-1930*, eds. Marsha Morton and Barbara Larsen (London: Bloomsbury Press, forthcoming 2021) 25–49.

Boe, Roy A. "Edvard Munch's Murals for the University of Oslo", *Art Quarterly*, XXIII/3 (Autumn 1960): 233–246.

Bölsche, Wilhelm. *Ernst Haeckel: ein Lebensbild* (Dresden: C. Reisner, 1900).

Brändström, Andreas and Hilde Rognerud. *Munch! Nietzsche, Thiel och Nordens störste konstnär*, (exh. cat.), (Stockholm: Thielska Galleriet, 2013).

Brain, Robert Michael. "How Edvard Munch and August Strindberg Contracted Protoplasmania: Memory, Synaesthesia, and the Vibratory Organism in Fin-de-Siècle Europe", *Interdisciplinary Science Reviews*, 35/1 (March 2010): 7–38.

Brauer, Fae and Anthea Callen, eds. *Art, Sex and Eugenics: Corpus Delecti* (Hampshire, UK; Burlington, VT: Ashgate, 2008).

Collett, John Peter. *Historien om Universitetet i Oslo*, Oslo: Universitetsforlaget, 1999.

Cordulack, Shelly Wood. *Edvard Munch and the Physiology of Symbolism*, Fairleigh Dickinson University Press, 2002.

Dietrich and Dr Kammer. *Bericht über den V. Internationalen Kongress für Thalassotherapie* (5–9 June), Berlin: Kolberg, 1911.

Egeland, Alv and William J. Burke. *Kristian Birkeland: The First Space Scientist* (Dordrecht: Springer Verlag, 2005).

Evjen, Bjørg. "Measuring Heads: Physical Anthropological Research in North Norway", *Acta Borealia*, 2 (1997): 3–20.

Finsen, Niels. *La Photothérapie* (Paris: Georges Carré et C. Naud, 1899).

Frey Manuel. *Der reinliche Bürger: Entstehung und Verbreitung bürgerlicher Tugenden in Deutschland, 1960–1980* (Göttingen: Vandenhoeck & Ruprecht, 1997).

Friedman, Robert Marc. "Civilization and National Honour: The Rise of Norwegian Geophysical and Cosmic Science", *Making Sense of Space: The History of Norwegian Space Activities*, ed. John Peter Collett (Oslo and Stockholm: Scandinavian University Press, 1995) 3–39, 398–401.

———. "Making the Aurora Norwegian: Science and Image in the Making of a Tradition", *Interdisciplinary Science Reviews*, 35/1 (March 2010): 51–68.

Gambrell, Lynn. *Exploring the Invisible: Art, Science, and the Spiritual* (Princeton: Princeton University Press, 2002).

Godinez, Irene Berg. "Friluftsbad Christiania – frå reinsomhed til sport og rekreasjon", *Byminner* 97/2 (1997).

Haeckel, Ernst. *Die Welträthsel (The Riddle of the Universe at the Close of the Nineteenth Century)* [1899] (New York: Harper and Brothers Publishers, 1905).

Halse, Sven."Vitalisme - fænomen og begreb", *Kritik* 171 (2004): 1–7.

———. "Vitalismen: flere tilløb til en begrebsdiskussion", *Kritik* 38 (2006): 119–125.

Hau, Michael. *The Cult of Health and Beauty in Germany: A Social History, 1890–1930* (Chicago and London: The University of Chicago Press, 2003).

Hestmark, Geir. "Nansen – 'Vitenskaperen,'" *Nansen ved to århundreskifter*, eds. Olav Christiensen and Audhild Skoglund (Oslo: Aschehoug, 1996) 128–136.

Hope, Nicholas. *German and Scandinavian Protestantism 1700–1918* (Oxford: Clarendon Press, 1995).

Hubbard, William H., Jan Eivind Myhre, Trond Nordby, and Sølvi Sogner. *Making a Historical Culture: Historiography in Norway* (Oslo, Copenhagen, Stockholm, and Boston: Scandinavian University Press, 1995).

Hueppe, Ferdinand. *Handbuch der Hygiene* (Berlin: Verlag von August Hirschwald, 1899).

Hvidberg-Hansen, Gertrud and Gertrud Oelsner, eds. *The Spirit of Vitalism. Health, Beauty and Strength in Danish Art, 1890–1940.* (Copenhagen: Museum Tusculanum Press, 2011).

Kemperinck, Maria G. and Leonieke Vermeer, eds. *Utopianism and the Sciences, 1880-1930, Groningen Studies in Cultural Change* (Leuven: Peeters, 2010).

Keshavjee, Serena."'La Vie Renaissant de le Mort': Albert Besnard's 'Non-Miraculous' History of Creation", *Picturing Evolution and Extinction: Regeneration and Degeneration in Modern Visual Culture*, eds. Fae Brauer and Serena Keshavjee (Lady Stephenson Library, Newcastle upon Tyne: Cambridge Scholars Publishing, 2015) 61–82.

Körber, Lill-Ann. *Badende Männer: Der nackte männliche Körper in der skandinavischen Malerei und Fotografie des frühen 20. Jahrhunderts* (Bielefeld: transcript Verlag, 2014).

Koefoed, Holger. "Badeliv og vitalisme i kunsten", *Byminner*, 97/ 2 (1997): 60–66.

Konkurransen om den Kunstneriske Utsmykning av Universitetets Nye Festsal (Competition for the Artistic Decoration of the University's New Festival Hall), Kristiania, 1911.

Konow, Sten and Karl Fischer. *Norway: Official Publication for the Paris Exhibition, 1900* (Kristiania: Aktie-Bogtrykkerit, 1900).

Kyllingstad, Jon Røyne and Thor Inge Rørvik. *Universitet i Oslo, 1811–2011*, vol. 2, *1870–1911 Vitenskapenes universitetet*, ed. John Peter Collett (Oslo: Unipub, 2011).

Kyllingstad, Jon Røyne Kyllingstad. *Measuring the Master Race. Physical Anthropology in Norway, 1890–1945* (Cambridge, UK: Open Book Publishers, 2014).

Larson, Barbara. "Microbes and Maladies: Bacteriology and Health at the Fin de Siècle", *Lost Paradise: Symbolist Europe*, (exh. cat.), (The Montreal Museum of Arts, 1995) 385–393.

———. *The Dark Side of Nature: Science, Society, and the Fantastic in the Work of Odilon Redon* (University Park: The Pennsylvania University Press, 2005).

Larsen Øivind, Bård Alsvik, Magne Nylenna, and Elisabeth T. Swärd. *Helse og Nasjonbyggning* (Oslo: Gyldendal Norsk Forlag, 2005).

Lathe, Carla."The Group *Zum Schwarzen Ferkel*: A Study in Early Modernism, unpublished Ph.D. dissertation, University of East Anglia, 1972.

Lie, Einer and Hege Roll-Hansen. *Faktisk Talt: Statistikkens historie i Norge* (Oslo: Universitetetsforlaget, 2001).

Lowenthal, Wilhelm. *L'Enseignement actuel de l'Hygiène dans les facultés de médecine en Europe* (Paris: Librairie H. Le Soudier, 1887).

Mallon, Bill and Ian Buchanan, eds. *The 1908 Olympic Games: Results from All Competitors in All Events, with Commentary* (Jefferson, North Carolina: McFarland and Co., 2000).

Mallon, Bill and Ture Widlund. *The 1912 Olympic Games: Results from All Competitors in All Events, with Commentary* (Jefferson, North Carolina: McFarland and Co., 2001).

Micale, Mark. *The Mind of Modernism: Medicine, Psychology, and the Cultural Arts in Europe and America, 1880–1940* (Stanford University Press, 2003).

Mohr, Otto Lous Mohr. *Edvard Munchs Auladekorasjoner i lys av ukjente utkast og sakens akter* (Oslo: Gyldendal Norsk Forlag, 1960).

Morton, Marsha. "From Monera to Man: Ernst Haeckel, *Darwinismus*, and Nineteenth-Century German Art", *The Art of Evolution: Darwin, Darwinisms, and Visual Culture*, eds. Barbara Larson and Fae Brauer (Hanover: Dartmouth College Press and University Press of New England, 2009, 59–91.

Offizieller Katalog der internationalen Hygiene Austellung, Dresden, Mai bis Oktober, 1911 (Berlin: Rudolf Mosse. 1911).

Nansen, Fridtjof. *Farthest North, Being the Record of a Voyage of Exploration of the Ship "Fram" 1893–96 and of a Fifteen Months' Sleigh Journey by Dr. Nansen and Lieut. Johansen* (New York: Harper and Brothers, 1897).

Natale, Simone. "A Cosmology of Invisible Fluids: Wireless, X-Rays, and Psychical Research around 1900", *The Canadian Journal of Communication*, 36/2 (2011): 263–275.

Niemi, Einar, Jan Eivind Myhre, and Knut Kjeldstadli. *Norsk innvandringshistorie*. Volume 2: *i Nasjonalstatens Tid 1814-1940* (Oslo: Pax Forleg A/S, 2003).

Nissen, Hartvig. *ABC of the Swedish System of Educational Gymnastics: A Practical Handbook for School Teachers and the Home* (Philadelphia and London: F. A. Davis, 1891).

Olmstad, Finn. *Norsk Idretts Historie: Forsvar, sport, klassekamp 1861–1939* (Oslo: Aschehoug, 1987).

Pettersen, Petra. "Munchs Aula-Dekorasjoner", Gerd Woll et al, *Samlede Malerier: Catalogue Raisonne*, Vol. III, 1909–1920, (Oslo: Munch-Museet, Kaare Berntsen, and Cappelen Damm AS 2008) 829–851.

Richardson, Benjamin Ward. *The Health of Nations: A Review of the Works of Edwin Chadwick* (London: Longmans, Green, and Co., 1887).

Rousseau, Pascal. "Radiation: Metabolizing the 'New Rays", *Edvard Munch: The Modern Eye*, (exh. cat.), eds. Angela Lampe and Clément Chéroux (Centre Pompidou and Tate Modern, 2011) 160–169.

Skirstad, Berit and Kristin Felde. "Sports Policy in Norway", *National Sports Policies: An International Handbook*, eds. Laurance Chalip, Arthur Johnson, and Lisa Stachura (Westport, CT.: The Greenwood Press, 1996).

Stavenow-Hidemark, Elisabet. "Hygienism kring sekelskiftet." *Fataburen: Nordiska Museets och Skansens Årsbok* (Stockholm: Nordiska Museets förlag, 1970) 47–54.

Shaw, Jennifer. *Dream States. Puvis de Chavannes, Modernism, and the Fantasy of France* (New Haven: Yale University Press, 2002).

Slagstad, Rune. *Sporten: En Idéhistorisk Studie* (Oslo: Pax Forlag, 2008) 239–266.

Sørensen, Gunnar. "Vitalismens år?", *Cras*, (1981): 26–42.

Tøjner, Poul Erik. *Munch in His Own Words* (Munich: Prestel Verlag, 2001).

Treitel, Corinna. *A Science for the Soul* (Baltimore and London: Johns Hopkins University Press, 2004).

Voss, Julia Voss. "Monkey, Apes and Evolutionary Theory: From Human Descent to King Kong", *Endless Forms: Charles Darwin, Natural Science and the Visual Arts*, (exh. cat.), eds. Diana Donald and Jane Munro (New Haven: Fitzwilliam Museum, Cambridge and Yale Center for British Art, 2009) 215–234.

Wagner, Monika. *Allegorie und Geschichte. Ausstattungsprogramme öffentlicher Gebäude des 19. Jahrhunderts in Deutschland* (Wasmuth: Tübinger Studien zur Archäologie und Kunstgeschichte, 1989).

Weyl, Theodor, ed. *Handbuch der Hygiene* (Jena: Verlag von Gustav Fischer, 1883–1904).

Woloshyn, Tania. "Regenerative Tanning: Pigmentation, Neo-Lamarckisn Eugenics and the Visual Culture of the Cure de Soliel", *Picturing Evolution and Extinction: Regeneration and Degeneration in Modern Visual Culture*, ed. Fae Brauer and Serena Keshavjeee (Lady Stephenson Library, Newcastle upon Tyne: Cambridge Scholars Publishing, 2015) 193–215.

———. *Soaking Up the Rays, Light Therapy and Visual Culture in Britain, c. 1890–1940* (Manchester: Manchester University Press, 2017).

Yarborough, Tina. "Exhibition Strategies and Wartime Politics in the Art and Career of Edvard Munch, 1914–1921", unpublished Ph.D. dissertation, University of Chicago, 1995.

Ydstie, Ingebjørg and Karen Lerheim, eds. *Livskraft: Vitalismen som kunstnerisk impuls 1900–1930*, (exh. cat.) (Oslo: Munch Museum, 2006).

Chapter Three

Adam, Paul. *La Morale des Sports* (Paris: La Librairie mondiale, 1907); "Athletic Education", *The International Journal of the History of Sport*, 23/3-4 (20 August 2006).

Agathon (Alfred de Tarde, Henri Massis).« Les jeunes gens d'aujourd'hui », *L'Opinion* (13 April-29 June 1912); repb. Agathon, *Les jeunes gens d'aujourd'hui* (Paris: Plon-Nourrit et Cie, 1913).

Antliff, Mark. *Inventing Bergson. Cultural Politics and the Parisian Avant-Garde* (Princeton, NJ: Princeton University Press, 1993).

Baudin, Pierre. « Nos enfants. A quoi rêvent-ils, Que rêvons-nous pour eux? » (national survey directed by Henri Mazel), *Revue des Français*, 6 (25 June 1912).

Bazaillas, Albert. *Musique et inconscience. Introduction à la psychologie de l'inconscient* (Paris: Alcan, 1908).

Bénéton, Philippe. « La génération de 1912-1914. Image, mythe et réalité? », *Revue Française de Science Politique*, XXI/5 (Octobre 1971): 981–1009.

Berget, Alphonse. « Heure », *Larousse mensuel*, 71 (January 1913).

Bertaut, Jules. « La jeunesse d'aujourd'hui », *La Gaulois*, 1 (15 juin 1912).

Bourdeau, Jean. *La philosophie affective* (Paris: Alcan, 1912).

Brauer, Fae. "L'Art Eugénique: Biopower and the Biocultures of Regeneration", *Genetics and French Culture*, eds. Louise Lyle, Douglas Morrey; *L'Esprit Créateur*, 52/2 (2012): 42–58.

———. "Eroticizing Lamarckian Eugenics: The Body Stripped Bare during French Sexual Neoregulation", *Art, Sex and Eugenics. Corpus Delecti*, eds. Fae Brauer, Anthea Callen (Hampshire: Ashgate, 2008).

Brauer, Fae and Serena Keshavjee. *Picturing Evolution and Extinction. Regeneration and Degeneration in Modern Visual Culture*, Cambridge Scholars Publishing, 2015.

Colomer, André. « Aux sources de l'héroïsme individualiste. M. Bergson et les Jeunes gens d'aujourd'hui », *L'Action d'art*, 2 (1 March 1913).

Cor, Raphaël. *Essais sur la sensibilité contemporaine* (Paris: H. Falque, 1912).

Cottington, David. *Cubism in the Shadow of War. The Avant-garde and Politics in Paris, 1905–1914* (New Haven and London: Yale University Press, 1998).

Coubertin, Baron Pierre de. *Les Batailles de l'éducation Physique: Un Campagne de vingt-et-uns ans (1887–1908)* (Paris: Librairie de l'Éducation Physique, 1909).

———. «L'avenir des sports au point de vue psychologique, international et démocratique», repb. « Psychologie, internationalisme, démocratie », *Revue Olympique*, (July 1910).

———. « La psychologie du costume sportif » (February 1909).

———. *Essais de psychologie sportive* (1913); (Grenoble: Jerôme Million, 1992).

———. « Le sport peut-il enrayer la névrose universelle? », *Essais de psychologie sportive*, 1913.

———. *Pédagogie sportive* (Paris: Les Éditions G. Crès et Cie, 1921) (Paris: 1972).

———. *Une campagne de vingt et un ans* (Paris: Librairie de l'Education Physique, 1909).

———. « Une nouvelle formule d'éducation physique », *Revue mensuelle du Touring Club de France* (20 March 1902) 146–151.

———. *L'Education des adolescents au XXème siècle. I. L'Education physique: la gymnastique utilitaire. Sauvetage-Défense-Locomotion* (Paris: Alcan, 1905).

Delaplace, Jean Paul. *L'Opinion* (17 August 1912).

Ehrenberg, Alain. « Aimez-vous les stades. Architecture de masse et mobilisation », *Recherches*, 43 (April 1980).

Gervais, Thierry. « L'invention du magazine. La photographie mise en page dans *La Vie au grand air* (1888-1914) », *Études photographiques*, 20 (June 2007): 50–67.

Grellety, Dr Lucien. *Névrosés et décadents* (Mâcon: Protat Frères, 1913).

Guasco, J. R. *Le Collège d'athlètes* (Paris: Audin, 1913).

Hébert, George. *L'éducation physique ou l'entraînement complet par la méthode naturelle* (Paris: Vuibert, 1912).

Henriot, Émile. « À quoi rêvent les jeunes gens? », *Le Temps*, 23 (24 April 1912; 27 May; 4 June 1912).

Kasson, John F. *Amusing the Million: Coney Island at the Turn of Century* (New York, Hill & Wang, 1978).

Lalo, Charles. *Les Sentiments esthétiques* (Paris: Alcan, 1910).

Laudet, Fernand. « Enquête sur la jeunesse », *La Revue hebdomadaire* (23 March–20 July 1912).

Le Tainturier-Fradin. *L'activité sportive, germe d'une philosophie pratique*, 1913, quoted in *Recherches*, 43, (April 1980).

Legrogneux. « Complications footballkaniques. Extrait d'un correspondant de guerre en l'an 1950 », *Fantasio*, 157 (1 February 1913).

Lubar, Robert. "Cubism, Classicism and Ideology: The 1912 Exposicio d'Art Cubista in Barcelona and French Cubist Criticism", eds. Elisabeth Cowling, Jennifer Mundy, *On Classic Ground* (London: Tate Gallery, 1990) 309–324.

Lichtenstein, Jacqueline. *La couleur éloquente. Rhétorique et peinture à l'âge classique* (Paris: Flammarion, 2003).

Montherlant, Henry de. "Football et rugby Olympique", *Demain* (July 1924); repb. Françoise Lionel et Serge Laget, *Rugby en toutes lettres. Anthologie* (Biarritz: Atlantica, 1999).

Mosso, Angelo. *Exercices physiques et le développement intellectuel* (Paris: Alcan, 1904).
Mourey, Gabriel. « L'art et le muscle », *Gil Blas* (19 janvier 1913): 1, in response to the survey, "Art and Muscle".
Nordau, Max. *Dégénérescence*, Tome Premier: *Fin de siècle: Le Mysticisme* (Paris: Félix Alcan, Éditeur, 1894).
Nye, Robert A. "Degeneration, Neurasthenia and the Culture of Sport in Belle Époque France", *Journal of Contemporary History*, 17/1 (January 1982): 51–68.
Ory, Pascal. *L'exposition universelle* (Bruxelles: Complexe, 1989).
Pidoux, Vincent. « Sur les traces d'une théorie énergétique de l'esprit. Angelo Mosso et le travail des émotions », *PSN (Psychiatrie, Sciences Humaines, Neurosciences)*, 13/2 (2015): 43–69.
Rozet, Georges. « Préparons la revanche de Stockholm. Il faut créer un Collège d'athlètes », *L'Opinion* (17 August 1912).
Rabaud, Etienne. « Origine et transformation de la notion de dégénéré », *Revue de l'école d'anthropologie*, 17 (1907): 36–46.
Rabinbach, Anson. *The Human Motor: Energy, Fatigue and the Origins of Modernity* (Berkeley: University of California Press, 1992).
Rey, Etienne. *La renaissance de l'orgueil français* (Paris: B. Grassat, 1912).
Rousseau, Pascal. *Robert Delaunay: l'invention du pop* (Paris: Éditions Hazan, 2019).
Rozet, Georges. *La défense et Illustration de la Race Française* (Paris: Felix Alcan, 1911).
Schneider, William. "Toward the Improvement of the Human Race: The History of Eugenics in France", *The Journal of Modern History*, 54 (June 1982): 268–291.
———. *Quality and Quantity. The Quest for Biological Regeneration in Twentieth-Century France* (Cambridge: Cambridge University Press, 1990).
Stiegler, Barbara. *Nietzsche et la biologie* (Paris: PUF, 2001): 32–43.
Toulouse, Dr. "L'action est le but supérieur de la vie", *Demain*, 1 (April 1912): 1–7.
Weber, Eugen. "Pierre de Coubertin and the Introduction of Organized Sport", *My France. Politics, Culture, Myth* (Cambridge: Harvard University Press, 1991).
Winock, Michel. *Décadence fin de siècle* (Paris: Gallimard, 2017).
Wohl, Robert. *The Generation of 1914* (Cambridge: Harvard University Press, 1979).

Chapter Four

Editor, "Notes by the Way", *Light*, February 27 (1886): 97–98.
Alexandre-Bisson, Juliette. *Les phénomènes dits de matérialisation: étude expérimentale* (Paris: Alcan, 1921).
Alvarado, Carlos S. "Human Radiations: Concepts of Force in Mesmerism, Spiritualism and Psychical Research", *Journal of the Society for Psychical Research*, 70.3/884 (2006): 138–162.
———. "Musings on Materializations: Eric J. Dingwall on 'The Plasma Theory'", *Journal of Scientific Exploration*, 33/1 (2019): 73–113.
Beale, Lionel S. *Protoplasm: or, Life, Force, and Matter* (London: J. Churchill and Sons, 1870).
———. *Bioplasm: An Introduction to the Study of Physiology and Medicine* (London: J. and A. Churchill, 1872).
Beckman, Karen Redrobe. *Vanishing Women: Magic Film and Feminism* (Durham, N.C.: Duke University Press, 2003).
Bergson, Henri. *Creative Evolution*. London: Macmillan, 1922. Originally published as *L'évolution créatrice* (Paris: Alcan, 1907).
Bonfante-Warren, Alexandra, Melissa E. Buron, Donato Esposito, and Krystyna Matyjaszkiewicz, eds. *James Tissot* (San Francisco: Fine Arts Museums of San Francisco, 2019).
Botar, Oliver A. I. "Prolegomena to the Study of Biomorphic Modernism: Biocentrism, Lászlo Moholy-Nagy's 'New Vision' and Ernõ Kállai's Bioromantik." PhD Dissertation, University of Toronto, 1998.

———. “New Vision and the Aestheticization of Scientific Photography in Weimar Germany.’ Science in Context”, *Science in Context*, 17/4 (2004): 525–556. https://doi.org/10.1017/s0269889704000250.

Bowler, Peter J. *Reconciling Science and Religion: The Debate in Early-Twentieth Century Britain* (Chicago: The University of Chicago Press, 2001).

———. *The Non-Darwinian Revolution: Reinterpreting a Historical Myth* (Baltimore: Johns Hopkins University Press, 1988).

Brain, Robert. “Materialising the Medium: Ectoplasm and the Quest for Supra-Normal Biology in Fin-De-Siècle Science and Art.” *Vibratory Modernism*, eds. Anthony Enns and Shelley Trower (London: Palgrave Macmillan, 2013) 112–141.

———. “Protoplasmia: Huxley, Haeckel and the Vibratory Organism in Late Nineteenth Century Science and Ar.” *The Art of Evolution: Darwin, Darwinisms and Visual Culture*, eds. Fae Brauer and Barbara Larson (Hanover, NH: Dartmouth College Press, 2016) 92–123.

Brauer, Fae. “Magnetic Modernism: František Kupka’s Mesmeric Abstraction and Anarcho-Cosmic Utopia.” *Utopia: The Avant-Garde, Modernism and (Im)Possible Life*, eds. David Ayers, Benedikt Hjartarson, Tomi Huttunen, and Harri Veivo, Vol. 4 of *European Avant-Garde and Modernism Studi.* (Berlin, Boston: De Gruyter, 2015) 123–152.

———. “Scientistic Magnetism and Hauntological Metarealism: The Phantasmatic Doubles of Duchamp and Durville.” *Realisms of the Avant-Garde*, eds. David Ayers, Moritz Baßler, Sascha Bru, Ursula Frohne, and Benedikt Hjartarson, Vol. 6 of *European Avant-Garde and Modernism Studies* (Berlin: De Gruyter, 2020)42–75.

Brower, M. Brady. *Unruly Spirits: The Science of Psychic Phenomena in Modern France* (Urbana: University of Illinois Press, 2010).

Clarke, Bruce, and Linda Dalrymple Henderson, eds. *From Energy to Information: Representation in Science and Technology, Art, and Literature* (Stanford: Stanford University Press, 2002).

Crookes, William. “The Last of ‘Katie King’. The Photographs of ‘Katie King’ by the Aid of the Electric Light”, *The Spiritualist*, (vol. 4, no. 23, 5 June 1874): 270–271.

Crookes, W., and C.G. Helleberg. *Remarkable Spirit Manifestations* (Cincinnati: C.G. Helleberg, 1891).

Davey, William. *The Illustrated Practical Mesmerist: Curative and Scientific* (London: J. Burns, 1889).

Delgado, Anne L. “Bawdy Technologies and the Birth of Ectoplasm”, *Genders* 1998–2013 (vol. 54, 1 September 2011): https://www.colorado.edu/gendersarchive1998-2013/2011/09/01/bawdy-technologies-and-birth-ectoplasm

Dingwall, Eric J. “The Plasma Theory”, *Journal of the Society for Psychical Research*, 15/ 5 (1921): 207–219.

———. “Report on a Series of Sittings with Eva C”, *Proceedings of the Society for Psychical Research*, 32 (1922): 209–343.

Doyle, Arthur Conan. *The History of Spiritualism* (New York: Arno Press, 1975).

Ellenberger, Henri. *The Discovery of the Unconscious: The History and Evolution of Dynamic Psychiatry* (New York: Basic Books, 1970).

Enns, Anthony. “Vibratory Photography.” *Vibratory Modernism*, eds. Anthony Enns and Shelley Trower (London: Palgrave Macmillan, 2013) 117–197.

Farmer, John S. *Twixt Two Worlds: A Narrative of the Life and Work of William Eglinton* (London: Psychological Press, 1886).

Fischer, Andreas. “The Reciprocal Adaptation of Optics and Phenomena: The Photographic Recording of Materializations.” *The Perfect Medium: Photography and the Occult*, eds. Chéroux Clément and Andreas Fischer (New Haven: Yale University Press, 2005) 171–216.

Flammarion, Camille. *The Unknown* (New York: Harper and Brothers, 1900).

———. *Mysterious Psychic Forces: An Account of the Author's Investigations in Psychical Research, Together with Those of Other European Savants* (Boston: Small, Maynard and Co., 1907).

Flint, Kate. "'More Rapid than the Lightning's Flash': Photography, Suddenness, and the Afterlife of Romantic Illumination", *European Romantic Review*, 24/3 (2013): 369–383. https://doi.org/10.1080/10509585.2013.787242.

Geison, Gerald L. "The Protoplasmic Theory of Life and the Vitalist-Mechanist Debate", *Isis*,60/3 (1969): 273–292. https://doi.org/10.1086/350498.

Geley, Gustave. *From the Unconscious to the Conscious*, trans. Stanley De Brath. (London: William Collins, Sons and Co., 1920). Originally published as *De l'inconscient au conscient* (Paris: Félix Alcan, 1919).

———. "Ectoplasm", *American Society for Psychical Research*, (vol. XVIII, no. 4, April 1924): 272–279.

Gettings, Fred. *Ghosts in Photographs: The Extraordinary Story of Spirit Photography* (Montreal: Optimum, 1978).

Granger, Michael. "D'Où Vient Le Mot 'Ectoplasme' Dans Son Acception Spirite et Métapsychique?", *Revue Spirite*,157/2 (2014): 15–17.

Gunning, Tom. "Phantom Images and Modern Manifestations: Spirit Photography, Magic Theater, Trick Films and Photography's Uncanny." *Cinematic Ghosts: Haunting and Spectrality from Silent Cinema to the Digital Era*, ed. Murray Leeder (London: Bloomsbury, 2015) 17–38.

Harvey, John. *Photography and Spirit* (London: Reaktion Books, 2007).

Houdini, Harry. *A Magician Among the Spirits* (New York: Harper and Brothers, 1924).

Huxley, Thomas Henry. *On the Physical Basis of Life*. (New Haven: College Courant, 1869).

———. "On the Physical Basis of Life", *Collected Essays*, vol. I (1868): http://aleph0.clarku.edu/huxley/CE1/PhysB.html.

Jolly, Martyn. *Faces of the Living Dead: The Belief in Spirit Photography* (London: British Library Publishing Division, 2006).

Jones, Esyllt. "Spectral Influenza: T. G. and Lillian Hamilton, Interwar Spiritualism, and Pandemic Disease." *Epidemic Encounters: New Interpretations of Pandemic Influenza in Canada, 1918–1920*, ed. Magdalena Fahrni and Esyllt Jones (Vancouver: UBC Press, 2012) 193–221.

Juler, Edward. "Henry Moore: Sculptural Process and Public Identity." Life Forms: Henry Moore, Morphology and Biologism in the Interwar Years. Tate Research Publication, 2015. https://www.tate.org.uk/art/research-publications/henry-moore/edward-juler-life-forms-henry-moore-morphology-and-biologism-in-the-interwar-years-r1151314.

Lachapelle, Sofie. *Investigating the Supernatural: From Spiritism and Occultism to Psychical Research and Metapsychics in France, 1853–1931* (Baltimore: Johns Hopkins University Press, 2011).

Lodge, Oliver J. "Experience of Unusual Physical Phenomena Occurring in the Presence of an Entranced Person (Eusapia Paladino)", *Journal of the Society for Psychical Research*, 6 (1894): 306–336.

Maeterlinck, Maurice. *The Unknown Guest*, trans. Alexander Teixeira de Mattos (London: Methuen and Co., 1914).

———. *The Great Secret* (New York: The Century Co., 1922).

Matheson, Neil. "Ectoplasm and Photography: Mediumistic Performances for Camera." *The Machine and the Ghost: Technology and Spiritualism in Nineteenth to Twenty-First-Century Art and Culture*, eds. Neil Matheson and Sas Mays (Manchester: Manchester University Press, 2013) 78–102.

Morton, Marsha. "From Monera to Man: Ernst Haeckel, Darwinismus, and Nineteenth Century German Art." *The Art of Evolution: Darwin, Darwinisms, and Visual Culture*, eds. Fae Brauer and Barbara Larson (Hanover, NH: Dartmouth College Press, 2009) 59–91.

Natale, Simone. "The Medium on the Stage: Trance and Performance in Nineteenth Century Spiritualism", *Early Popular Visual Culture*, 9/3 (2011): 239–255. https://doi.org/10.1080/17460654.2011.601166.

Noakes, Richard. *Physics and Psychics: The Occult and the Sciences in Modern Britain* (Cambridge: Cambridge University Press, 2019).

Normandin, Sebastien. "Visions of Vitalism: Medicine, Philosophy and the Soul in Nineteenth Century France." PhD Dissertation, McGill University, 2013.

Normandin, Sebastien, and Charles T. Wolfe. *Vitalism and the Scientific Image in Post-Enlightenment Life Science, 1800–2010* (New York: Springer, 2013).

Pethes, Nicolas. "Psychicones: Visual Traces of the Soul in Late Nineteenth-Century Fluidic Photography", *Medical History*,60/3 (2016): 325–341. https://doi.org/10.1017/mdh.2016.26.

Raulin-Cerceau, Florence. "The Concept of Chemical Evolution Before Oparin." In *Genesis – In the Beginning: Precursors of Life, Chemical Models and Early Biological Evolution*, ed. Joseph Seckbach, Vol. 22 of *Cellular Origin, Life in Extreme Habitats and Astrobiology* (Dordrecht: Springer, 2012) 894–895.

Rehbock, Philip F. "Huxley, Haeckel, and the Oceanographers: The Case of Bathybius Haeckelii", *Isis*, 66/4 (1975): 504–533. https://doi.org/10.1086/351511.

Reichenbach, Karl. *Physico-Physiological Researches on the Dynamics of Magnetism, Electricity, Heat, Light, Crystallization, and Chemism, in Their Relations to Vital Force.* trans. by H. John Ashburner (London: H. Baillière, 1851).

———. *Researches on Magnetism, Electricity, Heat, Light, Crystallization and Chemical Attraction in Relation to the Vital Force*, trans. William Gregory. (New Hyde Park: University Books, 1974). First published in 1850 by Taylor Walton and Maberly (London).

Richet, Charles. "Concerning the Phenomenon Called Materialisation", *Annals of Psychical Science*,2 (1905): 207–289.

———. *Thirty Years of Psychical Research: Being a Treatise on Metapsychic*, trans. Stanley de Brath. (New York: MacMillan, 1923). Originally published as *Traité de métapsychique* (Paris: Alcan, 1922).

Robertson, Beth A. *Science of the Séance: Transnational Networks and Gendered Bodies in the Study of Psychic Phenomena, 1918–40.* (Vancouver: UBC Press, 2017).

Rochas, Albert de. *L'extériorisation De La sensibilité: étude expérimentale Et Historique* (Paris: Chamuel, 1896).

Sausman, Justin. "'It's Organisms That Die, Not Life': Henri Bergson, Psychic Research, and the Contemporary Uses of Vitalism." *The Machine and the Ghost: Technology and Spiritualism in Nineteenth to Twenty-First Century Art and Culture*, eds. Neil Matheson and Sas Mays, (Manchester: Manchester University Press, 2013) 16–36.

Schlun, Betsy van. *Science and the Imagination: Mesmerism, Media, and the Mind in Nineteenth-Century English and American Literature* (Berlin: Galda and Wilch Verlag, 2007).

Schoonover, Karl. "Ectoplasms, Evanescence, and Photography", *Art Journal*, 62/3 (2003): 30–43. https://doi.org/10.2307/3558519.

Schrenck-Notzing, Albert von. *Der Kampf Um Die Materialisations-Phänomene* (Munich: Ernst Reinhardt, 1914).

———. *Phenomena of Materialisation: A Contribution to the Investigation of Mediumistic Teleplastics*, trans. E. E. Fournier d'Albe. (London: Kegan Paul, Trench, Trübner and Co., 1920). Originally published as *Der Kampf um die Materialisations-Phänomene* (Munich: Ernst Reinhardt, 1914).

Sharp, Lynn L. *Secular Spirituality: Reincarnation and Spiritism in Nineteenth-Century France* (Lanham: Lexington Books, 2006).

Sommer, Andreas. "Policing Epistemic Deviance: Albert Von Schrenck-Notzing and Albert Moll", *Medical History*, 56/2 (2012): 255–276. https://doi.org/10.1017/mdh.2011.36.

Steinorth, Karl, ed. *International Ausstellung Des Deutschen Werkbunds: Film Und Foto* (Stuttgart: Deutsche Verlags-Anstalt, 1979).

Tucker, Jennifer. *The Social Photographic Eye: Brought to Light: Photography and the Invisible, 1840–1900*, ed. Corey Keller (San Francisco: San Francisco Museum of Modern Art, 2009).

Willburn, Sarah. "Viewing History and Fantasy through Victorian Spirit Photography." *The Ashgate Research Companion to Nineteenth Century Spiritualism and the Occult*, eds. Tania Kontou and Sarah Willburn (Farnham: Ashgate, 2012) 359–381.

Wilson, Leigh. *Modernism and Magic: Experiments with Spiritualism, Theosophy and the Occult* (Edinburgh: Edinburgh University Press, 2013).

Wolffram, Heather. "In the Laboratory of the Ghost-Baron: Parapsychology in Germany in the Early 20th Century", *Endeavour*, 33/4 (2009): 152–157. https://doi.org/10.1016/j.endeavour.2009.10.001.

Chapter Five

Antliff, Mark. *Inventing Bergson: Cultural Politics and the Parisian Avant-Garde* (Princeton University Press, 1993).

Badiou, Alain. "Of Life as a Name of Being, or Deleuze's Vitalist Ontology", *Gilles Deleuze. Immanence et Vie*, eds. E. Alliez et al. (Paris: Presse Universitaire de France, 1998).

Baraduc, Hippolyte. *La Force vitale. Extrait de la Chronique Médicale*, 15 avril et 1er mai 1897 (Clermont (Oise): Imprimerie Daix Frère, 1897).

———. *La Force vitale: Notre corps vital fluidique, sa formule biométrique* (Paris: P. Ollendorff, 1897).

Beard, George. *American Nervousness, Its Causes and Consequences. A Supplement to Nervous Exhaustion (Neurasthenia)* (New York: G. P. Putnam Sons, 1881).

Bergson, Henri. *Essai sur les données immédiates de la conscience*, Thèse pour le Doctorat (Paris: Ancienne Librairie Germer-Ballière et Cie; Félix Alcan, Editeur, 1889).

———. *Matière et mémoire: Essai sur la relation du corps à l'esprit* (1896); *Matter and Memory: Essay on the Relation of Body and Spirit*, trans. Nancy Margaret Paul and W. Scott Palmer (London: George Allen and Unwin, 1911).

———. *Time and Free Will: An Essay on the Immediate Data of Consciousness*. trans. F. L. Pogson (London: George Allen and Unwin, 1910).

———. *L'évolution créatrice* (Paris: Presses Universitaires de France/Quadrige, 1907, 1941).

———. *«Le rêve», «"Fantômes de vivants" et "recherche psychique"»* (Paris: Presses Universitaires de France, 2013).

———. "Phantasms of the Living" and "Psychical Research": Presidential Address to the Society for Psychical Research, London, 23 May 1913; *Mind-Energy*, trans. H. Wildon Carr, ed. Keith Ansell Pearson and Michael Kolkman (Basingstoke: Palgrave Macmillan, 2007).

Besant, Annie. *L'Homme et ses Corps*, translation de l'anglais par F. B. (Paris: Publications théosophiques, 1902).

———. *Man and His Bodies* (London: Theosophical Publishing Society, 1896.

Blanchot, Maurice. "Bergson and Symbolism", *Yale French Studies*, 4 (1949): 63–66.

Blavatsky, Helena Petrovna. *Isis Unveiled: A Master-Key to the Mysteries of Ancient and Modern Science and Theology* (New York: J. W. Bouton, 1877).

Brauer, Fae. "Scientistic Magnetism and Hauntological Metarealism: The Phantasmatic Doubles of Duchamp and Durville", *Realisms of the Avant-Garde*, eds. David Ayers et al (Berlin/Boston: Walter de Gruyter, 2020) 42–75.

———. "Capturing Unconsciousness: The New Psychology, Hypnosis and the Culture of Hysteria", *A Companion to Nineteenth Century Art*, ed. Michelle Facos (London et al: John Wiley & Sons, Inc., 2019) 243–262.

———. "Dealing with Cubism: Daniel-Henry Kahnweiler's Perilous Internationalism", *Dealing Art on Both Sides of the Atlantic, 1860–1940*, ed. Lynn Catterton (Leiden/Boston/Tokyo: Brill International Publishing, 2017).

———. "Magnetic Modernism. František Kupka's Mesmeric Abstraction and Anarcho-Cosmic Utopia", *Utopia. The Avant-Garde, Modernism and (Im)possible Life*, eds. David *Ayers* et al. (Berlin/Boston: De Gruyter, 2015) 135–163.

———. "Becoming Simian: Devolution as Evolution in Transformist Modernism", *Picturing Evolution and Extinction: Regeneration and Degeneration in Modern Visual Culture*, eds. Fae Brauer and Serena Keshavjee (Lady Stephenson Library, Newcastle upon Tyne: Cambridge Scholars Publishing, 2015) 127–156.

———. "Becoming Simian: Darwin, Picasso, Nolan and Creative Evolution", *Studies in History and Philosophy of Science*, (2023, forthcoming).

———. *Rivals and Conspirators: The Paris Salons and the Modern Art Centre* (Newcastle upon Tyne: Cambridge Scholars Publishing, 2013).

———. "Representing 'Le Moteur Humain': Chronometry, Chronophotography, 'The Art of Work' and the 'Taylored Body'", *Visual Resources: An International Journal of Documentation*, XIX/2 (June 2003): 83–106.

Brunelleschi, Umberto. La Foire aux Croûtes", *L'Assiette au Beurre*, No. 60, Hors Serie, 24 May 1905.

Congrès Magnétique International pour l'étude des applications du magnétisme humain, Séance du lundi, du 21 octobre 1889 (Paris: Communication de Docteur Huguet, La Faculté de Médecine de Paris: 1889).

Conrad, Joseph. *The Secret Agent: A Simple Tale* (London: Methuen & Co., 1907).

Daix, Pierre, and Joan Rosselet. *Picasso, The Cubist Years, 1907–1916: A Catalogue Raisonée of the Paintings and Related Works* (London: Thames and Hudson, 1979).

Dalrymple Henderson, Linda. "X Rays and the Quest for Invisible Reality in the Art of Kupka, Duchamp, and the Cubists", *Art Journal*, 47/44 (1988): 323–340.

———. "L'éther de l'espace, medium de l'art, de la science et de l'occultisme", *Repenser le medium* (Dijon: Les Presses de Réel, 2021) 71–103.

Deleuze, Gilles. *The Logic of Sense*, trans. Mark Lester with Charles Stivale (New York: Columbia University Press, 1990).

———. "The Logic of Sense", ed. *Constantin V. Boundos*; trans. Mark Lester with Charles Stivale (New York: Columbia University Press, 1990).

Deleuze, Gilles and Félix Guattari. *Anti-Oedipus*, trans. Robert Hurley, Mark Seem and Helen R. Lane (London and New York: Continuum, 2004).

Durville, Hector. *Le Magnétisme considéré comme agent lumineux. Extrait du "Traité expérimental de magnétisme"* (Paris: Librairie du Magnétisme, 1896).

———. *Le Magnétisme humain considéré comme agent physique, mémoire lu au congrès magnétique international* (Paris: Librairie du Magnétisme, 1890).

———. *Traité expérimental de magnétisme. Cours professé à l'École pratique de magnétisme et de massage* (Paris: Librairie du Magnétisme, 1895).

———. *Magnetisme Personnel ou Psychique* (Paris: Librairie du Magnétisme, 1890; 1895; 1896).

———. *Lois physiques du magnétisme. Polarité humaine. Traité expérimental et thérapeutique de magnétisme. Cours professé à la clinique du magnétisme en 1885–1886* (Paris: Librairie du Magnétisme, 1896).

———. *Magnétisme personnel. Éducation de la pensée. Développement de la Volonté. Pour être Heureuse, Fort, Bien Portant et Réussir en tout* (Paris: Librairie du Magnétisme, 1899).

———. *Le Fantôme des Vivants, anatomie et physiologie de l'âme. Recherches expérimentales sur le dédoublement des corps de l'homme* (Paris: Librairie du Magnétisme, 1909).

Frizot, Michel. "The All-Powerful Eye: The Forms of the Invisible", *A New History of Photography*, ed. Michel Frizot; trans. Susan Bennett el al (Cologne: Könemann, 1998).

Garnier, François. *Correspondance de Max Jacob*, Tome I, dirigée par André Brissaud (Paris: Éditions de Paris, 1953).

Gleizes, Albert and Jean Metzinger. *Du Cubisme* (Paris: Eugène Figuière, Éditeurs, 1912).

Jacob, Max. *Saint Matorel; illustré d'eau-fortes par Pablo Picasso* (Paris: Henry Kahnweiler, Éditeur, 1911).

———. *Le Siège de Jérusalem: Grande tentation céleste de Saint Matorel* (Paris: Eugène Delatre; Henry Kahnweiler, Éditeur, 1913).

———. *Récit de ma conversion* (dictated to Paul Frizot, Abbé Louis Foucher, 1939); printed *La Vie intellectuelle*, 1951.

———. *Max Jacob: Oeuvres*, Antonio Rodriguez (Paris: Quarto, Gallimard, 2012).

Jones, Patrick. "Strange Distance: Bergson and Symbolism", 'L'anomalie en question', *TRANS – Revue de littérature génerale et compare*, 26 (2021): Unpaginated.

Josephson Storm, Jason Ā. *The Myth of Disenchantment: Magic, Modernity and the Birth of the Human Sciences* (Chicago and London: The University of Chicago Press, 2017).

Kern, Stephen. *The Culture of Time and Space, 1880-1918* (Cambridge, Mass.; London, England: Harvard University Press, 1983; 2003).

Lancelin, Charles. "Méthode de Dédoublement personnel", *Le Monde Physique: Organe Mensuel de "l'Institut de Recherches Psychiques de France" pour l'étude expérimentale des Phénomènes spirites, magnétiques, hypnotiques et occultists"* No. 1 (March 1911): 14–19.

Le Bon, Gustave. *L'Évolution de la matière* (Paris: Flammarion, 1905).

Lefranc, Léon. "Le Corps Astral du Vivant. Forme – Matière – Couleur", *Le Monde Psychique*, 1 (1911): 70–81.

———. "Recherches Expérimentales sur le deuxième corps invisible du l'homme vivant (Le corps astral) à l'aide de magnétisme", *Le Monde Psychique*, 3 (1 March 1911): 8–13.

———. "Le Corps Astral du Vivant", Deuxième Partie, Chapître III, Essai du physiologie du corps Astral du vivant. *FORME – MATIERE – COULEUR*, 3 (1 March 1911): 72–80.

Malraux, André. *La tête d'obsidienne* (Paris: Gallimard, 1974).

Merleau-Ponty, M. *Phénoménologie de la perception* (Paris: Librairie Gallimard, 1945).

Miller, Arthur L. *Einstein, Picasso: Space, Time and the Beauty That Causes Havoc* (London: Hachette, 2008).

Mizuta Lippit, Akira. "*from* Modes of Avisuality: Psychoanalysis – X-ray – Cinema", Chapter 15, *The Spectralities Reader: Ghosts and Haunting in Contemporary Cultural Theory*, eds. Maria del Pilar Blanco and Esther Peeren (London: Bloomsbury Publishing, 2013) 258.

Olivier, Fernande. *Picasso et ses amis* (Librairie Stock, Delamain et Boutelleau, 1933).

Paul, Harry. "The Debate over the Bankruptcy of Science in 1895", *French Historical Studies*, 5/3 (Spring, 1968): 299–327.

Péladan, Joséphin. "Le Radium et les miracles", *La Nouvelle revue*, XXVI (1 January 1904): 35–47.

———. "Le Radium et l'hyperphysique", *Le Mercure de France* (50, Bo. 174, June 1904) 608–637.

Penrose, Roland. "Picasso's Portrait of Kahnweiler", *The Burlington Magazine*, "Modern Art (1908–25)", 116/852 (March 1974): 124–133.

Quinn, Susan. *Marie Curie: A Life* (Plunkett Lake Press, 2019).

Read, Peter. *Picasso and Apollinaire: The Persistence of Memory* (Berkeley/Los Angeles: University of California Press, 2008).

Richardson, John, with Marilyn McCully. *A Life of Picasso: Volume I: 1881–1906* (New York: Random House, 1991).

Rochas d'Aiglun, Lt.-Colonel de. *Les Sentiments, La Musique et Le Geste* (Grenoble: Librairie Dauphinoise, 1900).

Rousseau, Pascal. "The Magic of Images: Hallucination and Magnetic Reverie in the Work of Henri Rousseau", *Henri Rousseau: Jungles in Paris* (London: Tate Publishing, 2005) 191–203.

Rousselot, Jean. *Max Jacob au sérieux essai* (Charlieu: La Bartavelle, 1994).
Salmon, André. *La jeune peinture française* (Paris: Société des Trente, Albert Messein, 1912).
———. *André Salmon on French Modern Art*, trans.; annot. Beth Gersh-Nešić (New York: Cambridge University Press, 2006).
Sauvebois, Gaston. "Le Cubisme", *Le Gil Blas* (5 October 1911).
Schneider, Judith Morganroth. "Max Jacob on Poetry", *The Modern Language Review*, 69/2 (April 1974): 290–296.
Schwarz, Arturo. *Marcel Duchamp* (New York: Harry No. Abrams, Inc., Publishers, 1975).
Thau, Annette. "The Esthetic Reflections of Max Jacob", *The French Review*, XLV/4 (March 1972): 800–812.
Tinterow, Gary. "Vollard and Picasso", *Cézanne to Picasso: Ambroise Vollard, Patron of the Avante Garde* (New York: Metropolitan Museum of Art, 2006).
Vernéde, Louis. "Le Bergsonisme ou une Philosophie de la Mobilité", *La Phalange*, 72 (20 June 1912).

Chapter Six

Allard, Roger. "Les Beaux Arts." *La Revue indépendante*, 3 (Aug. 1911): 134.
Antliff, Mark. "The Fourth Dimension and Futurism: A Politicized Space." *Art Bulletin*, 82/4 (December 2000): 722–733.
———. *Inventing Bergson: Cultural Politics and the Parisian Avant-Garde* (Princeton, NJ: Princeton University Press, 1993).
Apollinaire, Guillaume. *The Cubist Painters*. Translated and commentary by Peter Read. (Berkeley: University of California Press, 2004). Originally published as *Les peintres cubists: Méditations esthétiques*. Paris: Éditions Athena, 1912.
Apollonio, Umbro, ed. *Futurist Manifestos*. Afterword by Richard Humphreys. trans. Robert Brain, R. W. Flint, J. C. Higgitt, and Caroline Tisdall. (Boston: MFA Publications, 2001).
Berghaus, Günter. *Futurism and Politics: Between Anarchist Rebellion and Fascist Reaction, 1909–1944*. (Providence, RI: Berghahn Books, 1996).
Bergson, Henri. *Creative Evolution*. (New York: Cosimo Classics, 2005). Originally published as *L'évolution créatrice*. Paris: Félix Alcan, 1908.
———. *Matter and Memory*, trans. Nancy Margaret Paul and W. Scottt Palmer. (Mineola, NY: Dover, 2004). Originally published as *Matière et mémoire: essai sur la relation du corps à l'esprit*. Paris: Félix Alcan, 1896.
———. *Time and Free Will: An Essay on the Immediate Data of Consciousness*, trans. F. L. Pogson. (Mineola, NY: Dover, 2001). Originally published as *Essai sur les données immédiates de la conscience*. Paris: Félix Alcan, 1889.
Boccioni, Umberto. *Gli scritti editi e inediti*, ed. Zeno Birolli. (Milan: Feltrinelli, 1971).
———. *Lettere futuriste*, ed. Federica Rovati. (Rovereto, Italy: Egon and Museo di Arte Moderno e Contemporanea di Trento e Rovereto, 2009).
———. Notes on Henri Bergson's *Matière et memoire* [*Matter and Memory*]. *Boccioni papers*, 1899–1986, J. Paul Getty Research Institute, Los Angeles.
———. *Pittura e scultura futuriste*, ed. Zeno Birolli. (Milan: Abscondita, 2006).
Bois, Yve-Alain. "On Matisse: The Blinding; for Leo Steinberg", trans. Greg Sims. *October* 68 (Spring 1994): 62–121.
Boulouch, Nathalie. *Les Autochromes Lumière: La couleur inventée*. Preface by Maurice Trarieux-Lumière. (Lyon, France: Alain Scheibli Editions, 1995).
Chevreul, Michel Eugène. *The Laws of Contrast of Color*. trans. John Spanton. (London: Routledge, 1857).
Coen, Ester. *Umberto Boccioni: A Retrospective*. (New York: Harry Abrams and the Museum of Modern Art, 1988).

Corra, Bruno, and Arnaldo Ginna. *Il pastore, il gregge e la zampogna*. Bologna: Libreria Beltrami, 1912.

Delaunay, Robert, and Sonia Delaunay. *The New Art of Color*, ed. Arthur Cohen; trans. David Shapiro and Arthur Cohen. (New York: Viking Press, 1978).

Dorazio, Virginia Dortch. *Giacomo Balla: An Album of His Life and Work*. Introduction by Giuseppe Ungaretti. (New York: Wittenborn EC, 1969).

Freud, Sigmund. *Standard Edition of the Complete Psychological Works of Sigmund Freud. Vol. 11, Five Lectures on Psycho-Analysis, Leonardo da Vinci and Other Works*, ed. and trans. James Strachey, in collaboration with Anna Freud. (London: Hogarth and the Institute of Psychoanalysis, 1957).

Gage, John. *Color and Culture: Practice and Meaning from Antiquity to Abstraction*. London: Thames and Hudson, 1993).

Greene, Vivien, ed. *Divisionism/Neo-Impressionism: Arcadia and Anarchy*. Exhibition catalog. (New York: Solomon R. Guggenheim Museum, 2007).

———. ed. *Italian Futurism, 1909–1944: Reconstructing the Universe*. (New York: Solomon R. Guggenheim Museum, 2014).

———. "*Pittura ideista*: The Spiritual in Divisionist Painting", *California Italian Studies*,5/1 (2014): 1–16.

Higgins, Scott. *Harnessing the Technicolor Rainbow*. (Austin, TX: University of Texas, 2007).

Kandinsky, Wassily. *Concerning the Spiritual in Art*. (Boston: MFA Publications, 2006).

Kubnick, Henri. *Les Frères Lumière*. (Paris: Librarie Plon, 1938).

Lista, Giovanni. *Balla*. (Modena, Italy: Edizioni Galleria Fonte d'Abisso, 1982).

———. *Balla: La modernità Futurista*. (Milan: Skira, 2009).

———. *Cinema e fotografia futurista*. (Milan: Skira, 2001).

Mainz, Valerie, and Griselda Pollock, eds. *Work and the Image, Vol II: Work In Modern Times–Visual Mediations and Social Processes*. (Aldershot, UK: Ashgate, November 2000).

Marinetti, F. T. *Critical Writings*, ed. Günter Berghaus. (New York: Farrar, Straus and Giroux, 2006).

———. *Guerra sola igiene del mundo*. (Milan: Ed. Futuriste di "Poesia", 1915).

Mather, David S. *Futurist Conditions: Imagining Time in Italian Futurism*. (London: Bloomsbury Academic, 2020).

———. "Mobilizing Desire in Umberto Boccioni's 'Small Dress Shoe + Urine'", *Getty Research Journal* 6 (2014): Los Angeles: Getty Research Institute; 185–194.

Matisse, Henri. *Matisse on Art*, ed. Jack Flam. Revised edition. (Berkeley: University of California Press, 1995).

Mattioli Rossi, Laura, ed. *Boccioni's Materia: A Futurist Masterpiece and the Avant-garde in Milan and Paris*. (New York: Solomon R. Guggenheim Foundation, 2003).

Nowotny, Robert. *The Way of All Flesh Tones: A History of Color Motion Picture Processes, 1895–1929*. (New York: Garland, 1983).

Poggi, Christine. "*Folla/Follia*: Futurism and the Crowd", *Critical Inquiry* 28 (Spring 2002): 709–748.

———. *Inventing Futurism*. (Princeton, NJ: Princeton University Press, 2009).

Pruraux, Henri des. "Il sogetto nella pittura", *La Voce* 4/44 (October 31, 1912): 921.

Ramsaye, Terry. "Color Photography and the Motion Picture." *Photoplay* 15/4 (March 1919): 84–86.

Romiti, Giulielmo. "Anatomia generale." *Trattato di Anatomia Umana, Vol. 1*. (Milan: F. Vallardi, 1912).

Severini, Gino. *Écrits sur l'art*. (Paris: Editions Cercle d'Art, 1987).

Smith, Terry, ed. *In Visible Touch: Modernism and Masculinity*. (Chicago: University of Chicago, 1997).

Taylor, Richard, and Ian Christie, eds. *The Film Factory: Russian and Soviet Cinema in Documents*, trans. Richard Taylor. (Cambridge, MA: Harvard University Press, 1988).
Urban, Charles. *A Yank in Britain: The Lost Memoirs of Charles Urban, Film Pioneer*, ed. Luke McKernan. (East Sussex, England: Projection Box, 1999).

Chapter Seven

Baljeu, J. *Theo Van Doesburg*. (London: Studio Vista, 1974).
Ball, H. *Flight Out of Time: A Dada Diary*, edited and introduced by John Elderfield. (Berkeley: University of California Press, 1974).
Cabanne, P. *Dialogues with Marcel Duchamp*. London: Thames and Hudson, 1971).
Cage, J. *Silence: Lectures and Writings*. (London: Calder and Boyars, 1986).
Huelsenbeck, R. *Dada Almanac*. (London: Atlas Press, 1920, 1998).
Jean, M. ed. *Jean Arp: Collected French Writings*. (London: Calder and Boyars, 1974).
Naumann, F. ed. *New York Dada 1915–1923*. (New York: Harry N. Abrams 1994).
Richter, H. *Dada: Art and Anti-Art*. (London: Thames and Hudson, 1964, 2016).
Tupitsyn, M. ed. *Russian Dada 1914–1924*. (Madrid: Museo Nacional Centre de Arte Reina Sofia; Cambridge, MA: The MIT Press, 2018).
Tzara, T. *Seven Dada Manifestos and Lampisteries*. (New York: Calder, London and Riverrun Press, 1977).

Chapter Eight

Ades, Dawn. *Dada and Surrealism Reviewed*. (London: Arts Council of Great Britain, 1978).
———. *Dalí*. (London: Thames and Hudson, 1982).
Ansell-Pearson, Keith. *"Bergson and Creative Evolution/Involution: Exposing the transcendental illusion of organic life"*. *The New Bergson*, ed. John Mullarkey (Manchester and New York: The University of Manchester Press, 1999) 146–167.
———. *Thinking Beyond the Human Condition*. (London and New York: Bloomsbury, 2018).
Bergson, Henri. *Creative Evolution*, 1907; trans. Arthur Mitchell. (Mineola, New York, Dover Publications, Inc., 1998).
———. *Key Writings*, eds. Ansell-Pearson, Keith and John Ó Maoilearca. (London, New York: Bloomsbury, 2002).
———. *Matter and Memory*, trans. Nancy Margaret Paul and W. Scott Palmer. (London: George Allen & Unwin Press, Inc., 1970).
Breton, André. *Manifestoes of Surrealism*, trans. Richard Seaver and Helen R. Lane. (Ann Arbor: The University of Michigan Press, 1972).
———. *What is Surrealism? Selected Writings*, ed. Franklin Rosemont (New York: Pathfinder, 1978).
Caillois, Roger. *The Edge of Surrealism: A Roger Caillois Reader*, ed. Claudine Frank, trans. Claudine Frank and Camille Naish (Durham and London: Duke University Press, 2003).
Cohen, Richard A. "Philo, Spinoza, Bergson: the rise of an ecological age". In *The New Bergson*, ed. John Mullarkey (Manchester and New York: The University of Manchester Press, 1999) 18–31.
Deleuze, Gilles. *Bergsonism*, trans. Hugh Tomlinson and Barbara Habberjam (New York: Zone Books, 1991).
Fijalkowski, Krzysztof and Michael Richardson, eds. *Surrealism: Key Concepts*. (London and New York: Routledge, 2016).
Grosz, Elizabeth. *Becoming Undone: Darwinian Reflections on Life, Politics, and Art*. (Durham and London: Duke University Press, 2011).
Gunter, Peter A.Y. "Bergson and the War Against Nature". *The New Bergson*, ed. John Mullarkey (Manchester and New York: The University of Manchester Press, 1999) 168–182.

Helin, Jenny, Tor Hernes, Daniel Hjorth and Robin Holt. *The Oxford Handbook of Process Philosophy and Organisation Studies* (Oxford: Oxford University Press, 2014).

Kadri, Raihan. *Reimagining Life: Philosophical Pessimism and the Revolution of Surrealism* (Plymouth: Farleigh Dickinson University Press, 2011).

Kerslake, Christian. *Deleuze and the Unconscious.* (London and New York: Continuum, 2007).

Lefebvre, Alexander and Melanie White, eds. *Bergson, Politics, and Religion.* (Durham and London: Duke University Press, 2012).

Lehan, Stephen. "Bergson and the Discourse of the Moderns". *The Crisis in Modernity: Vitalist Controversy*, eds. Frederick Burwick and Paul Douglass (Cambridge: The Cambridge University Press, 1992) 306–329.

Linstead, Steven. "Henry Bergson (1859-1941)", *The Oxford Handbook of Process Philosophy and Organisation Studies*, eds. Jenny Helin, Tor Hernes, Daniel Hjorth and Robin Holt, (Oxford: Oxford University Press, 2014) 218–235.

Marrati, Paola. "James, Bergson, and an Open Universe", *Bergson, Politics and Religion*, eds. Alexandre Lefebvre and Melanie White (Durham & London: Duke University Press, 2012) 299–312.

Matz, Jesse. *Modernist Time Ecology* (Baltimore: Johns Hopkins University Press, 2018).

Merleau-Ponty, Maurice. *Nature: Course Notes from the Collège de France,* trans. Robert Vallier (Evanston, Illinois: Northwestern University Press, 2003).

Roberts, Donna. "Surrealism: The Ecological Imperative". *Surrealism: Key Concepts*, eds. Fijalkowski and Richardson (London and New York: Routledge, 2016) 216–227.

———. "Nature". *The International Encyclopedia of Surrealism*, Vol. 1, eds. Michael Richardson, Dawn Ades, Krzysztof Fijalkowski et al. (London: Bloomsbury, 2019) 261.

Torry, Malcolm. *Actology: Action, Change, and Diversity in the Western Philosophical Tradition.* (Eugen, Oregon: Resource Publications, 2020).

Chapter Nine

Andreu, Pierre. *Notre maître, Georges Sorel* (Paris: Editions Bernard Grasset, 1953).

Antliff, Mark. *Avant-Garde Fascism: The Mobilization of Myth, Art and Culture in France, 1909-1939* (Durham: Duke University Press, 2007).

———. "Bad Anarchism: Aestheticized Mythmaking and the Legacy of Georges Sorel", *Anarchist Developments in Cultural Studies*, 2 (2011): 155–187.

———. *Inventing Bergson: Cultural Politics and the Parisian Avant-Garde* (Princeton: Princeton University Press, 1993).

Barden, Garrett. "Method in Philosophy", *The New Bergson*,ed. John Mullarkey (New York: St. Martin's Press, 1999), 32–40.

Bergson, Henri. *An Introduction to Metaphysics*, trans. T.E. Hulme, Introduction by John Mullarkey (London: Palgrave MacMillan, 1903, 2007).

———. *Creative Evolution.* Trans Arthur Mitchell (New York: Henry Holt and Company, 1911, rpt. 1931).

———. *The Creative Mind*, trans. Mabelle L. Andison (New York: Philosophical Library, 1946).

———. *Matter and Memory*, trans. N.M. Paul and W.S. Palmer (New York: Humanities Press, 1896, 1978).

———. *Mélanges* (Paris: Presses Universitaires de France, 1972).

———. *Time and Free Will*, trans. F.L. Pogson, 1910 (rpt. New York: Harper & Row, 1889, 1960).

Berth, Eduoard. "Anarchisme individualiste, Marxisme orthodox, syndicalisme révolutionaire", *Le Mouvement socialiste*, (1 May 1905): 6–35.

———. "Marx Inédit!", *Le Mouvement socialiste*, (1 November, 1904): 98–99.

———. *Les Méfaits des Intellectuels* (Paris: Rivière, 1914).
Clark, John. *Max Stirner's Egoism* (London: Freedom Press, 1976).
Colomer, André. "L'Art, l'anarchie, et l'ame Chrétienne", *L'Action d'art*,(15 April 1913): 1–2.
———. "La Bande", *L'Action d'art*, (10 November 1913): 2.
———. "Illusions sociales et delusions scientist", *L'Action d'art*, (25 August 1913): 2.
———. "M Bergson et les Jeunes Gens d'Aujourd'hui", *L'Action d'art*,(1 March, 1913): 1.
———. "La Science et l'Intuition: Leurs rôles dans l'individualisme", *L'Action d'art*, (10 May 1913): 3.
Golman, Lawrence. "Education as Politics: university Adult Education in England since 1870", *Oxford Review of Education*,(March-June 1999): 89–101.
Grogin, R.C. *The Bergsonian Controversy in France 1900–1914* (Calgary: University of Calgary Press, 1988).
Kropotkin, Peter. "La Croisade contre la Science de M. Bergson", *Les Temps Nouveaux*, (15 October, 1913): 2–4.
———. *Anarchism and Modern Science* (London: Freedom Press, 1912)
"Librarie d'Action d'art", *L'Action d'art*, (25 July 1913): 4.
Flers, Marion de. "*Le Mouvement socialiste* (1899–1914)", *Cahiers du Georges Sorel*,5 (1987): 49–76.
Marx, Karl, with Friedrich Engels, and V.I. Lenin. *Anarchism and Anarcho-Syndicalism* (New York: International Publishers, 1972).
Maximoff, G.P. *The Guillotine at Work: Twenty Years of Terror in Russia* (Sanday: Cienfuegos Press, 1979).
Miles, Andy. "Workers' Education: The Communist Party and the Plebs League in the 1920's", *History Workshop*, (Autumn, 1984): 102–114.
Mullarkey, John. *Bergson and Philosophy* (Notre Dame: University of Notre Dame Press, 2000).
Papinikolas, Theresa. *Anarchism and the Advent of Paris Dada: Art and Criticism, 1914-1924* (London: Ashgate, 2010).
Paul, Eden and Cedar. *Creative Revolution: A Study of Communist Ergatocracy* (London: Plebs League, 1920).
Rée, Jonathan. *Proletarian Philosophers: Problems in Socialist Culture in Britain, 1900-1940* (Oxford: Clarendon Press, 1984).
Roth, Jack. *The Cult of Violence: Sorel and the Sorelians* (Berkeley: University of California Press, 1981).
Ruhe, Algot. *Henri Bergson: An Account of his Life and Philosophy* (London: MacMillan, 1914).
Sonn, Richard. *Sex, Violence and the Avant-Garde: Anarchism in Interwar France* (University Park: The Pennsylvania State University Press, 2010).
Sorel, Georges. *The Illusions of Progress*, trans. John and Charlotte Stanley (Berkeley: University of California Press, 1908, 1969).
———. *Reflections on Violence*. Trans, T.E. Hulme (New York: Peter Smith, 1915, rpt. 1941, 1908).
Soulez, Philippe. *Bergson politique* (Paris: Presses Universitaires de France, 1989).
Stanley, John. *Sociology of Virtue: The Political and Social Thought of Georges Sorel* (Berkeley: University of California Press, 1981).
Stirner, Max. *The Ego and its Own*. Trans, Steven Byington (London: Rebel Press, 1907, 1993).
Vernon, Richard. *Commitment and Change: Georges Sorel and the Idea of Revolution* (Toronto: University of Toronto Press, 1978).
Ward, Paul. *Red Flag and Union Jack: Englishness, Patriotism and the British Left, 1881-1924* (Bury St. Edmunds: The Boydell Press, 1998).
Worms, Frédéric. "*Matter and Memory* on Mind and Body", *The New Bergson*, ed. John Mullarkey (New York: St. Martins Press, 1999) 88–98.

Chapter Ten

Alexievich, S., K. Hammond, and L. Lezhneva. "'I Am Loth to Recall': Russian Women Soldiers in World War II", *Women's Studies Quarterly*, 23/3/4 "Rethinking Women's Peace Studies" (Fall/Winter, 1995): 78–84.

Anzharov, A. cinematographer, Kohts, R.A. editor, *Iz kinodocumentov darvinovskogo muzeia*, short film about the Darwin Museum, undated but c.1948/49-mid 1950s: https://www.youtube.com/watch?v=7n8ymciNm-c.

Ashwin, S. "Introduction: Gender, State and Society in Soviet and post-Soviet Russia", *Gender, State and Society in Soviet and post-Soviet Russia*, ed. S. Ashwin (London and New York: Routledge, 2000) 1–29.

"Canines", in, G. Lindsey-Campbell, ed., *The Oxford Handbook of Animals in Classical Thought and Life*, Oxford and New York: Oxford University Press, 286–287.

Crocq, M-A. and L. Crocq. "From Shell Shock and War Neurosis to Posttraumatic Stress Disorder: A History of Psychotraumatology", *Dialogues in Clinical Neuroscience*, 2/1 (March 2000): 46–55. https://www.ncbi.nlm.nih.gov/pmc/articles/PMC3181586/

Darwin, C.R. *On the Origin of Species by Means of Natural Selection, or the Preservation of Preferred Races in the Struggle for Life*, London: John Murray, 1861 (3rd edition): http://darwin-online.org.uk/content/frameset?pageseq=1&itemID=F381&viewtype=text.

Davis, P.K. *100 Decisive Battles from Ancient Times to the Present* (London and New York etc.: Oxford University Press, 2001).

Duffin, J. "Cadavers and Patients: Lannec's Vital Principle and the Historical Diagnosis of Vitalism", in G. Cimino and F. Duchesne, eds, *Vitalisms From Haller to the Cell Theory* (Fierenze: Leo S. Olsicki, 1997): 205–224.

Engels, F. "The Part Played by Labour in the Origin of Man from Apes", written May-June 1876, first published, in *Die Neue Zeit 1895-06*, trans. C. Dutt (Moscow: Progress Publishers, 1934. Marxists.org archive, https://www.marxists.org/archive/marx/works/1876/part-played-labour/.

———. *The Dialectics of Nature* (Moscow: Progress Publishers, 1934) https://marxists.catbull.com/archive/marx/works/1883/don/ch07b.htm.

Graham, H.F. "The Significant Role of the Study of Ancient History in the Soviet Union", *The Classical World*, 61/3 (1967): 85–97.

Hosking, G. *A History of the Soviet Union 1917–1991 Final Edition* (London: Fontana Press, 1992).

Hunt, P. "Battle of the Trebbia River: Roman-Carthaginian History", *Encyclopaedia Britannica*, https://www.britannica.com/event/Battle-of-the-Trebbia-River.

Joravsky, D. *The Lysenko Affair* (Chicago: University of Chicago Press, 1986).

Kalacheva, I.P. "60 let velikoi pobedy: darvinovskii muzei v gody velikoe otechestvennoe voini", *Trudy gosudarstvennogo darvinovskogo muzeia*, Issue VIII (Moscow: Federal'noe agenstvo po kul'ture I kinematografii, department kul'tury goroda Moskvy, gosudarstvennyi Darvinovskii muzei, 2007): 98–120.

———. "Gosudarstvennyi darvinovskii muzei v gody otechestvennoi voiny 1941-1945", *Darvinovskii muzei: 100 let so dnia osnovanie, 1907–2007*, ed. A. Trofimov (Moscow: Izdatel'skaia programma "Interrossa", 2007): 37–42.

Kalacheva, I.P. and A.I. Kliukina. "100 let zhizni darvinovskogo muzeia", *Trudy gosudarstevennogo darvinovskogo muzeia*, XI (2007): 64–205.

Kandinsky, W. *Concerning the Spiritual in Art*, (1912), trans. M Sadlier 1914, with re-translation by F. Golfing, M. Harrison and F. Ostertag (New York: George Wittenborn Inc., 1970).

Kazennaya, E. and E. Divid. "Chronic Trauma and Dissociation in Russia: A Brief Overview of the History of Psychological Services and Psychotherapy in Russia", European Association for Trauma and Dissociation (ESDT) (no page numbers), https://www.estd.org/countries/russia.

Kol'tsov, N.K. "Review of Professor Hans Driesch, *Vitalism: Its History and System*, authorised translation by Professor A.G. Gurevich, Nauka Publishing House: Moscow 1915", *Priroda*, 12 (1914): 1528–1530.

Kohts, R.A. cinematographer, *Opyt c shimpanze, raboty N. Ladygina-Kots*, short film from the collection of the State Darwin Museum, Moscow, no date: https://www.youtube.com/watch?v=8imskoHgAb0.

———. cinematographer, *Proshchal'noe privetstviia Aleksandra Feodorovicha Kots*, short film, c.1957, with English subtitles translated from the Russian by Sunny M. Boscoc: https://www.youtube.com/watch?v=jyBuSOn7h_w.

Kots, A.F. "Bor'ba V.A. Vagnera protiv idealizma i mekhanizma", AGDM, o.12430, f.17, ed.khr.97.

———. *Muzei evoliutsionnoi istorii moskovskikh vysshikh kursov za 1913-1914*, Moscow, 1914 (printed pamphlet, no publisher named). Arkhiv gosudarstvennyi darvinovskii muzei (AGDM) f.12430, o.32, ed.khr.165, 16pp.

Krementsov, N. "Darwinism, Marxism and Genetics in the Soviet Union", in D. Alexander and R. Numbers, eds, *Biology and Ideology from Descartes to Darwin* (Chicago and London: University of Chicago Press, 2010): 215–246.

———. "V.A. Vagner and the Origin of Russian Ethology", *International Journal of Comparative Psychology*, 6/1 (1992): 61–70. https://archive.org/stream/internationaljouv6no1newy/internationaljouv6no1newy_djvu.txt.

Krementsov, N. and W. deJong-Lambert. "Lysenkoism Redux: Introduction", in N. Krementsov and W. deJong-Lambert, eds, *The Lysenko Controversy as a Global Phenomenon*, vol, 1 (Cham, Switzerland: Palgrave MacMillan/Springer Nature, 2017): 1–34.

Krylova, A. "'Healers of Wounded Souls': The Crisis of Private Life in Soviet Literature, 1944–1946", *The Journal of Modern History*, 73 (2001): 307–331.

Ladygina-Kots, N.N. "Avtobiografiia", in *Gosudarstvennyi darvinovskii muzei* (Moscow: Mozhaiskii poligraficheski kombinat, 1993): 79.

———. *Infant Chimpanzee and Human Child*, ed. F. de Waal, trans. B. Vekker (Oxford, New York: Oxford University Press, 2002).

Levy, J., A. Goldstein, and R. Feldman. "The Neural Development of Empathy is Sensitive to Caregiving and Early Trauma", *Nature Communications*, 10 (2019): article no: 1005. https://www.nature.com/articles/s41467-019-09927-y

Lysenko, T.D. "The Situation in Biological Science'. Address given to the Vaskhnil Session, 31 July 1948, in T.D Lysenko, *Agrobiology. Essays on Problems of Genetics, Plant Breeding and Seed Growing* (Moscow: Foreign Languages Publishing House, 1954) 515–554.

Merridale, C. "The Collective Mind: Trauma and Shell-Shock in Twentieth Century Russia", *Journal of Contemporary History*, 35/1 (January 2000): 39–55.

McLeish, J. *Soviet Psychology: History, Theory, Content* (London: Methuen, 1975).

Nietlisbach, G., A. Maerker, W. Rossler, and H. Haker. "Are Empathic Abilities Impaired in Post-traumatic Stress Disorder?", *Psychological Reports*, 106/3 (2010): 832–844. https://www.ncbi.nlm.nih.gov/pubmed/20712172.

"Nimrud dogs" Cambridge University Museum of Archaeology and Anthropology: http://oracc.museum.upenn.edu/nimrud/livesofobjects/catordog/index.html

Phillips, S.D." 'There Are No Invalids in the USSR!' A Missing Soviet Chapter in the New Disability History'" *Disability Quarterly*, 29/3 (2009). https://dsq-sds.org/article/view/936/1111.

Popov, I. *Orthenogenesis Versus Darwinism* (Chem, Switzerland: Springer Nature AG. 2018).

Rudova, L. "Bergonism in Russsia: the Case of Bakhtin", *Neophilogus*, 80/2 (April 1996): 175–188.

Ruse, M. "Charles Darwin and Artificial Selection", *Journal of the History of Ideas*, 36/2 (April-June 1975): 339–350. https://link.springer.com/article/10.1007/BF00212098.

Samsonova, A. "Kavkazskaia Ovtcharka" (Caucasian Shepherd Dog), revised by Renee Sporre-Wiles, 2010, Federation Cynologique Internationale (AISRL), 02/03/20111/En, FC Standard no.328, https://www.fci.be/Nomenclature/Standards/328g02-en.pdf, 1-9. For some extra history and images, see also: American Kennel Club, "Caucasian Shepherd Dog", 2022, https://www.akc.org/dog-breeds/caucasian-shepherd-dog/.ab.1.0.0j0i22i30l9.5201.13684.

Simpson, P. "A Cold War Curiosity? The Soviet Collection at the Darwin Memorial Museum Down House, Kent", *Journal of the History of Collections*, 30/3 (2018): 487–509.

———. "Imag[in]ing Post-Revolutionary Evolution: The Taylorised Proletarian, 'Conditioning' and Soviet Darwinism in the 1920s", in B. Larson and F. Brauer, eds, *The Art of Evolution: Darwin, Darwinisms and Visual Culture* (Lebanon NH: Dartmouth College Press/University of New England Press, 2009): 226–261.

———. "Lysenko's 'Michurinism' and Art at the Moscow Darwin Museum 1935-1964", in W. deJong-Lambert and N. Krementsov, eds, *The Lysenko Controversy As a Global Phenomenon: Genetics and Agriculture in the Soviet Union and Beyond*, vol. 1 (Cham, Switzerland: Palgrave MacMillan/Springer Nature, 2017): 129–175.

"The Jennings Dog/the Duncombe Dog/The Dog of Alcibaiades", The British Museum, https://www.britishmuseum.org/research/collection_online/collection_object_details.aspx?objectId=467443&partId=1.

Todes, D.P. *Ivan Pavlov: A Life in Science* (New York: Oxford University Press, 2014).

———. *Pavlov's Physiology Factory Experiment, Interpretation, Laboratory Enterprises* (Baltimore, Maryland, USA: Johns Hopkins University Press, 2002).

Udal'tsova, V. "Konstantin Konstantinovich Flerov", *Gosudarstvennyi darvinovskii muzei. Sokrovishcha russkogo iskusstva* (Moscow: Belyi gorod, 2007): 51–56.

Vuchinich, A. *Darwinism in Russian Thought* (Berkeley, Los Angeles, Oxford: University of California Press, 1989).

Wanke, P. *Russian/Soviet Military Psychiatry 1905-1945* (Abingdon, Oxon., and New York: Routledge, 2005).

West, J. *Russian Symbolism: A Study of Vyacheslav Ivanov and the Russian Symbolist Aesthetic* (London: Methuen, 1970).

Wünsche, Isabel. *"Organic Visions and Biological Models in Russian Avant-garde Art", Biocentrism and Modernism*, eds. Botar, O.I.A and Wünsche, I. (Farnham Surrey, and Burlington, VT: Ashgate, 2011): 127–153.

———. *The Organic School of the Russian Avant Garde: Nature's Creative Principles* (London and New York: Routledge, 2018).

Zhdanov, A.A. "Soviet Literature – The Richest in Ideas, The Most Advanced Literature", *Soviet Writers' Congress 1934: The Debate on Socialist Realism and Modernism in the Soviet Union. Gorky, Radek, Bukharin, Zhdanov and others*, ed. H.G. Scott (London: Lawrence and Wishart, 1977): 15–24.

Index

Note: Titles of books and artworks appear at the end of entries for corresponding authors and artists. Under 'Bergson', Bergsonian concepts appear in inverted commas, e.g. 'élan vital'

For Product Safety Concerns and Information please contact our EU representative GPSR@taylorandfrancis.com Taylor & Francis Verlag GmbH, Kaufingerstraße 24, 80331 München, Germany

Batch number: 10777126

Printed by Printforce, the Netherlands